SYSTEMS BIOLOGY
Principles, Methods, and Concepts

SYSTEMS BIOLOGY
Principles, Methods, and Concepts

Edited by
Andrzej K. Konopka

CRC Press
Taylor & Francis Group
Boca Raton   London   New York

CRC Press is an imprint of the
Taylor & Francis Group, an informa business

CRC Press
Taylor & Francis Group
6000 Broken Sound Parkway NW, Suite 300
Boca Raton, FL 33487-2742

First issued in paperback 2019

© 2007 by Taylor & Francis Group, LLC
CRC Press is an imprint of Taylor & Francis Group, an Informa business

No claim to original U.S. Government works

ISBN-13: 978-0-8247-2520-4 (hbk)
ISBN-13: 978-0-367-38986-4 (pbk)

Visit the Taylor & Francis Web site at
http://www.taylorandfrancis.com

and the CRC Press Web site at
http://www.crcpress.com

PREFACE

This text presents biology as an autonomous science from the perspective of fundamental modeling techniques. It is designed as a desk reference for practitioners of diverse fields of life sciences, as well as for these intellectually mature individuals who would, themselves, like to practice the art of systems biology in the future.

Albeit systems biology exists for well over two millennia, it has enjoyed a spectacular rejuvenation in the recent years. The computer has indeed been a major tool for systems scientists, including systems biologists, since at least the 1960s. This is probably a reason why most conceptual foundations of today's systems biology appear to be expressed with the help of terminology borrowed from logic (Chapters 1, 7, and 8), linguistic (Chapters 1, 7, and 8), theory of knowledge (Chapters 1, 2, 4, 5, 7, and 8), computer science (Chapter 1, 2, and 8), general systems theory (Chapters 1, 6, and 7), and dynamical systems (Chapters 2, 3, 5, 6, and 9). Because of the diversity of flavors of possible applications, the general modeling methods are presented from several different perspectives such as for instance biochemistry (Chapter 2), thermodynamics (Chapters 3 and 9), engineering (Chapters 7 and 8), and ecology (Chapter 5).

Each chapter has been carefully reviewed and edited such that it will most likely provide the reader with a factually and methodologically rigorous state-of-the-art tutorial, survey, and review of modeling convoluted (complex) organic systems. Of course it would be naïve to guarantee that all mistakes or misstatements are eliminated by the editing (most probably are). If any errors are left, I will certainly feel responsible for them and therefore I would greatly appreciate it if the readers could point them out to me by writing an e-mail or a "snail" mail.

At this point, I would like to extend my most sincere thanks to Jean-Loup Risler (University of Evry) and Laurie F. Fleischman (BioLingua™ Research, Inc.) for multiple reading, reviewing, and help with editing the scientific content of the chapters of this handbook. The task of meritoriously revising the final drafts of this book would have simply been impossible without their selfless contribution. I would also like to thank our outstanding acquisition editor Anita Lekhwani who not only initiated this project but also relentlessly motivated it until its fruition, with her enthusiasm and spectacular resolve. Finally, I also owe thanks to my organization, BioLingua™ Research, Inc., and its board of directors for their enthusiastic support of my involvement in this project, as well as for releasing me from the fund-raising duties and teaching assignments during almost the whole of 2005.

Andrzej K. Konopka
Gaithersburg, MD

Contributors

B. Andresen
Niels Bohr Institute
Copenhagen, Denmark

S. W. Kercel
New England Institute
University of New England
Portland, Maine

W. Klonowski
Institute of Biocybernetics and Biomedical
  Engineering and GBAF Medical Research
  Center
Polish Academy of Sciences
Warsaw, Poland

A. K. Konopka
BioLingua™ Research, Inc.
Gaithersburg, Maryland

J. Nulton
Department of Mathematics and Statistics
San Diego State University
San Diego, California

K. L. Ross
Department of Philosophy
Los Angeles Valley College
Van Nuys, California

A. Salamon
Department of Mathematics and Statistics
San Diego State University
San Diego, California

P. Salamon
Department of Mathematics and Statistics
San Diego State University
San Diego, California

J. H. Schwacke
Department of Biostatistics, Bioinformatics,
  and Epidemiology
Medical University of South Carolina
Charleston, South Carolina

R. E. Ulanowicz
Chesapeake Biological Laboratory
University of Maryland Center for
  Environmental Science
Solomons, Maryland

E. O. Voit
The Wallace H. Coulter Department of
  Biomedical Engineering
Georgia Tech and Emory University
Atlanta, Georgia

Contents

1 Basic Concepts of Systems Biology

Andrzej K. Konopka

CONTENTS

1.1 INTRODUCTION

Systems Biology is a well-established area of human activity that can be traced back at least to the writings of *Aristotle* around 350 B.C. [1]. The term "system" signifies here an entity that represents wholeness, which can be superficially divided (fractioned) into components (parts) but whose pivotal properties cannot be fully explained solely from the knowledge of the parts. The paradigms of life sciences were changing with time but the core fundamental concepts of systems biology are apparently resistant to these changes and remain pretty much the same today as they were in the time of Aristotle. That is particularly true of the holistic Aristotelian notion of the system as a functionally robust but otherwise complex whole. It also remains true of other basic methods and concepts of life sciences (such as, for instance, embryogenesis) described in Aristotle's work [1].*

I should mention here that because of *semantic* and *pragmatic* ambiguity of the popular credo of scientific holism [3,4], "system is more than a sum of its parts," one can distinguish several coexisting paradigms of systems biology that generate different shades of meanings of words "system," "more," "sum," and "parts." In contrast to physics-related research programs, there does not need to be a "winning paradigm" in biology. Several different schools of thoughts, methodologies, and belief systems can easily coexist in *life sciences* and it is reportedly a rare event that one well-formulated paradigm would prevail (i.e., "win") over the others [4].

Despite an impressive, time-resistant success of the foundations of biology as an autonomous science, today's academic methodologies of science clearly pertain to mechanics and arguably to a few other branches of physics while completely neglecting the cultures of chemistry [5] and, even more so, of biology [4–8]. As a consequence, it has been erroneously assumed by many that by setting appropriate habits of experimentation and theorizing (external controls), it must be always possible to eliminate the influence of the observer's mind (internal controls) on scientific observations in every imaginable field of science, not only mechanics. In reality, this assumption has been only partly effective even in physics proper and, without a doubt, research in biology and chemistry continues to rely not only on external but also on internal controls at the researcher's disposal. It is fair to say that the more convoluted (complex) a material system is the more important the properties (activities) of its observers (modelers) appear to be.

Within similar lines I wrote in another text a few years ago [9]:

> ... As most of us know, heuristic reasoning can be conclusive. Yet biologists face the problem of not being able to specify all the rules applied to derive conclusions. Nor are they able to list all assumptions on which those rules ought to operate. It seems that these inabilities are a reflection of the complexity of biological systems themselves. Biological phenomena are often represented by models that are still too complex to be described in a communicable manner. Further and further modeling is required until our observations can be communicated in a linguistically comprehensive way. The cascade of models gives us the advantage of creating "communicable reality" but does not help us to judge the evidence pertinent to "real" (i.e., not necessarily communicable) reality. To the contrary, the more advanced a model in a cascade is, the further is its "distance" (in terms of number of modeling steps) from the modeled system ....

We could perhaps add that because of the extraordinary complexity of living things and because of biologists' dependence on natural language, biology has been considered a "descriptive" science for centuries. It has therefore not been unusual for biologists to believe that the actions of observers are as important for the outcomes and interpretations of findings as the actual observations are assumed to be. An additional reason for the scientific merit of taking into account an observer is the fact that our human means of observation (such as senses) do not need to be appropriate for even detecting, not to mention their modeling, the complex natural phenomena we would be

---

* Most of Aristotelian foundations of modern biology are described in his work entitled *Physics*. It is therefore relevant to notice that what Aristotle called "Physics" should not be confused with today's name of the science called physics. For one thing, Aristotle's attempts to study physical and astronomical phenomena have reportedly been quite lousy and have long been forgotten. In contrast to that, his work on foundations of biology continues to inspire generations after generations of biologists ever since his written testimonials have been known. In the absence of better terminology, Aristotle called "Physics" all kinds of studies of Nature; particularly living things such as plants and animals that today would belong to biology. Incidentally, it is believed that the term "biology" did not exist until Jean-Baptiste Pierre Antoine de Monet, Chevalier de Lamarck (known to most of us as Lamarck) mentioned it for the first time in his famous introduction to his "Philosophie Zoologique" published in 1809 [2]. Since this first published mention of the word *biology* refers to a title of an unfinished manuscript that Lamarck begun writing between 1800 and 1801, it is believed that he invented the term *biology* around 1800. (Some other accounts credit Lamarck and Treviranus for independently coining the term *biology* in 1802). Anyway, Aristotle could not possibly be aware of the work undertaken 2200+ years after his death and therefore he included today's biology in his "Physics."

potentially interested in observing if we were able to. In this situation we rely on models that (again) may be very distant from the actual natural objects, phenomena, and processes we should be observing. This in turn brings us to yet another good reason for taking into account an observer: the attributes of models we are capable of observing may or may not be adequate to the actual natural systems we should be observing.

It follows from the foregoing comments that, in an abstract characterization, systems biology can be seen as studies of properties of pairs, each consisting of an *observer* and a subject of observation (an *observed*) [10].

1.2 GENERAL SYSTEMS THEORY

General systems theory [11–13] (whose variants are also known under the names of *cybernetics, systems research, or systems thinking*) is an academic culture that promotes the *functional organization* as a primary concept instrumental for systems definition and description. In most general settings, it is based on the assumptions that:

1. Systems behavior (functioning) as a whole is governed by general rules (or laws).
2. The rules do not need to be expressed in terms of properties of the components (parts) of the system and ideally would not have much *relevance* to the laws followed by the parts.

Traditionally, the general systems theory has been known in four complementary practical incarnations (paradigms) [10] that explored different ways of overcoming misgivings of reductionist methodologies (see section 1.4 of this chapter), as well as approaching a working definition of complexity:

1. Chaos theory [14]
2. Cellular automata [15]
3. Catastrophe theory [16]
4. Hierarchical systems [17]

The first three of the foregoing incarnations of general systems theory (chaos, cellular automata, and catastrophes) are all variants of dynamic systems, each of which in turn can be represented by a finite set of differential equations. The fourth incarnation of systems biology paradigms, hierarchical systems, appears to constitute a focal point of methodology. On the one hand the postulate of existence of hierarchical systems and the concept of hierarchy constitute a condition of clear thinking (condition 3 in section 1.2.1) while, on the other hand, the hierarchical organization can be used to assume that functioning (action) at one level of hierarchy does not need to be reducible to functioning at another level.*

1.2.1 PRINCIPLES OF CLEAR THINKING

One aspect of the appearance of complexity is our complete dependence on the object language in which observer's causal questions (what? how? and why?) are formulated and the answers are communicated. The object language of scientific observations [4,5,10,12,18–21] is meant to be simple but the observed situations almost always appear to be complex, at least *a priori*. Therefore,

---

* Of course the opposite assumption, stating that properties at one level can be derived from behavior of things on another level, can be and sometimes is made in the spirit of causal reductionism (see discussion in section 1.4 of this chapter.) Our point here is that the framework of hierarchical systems allows one to choose between reductionist and nonreductionist principles of methodology. Whether the choice is really essential or not is debatable and the answer clearly depends on the specific nature of the problem.

the observer is forced to actively ask and answer questions that pertain to the following *principles of clear thinking* during the act of observing as well as before and after it:

1. Principle of *formal correctness* — effective reasoning in science should conform to the rules of logic (i.e., should be formally correct)
2. Principle of *causality* — every scientifically explainable object, process, or phenomenon (a fact) can be thought of as materially or logically entailed by a cause (also a fact, object, or process)
3. Principle of *hierarchical organization* (a postulate of the existence of hierarchy of things) — every scientifically explainable object, process, or fact belongs to a hierarchical organization (a hierarchical system) and, at the same time, contains things organized hierarchically.
4. Principle of *logicolinguistic fractionability* — every scientifically explainable system can either be represented as a subsystem (or a part) of a plausible (or well-defined) larger system or be partitioned into subsystems whose properties are believed to be plausible or well defined. To avoid the so-called *infinite regress,* one additionally assumes the existence of the minimal parts (logicolinguistic atoms) that cannot be broken into smaller parts.
5. Principle of *material adequacy* [21,22] — all concepts and linguistic constructs (particularly definitions) used to explain observations of natural systems must not only be true but must also correspond to factual reality to a maximum degree possible. Our explanations must be as relevant and as pertinent as possible to the actual observed process or phenomenon that we try to explain. (Trivial or irrelevant truths should not be acceptable as valid explanations).

The foregoing principles are arguably the rules for the observer's "mind only" because they cannot be effectively applied without semantic and pragmatic interpretation.* That is to say even if a hard-core reductionist believer in materialism would like to see the parts and hierarchies right there in nature (as some extreme materialists do), it would not really matter because such constructs are only possible within a meta-language of reasoning performed by the observer's structuring mind. Ergo, the need for semantic and pragmatic interpretation of the rules of clear thinking are one of the reasons for systems biology to be concerned with the scientifically viable methodology of modeling, as well as with the distinction between a natural system and its model.

1.3   THE CONCEPT OF TRUTH IN NONDEDUCTIVE SCIENCE

About 2 years ago, I wrote [22]:

> In everyday practice of science we assume that when an individual communicates results of an experiment he or she intends to tell the truth. We need also to assume that all scientists intend to be truthful to the best of their human capacity and therefore their scientific communications should not contain intentional lies. Nor should their messages be intended to misguide or otherwise mislead others into believing that something is true that in fact is not so. In other words we assume that individuals whose craft is science choose to be truthful and honest and do in fact pursue the virtue of truthfulness to the maximum of their capacity in every act of scientific communication.

Even with these assumptions in mind, the business of approximating truth is intricate and elusive. First of all, we need to follow pragmatic principles of clear thinking (see section 1.2.1),

---

* The principle of formal correctness (principle #1) is perhaps an exception here. There exist formal systems (such as, for instance, propositional calculus in logic) whose syntax is sufficient to account for their own semantics.

which often affect the nature of questions we are able to ask. That is to say, it cannot be taken for granted that the questions we are capable of asking are most adequate regarding our interest in the answer. Second, we are forced to use a subset of natural language to formulate our queries and to communicate with each other. Ergo, we often need a linguistic convention that would define a meta-language to specify the (object) language designed for communication.

1.3.1 Grand Theories of Truth

Historically, we know of at least three different grand theories of truth:

1. Correspondence theory: What we say or see is true if and only if it corresponds to the factual reality. Every pursuit of this doctrine requires, from the outset, that we take into account the possible discrepancies between real facts and their models.
2. Coherence theory: What we say or see is true if and only if it conforms to an existing system of knowledge without contradictions. The validity of this theory depends on our capacity to make sure that we have taken into account all the existing knowledge (an obvious overstatement of our potential).
3. Pragmatic (utility-based) theory: What we say or see is true if and only if it is being adequately used. One technical problem with this doctrine is due to our ability (or lack thereof) to determine adequacy of usage. Another problem is integrity (or lack thereof) of the users. That is to say, concepts and assertions can be used with or without regard to merits of such utilization as long as it (the utilization) brings social, reproductive, political, or any other merit-unrelated success to the users.

There also exist theories of truth that are blends of the foregoing three "pure" doctrines. For instance, we can often encounter in science the following tacit doctrines:

4. Truth by convention (naive epistemological idealism): We agree to believe that certain general assertions are true and do not require the evidence to demonstrate this fact. (One classical example is an axiomatic system in mathematics, such as, for instance, Euclidian geometry, where the axioms are assumed to be true theorems.)
5. Truth by nature of things (naive ontological realism): Whatever we believe about the observed relation between objects or phenomena or processes, the relation is true or false in objective reality and has nothing to do with the observer's ability (or lack thereof) to determine that.

In formal logic and most of mathematics (i.e., in deductive sciences), we employ a formal language (such as, for instance, predicate calculus or arithmetic) to determine if a given specific statement formulated in this language is true or false. All we need to do is to check correctness of use of the rules of syntax of the language. In this respect, the theories of coherence (theory 2 above) and convention (theory 4 in our list) are favorite doctrines of truth in deductive science. The coherence theory and conventionalism are also doctrines of choice in various disciplines of computer applications that require handling of large data sets (such as for instance database searches, data mining, and pattern acquisition). The predominant paradigm in these fields is the creation of standards for judging the quality of data or software. (Presumably, any standard is better than no standard at all.)

The schemes of arriving at possible truth in nondeductive science appear to differ from field to field. For instance, most of physics and astronomy seem to rely on simple induction — derivation of general rules of system's activity from instances of specific processes. On the other hand, chemistry and biology — in a very general view — seem to require more than a simple induction to validate scientific claims. This not-so-simple induction along with other complicated modes of inference is often called *pragmatic inference* [5,9] and will be discussed briefly in the next section. It is not clear which of the five theories of truth will be used by individual researchers. It looks as though truth by correspondence (theory 1) is a favorite doctrine in systems biology and other fields of science* because of the built-in distrust in our models. In fact, the correspondence theory requires that we verify the fidelity, accuracy, and/or precision with which the models represent nature (i.e., the factual reality).

1.3.2 DEDUCTION, INDUCTION, AND PRAGMATIC INFERENCE

Forming conclusions from premises is called *inference*. One can distinguish several types (modes) of systematic inference. Their nonexhaustive list includes:

1. Deduction (from general to specific)
2. Induction (from specific to general)
3. Pragmatic inference
   a. Abduction (finding the most plausible explanation available)
   b. Consilience (transcending different sequences of inferences)

Formal logic, as we know it today, can only deal with deductive inferences and to some extent with induction (such as inductive proofs of properties of integers). All other types of systematic inference that appear similar to an intricate (i.e., not simple) induction are often referred to collectively as pragmatic inference [9,23,24]. They cannot be easily formalized and thereby require language and domain-of-application–dependent conventions that would define their semantics (syntax alone will not suffice).**

Inference via deduction always leads to specific conclusions from general premises. For instance, the implication: "every man is mortal, I am a man, and therefore I am mortal" is an example of deduction. The general scheme of deduction is:

1. Every object A has a property B.
2. Therefore, a given specific individual object A has a property B.

Induction is a derivation of a general conclusion from specific premises. The scheme of inductive inference is:

1. Every individual object A observed thus far has a property B
2. Therefore every A has property B

The question of what the phrase "thus far" should exactly mean has been hotly debated for centuries and is known under the name the *Problem of Induction* [25,26]. If we see the night following the day, say 1000 times, does it mean that the night *always* follows the day? Or perhaps, in order to decide that it is always a case, we need to see that happening 10,000 times instead of 1000? What about 10,000,000 observations? Will we be more certain of the outcome than we would have been with only a 1000 events?

---

* Mechanics and some other mathematics-dependent fields of physics (such as quantum theory) may be the exceptions here. In these fields, a naïve belief in reality of models (theory 5) seems to prevail over correspondence to factual reality. This is one of the nongeneric situations addressed by Hertz–Rosen modeling relation (see section 4 of this chapter) in which the surrogate formal system is identified with the modeled natural system. In typical situations regarding modeling nature, the two systems (natural and surrogate) should be kept separate.

** That said, the vocabulary and syntax of predicate calculus and quantification can be used as stenographic shortcuts in meta-languages describing all kinds of inferential schemes, including those of pragmatic inference.

Ever since Kant [27], the general answer to these kind of questions is to use reason and experience together in determining the number (and kind) of observations and then transcend all we know plus the outcomes of these observations into a new tidbit of knowledge. That has been one aspect of Kant's and his followers' transcendental methodology. It has also been an underlying principle of heuristics involved in planning experiments in all fields of science since, at least, the Hellenic period in ancient Greece.

The question of how to effectively do the transcending of our existing knowledge in the light of new observations and new inferences is a subject of specific methodology of each field of science and sometimes a style of individual researcher. However, there exist a few general guidelines that can be summarized under the name of pragmatic inference (blend of few types of inference plus a few pragmatic rules of transcending knowledge). Two known examples of types of pragmatic inference, abduction and consilience, are described below.

Abduction, also known as inference to the best explanation, is a derivation of the most plausible premises (explanations) from known alleged consequences. The most general scheme of abduction is:

1. B is a set of observations
2. A, if true, has a good potential to be a valid explanation of B
3. No hypothesis other than A appears to explain B as well as A does.
4. Therefore, A is probably true

C.S. Peirce who coined this term [28] has noticed that abductive reasoning can and often does violate principles of formal logic and set theory. An example of typical abductive reasoning is:

Sentence 1: My bank has many $20 bills.
Sentence 2: I have a $20 bill in my pocket.

Therefore

Explanation: I obtained my $20 bill from my bank.

The obvious logical problem with this explanation is the fact that my bank does not need to be a unique source of $20 bills. We could make the explanation stronger with the help of additional evidence (such as, for instance, nonexistence of other than my banks in my geographic area or some peculiar additional characteristics of the specific $20 banknotes) but generally abductive reasoning tends to violate rules of logic.

Despite the logical weakness of abduction, it is a popular tool of providing plausible explanations. Perhaps the most dramatic examples of acceptance of logically flawed abductive conclusions are superficial statistical inferences such as, for example, this:

Statement 1: A large set (population) X contains a large number of red objects.
Statement 2: A not so large (smaller than X) set Y contains a significantly large number of red objects.

Therefore

Conclusion: Y is a subset (sample) of X (population) or, in other words, X contains Y.

It is obvious that without factual, empirical evidence for Y being a subset of X, the conclusion is false and a statistics done on Y does not need to correspond to statistical properties of X at all. In particular, the elements of Y do not need to constitute a sample drawn from the population X. Despite the misgivings of the foregoing erroneous sampling, the flawed "statistical" inference is commonly performed to demonstrate statistical support for specific experimental conclusions. In other words, abduction is and will be with us for the time to come but we need to follow Peirce's advice and exercise methodological prudence while using it, at least by making sure that sampling for our own statistical studies is done correctly.

Consilience, as we understand it today, is a synergistic combination of two or more (originally unrelated) sequences of inferences that can be transcended into a useful new concept or hypothesis. Statements or intermediary conclusions from one sequence can synergistically enlighten another (otherwise unrelated) chain of inferences and even (abductively) affect the conclusions.* The important aspect of consilience is a semantic and/or pragmatic interpretation (that may or may not be biased by further transcending by the observer's "structuring mind") of observations and syntax-based inferences. In many respects, consilience can be seen as abduction and, even more so, as a way of *knowledge integration* via creation of novel (or just useful) abstract entities. From this perspective, one can distinguish the following types of consilience:

1. Data integration via *naming* (or renaming) and adjusting *reference* — Objects, phenomena, and processes are bound together with concepts as well as relations between concepts. The result of such blend is then phrased in a subset of natural language, such as a jargon of a given discipline. An appropriately implemented metaphor, for instance, can facilitate understanding of a group of phenomena in light of an existing or a brand new conceptual framework alike. Similarly, a properly chosen analogy can catalyze understanding by bridging the phenomena that were unrelated to each other before.

2. Data *compression* via *generalization* — A set, usually a very large one, of individual factual observations can often be described in a brief manner with the use of knowledge from outside this data collection. The best examples here are trajectories of planets that, once discovered, compress the detailed knowledge of a huge number of otherwise unrelated positions of a celestial body in space. Similarly, chemical equations bring our understanding of chemical reactions to an exponentially more satisfactory level than a huge set of individual descriptions of experiments could afford. Numerous other examples, such as gas laws, include the use of compact mathematical equations to describe (model) factual situations that without such formalism would merit a lengthy and often inconclusive description.

3. Action (activity) *iconization* — Advanced use of symbols can, and often does, enhance or redirect understanding. All kinds of schemata (including the overwhelming presence of hierarchies in our definitions and reasoning) can symbolically represent a large body of our knowledge (new and old alike) without any linguistic description of what exactly constitutes this enlightened wisdom. Even a single symbol charged with a distillation of a huge body of shared cultural experience, such as a trash can on desktops of most personal computers, can be used to enhance or redirect understanding without any exhaustive explanation. The essence of iconization is abbreviating a conceptual mindset shared by all potential observers with common experience. The symbols used as icons must be interpreted in the same way by all their users or else the icons will not have

---

* The original concept of consilience has been coined by the "gentleman of science" William Whewell (1794–1866) in his 1840 treaty on the methodology of inductive sciences [29]. In this original version, he discussed only consilience of inductions as a source of unexpected discoveries that could not be attained by simple induction alone. However, the pragmatic principle of synergistic (or inhibitory) influences of sequences of inference need not be limited to inductive inference and for all practical purposes such limits do not exist. As a matter of fact, Whewell's understanding of the notion of induction was reportedly [30] much larger than the standard meaning of this concept would imply. Whewell's "induction" should be understood, in today's terms, as a knowledge integration with the help of old and new facts and concepts configured in an effectively knowledge-enriching manner. Consilience, then, is a lucky combination of such "inductions," which brings a quantum leap of understanding a phenomenon or a process or a class of processes.

their knowledge-summarizing power (and thereby will not be icons in other than an open-interpretation, artistic sense).*

1.4 REDUCTIONISM VS. HOLISM: A REAL ISSUE OR A MERE DETRACTION?

Some time ago I characterized the naive version of reductionism in today's "postmolecular" and "postgenome" biology as follows [10]:

> According to the common-sense, naïve, interpretation of reductionism life should be fully explained via exhaustive studies of systems of differential equations (dynamic systems) but very few, if any at all, biologists would see the point in following this research programme. An extreme version of this paradigm adopts an assumption that phenotype can ultimately be modeled by dynamic systems (i.e. systems of differential equations in which rates, forces, are time derivatives). In a little less extreme version of naive reductionism genes are assumed to prescribe (in an unknown way) the epigenetic rules (also unknown) that in turn control interactions (of unknown nature and number) of proteins (of unknown kind and function) with each other as well as with other (also unknown) ligands.

As we can see from this quite accurate description, naive reductionist agenda, which has been so successful in physics of the past three centuries, is not very convincing in biology in general (not even in the physics-friendly "postgenome" biology).

That said, I hasten to mention that the long-standing debate between reductionism and holism has been complicated by the intensity of political agendas of the most recent two millennia (including the 20th century). Not only are the general methodological programs of natural science difficult to express in language but also the unforgiving political pressures in different historical periods prevented their scholars and savants from having any opinion at all. The lucky ones who actually could express their opinions about their methodological worldviews often did not survive this experience or, if they have, they tended to revoke their original claims under the stimulating ambience of prison cells, torture chambers, or fear thereof. The last few decades of the 20th century went a bit easier on opinionated scientists talking of the glory of reductionism. However, it would be a mistake to say that rare instances of unemployment among (Western) life science faculties as well biologists' "vacations" in remote parts of former Soviet Siberia had nothing to do with attempts to encourage silence about the holistic (systemic) methodological attitudes. In fact, silencing (or enhancing) the scientists' general worldviews continues to go on all the time for the reasons that have nothing to do with science or even with the merits of the silenced (or enhanced) opinions themselves. In the face of all this methodological and ideological mess, a good question to ask is: What is the actual problem to be addressed with the help of learned opinions about the worldview?

As far as systems biology is concerned, the pivotal issue is reportedly a possibility of novelty of systems behavior regarding its ambiance (generalized material and symbolic environment). In principle, the occurrence of new, unexpected systemic characteristics (novelty) could be induced via either emergence or anticipation or both. Emergence is an apparently spontaneous generation of novel systemic properties that could not be predicted from the properties of components (parts) of a complex system alone. Anticipation is characteristic of the so-called anticipatory systems [31]** that contain predictive models of themselves as well as their ambiance and thereby can

---

* Specialists in a given field experience similar observations and use similar tools of interpretation. Because of that, they can effortlessly communicate their opinions to each other but are often misunderstood by nonspecialists.

** An anticipatory system is "… a system containing a predictive model of itself and/or its environment, which allows it to change state at an instant in accord with the model's prediction pertaining to a later instant." [31]. For the outside observer, the net effect of emergence should be the same as that of an anticipatory behavior (i.e., the apparent novelty) but the actual explanation of the phenomenon of novelty is entirely different.

unexpectedly, for the observer, change their current activity (behavior if you will) in response to the predicted future states.

The intense and often politically charged debate mentioned earlier in this chapter was to a great extent due to a fight between believers and nonbelievers in genuine novelty induced by emergence [32]. But were the reasons for this fight or debate scientifically meritorious? Anybody who has ever seen plants emerging from the seeds planted long ago does not have doubt that emergence is a fact. Anybody who ever made a chemical synthesis or even saw a chemical reaction occurring in a test tube does not have doubt that emergence is a fact. As a matter of fact, the phenomenon of emergence appears to be so common in nature that it should pass even the most severe test of systematic doubt. Why then is disbelief in spontaneous generation of novelty (such as emergence) possible? Well, because admitting the fact of emergence (and/or anticipation for that matter) sometimes contradicts other beliefs that their practitioners may find important for right as well as wrong reasons. One of the best characterizations of this situation is given by the dean of modern biology, Mayr [32], while describing properties of emergence:

> During its long history, the term "emergence" was adopted by authors with widely diverging philosophical views. It was particularly popular among vitalists, but for them, as is evident from writings of Bergson and others, it was a metaphysical principle. This last interpretation was shared by most of their opponents. J. B. S. Haldane (1932:113) remarked that "the doctrine of emergence ... is radically opposed to the spirit of science." The reason for this opposition to emergence is that emergence is characterized by three properties that appear at first sight to be in conflict with a straightforward mechanistic explanation: first that a genuine novelty is produced — that is, some feature or process that was previously nonexistent; second, that the characteristics of this novelty are qualitatively, not just quantitatively, unlike anything that existed before; third, that it was unpredictable before its emergence, not only in practice, but in principle, even on the basis of an ideal, complete knowledge of the state of the cosmos.

Reductionism is a collective label intended to mark several worldviews and general research strategies (sometimes distinctly different from each other). Selected aspects of these strategies sometimes indeed reside behind the actual day-to-day methodology of scientific research but sometimes are erroneously believed to motivate scientific methods.

One can roughly distinguish at least the following four variants of reductionism [10,33]:

R1. *Ontological Reductionism* — a worldview that the whole is strictly "nothing but" the sum of parts. One can clearly distinguish two extreme traditions here:

   R1a) A doctrine of *reductive materialism*: All properties of the system should be completely derivable from properties of its parts and therefore emergence or anticipation, as a generation of surprising novelty, could not happen at all.

   R1b) A doctrine of *reductive idealism* sometimes referred to as *physicalism*: All parts of the system fit perfectly the grand "plan" or "design" according to which nature is organized. There is no room for novelty as an attribute of nature. Any impression of novelty should be due to our ignorance about some behaviors of the parts but the systems characteristics should still be (ultimately) a combination of properties of the smallest components.* Ergo, if we happen not to satisfactorily explain the system from the properties of its alleged smallest parts, we would need to look for errors in identification of components or properties and then repeat the process of explaining over and over again. Not surprisingly, reductive idealism is almost identical to the strong version of holism (see the sequel to this section).

---

* Physicalism is *de facto* a mixture of ontological and causal reductionism because it is based on the assumption of bottom-up, upward, causation and the existence of the smallest components (atoms) that cannot be further divided into parts.

Ontological reductionism is probably extinct from science by now but it had played a constructive role as a belief system that fueled the development of mechanics, mechanical devices, and even medicine (anatomy, morphology, cytology) for several centuries.

R2. *Methodological Reductionism* — a research strategy that explores a concept of substituting a given system with a surrogate system. In this variant of reductionism, there should be no problem with accepting emergence as long as it could be explained by the properties of a surrogate system. The methodological version is pragmatically useful in science because it leads to a clear-cut distinction between a natural system (ontological entity) and its description (epistemic entity). This distinction is in fact reflected in a meta-language describing the *Hertz–Rosen modeling relation* [8,31,34] discussed in several places in this volume (including this chapter). In the case of convoluted (complex) systems, the boundary between modeled system and its model (the epistemic cut [35]) is not easy to find and, when found, often leads to paradoxes or inadequate theories. That is one of the reasons why a singe (largest) model cannot possibly be sufficient for an adequate description of convoluted (complex) systems. To approach a satisfactory systems description, a multitude (but an unknown number) of complementary models is needed.*

R3. *Epistemological Reductionism* — a view that the whole can (in principle) be a sum of parts subject to adequacy of definition of "parts" and "sum." Again, there should be, in principle, no problem with emergence as long as it could be derived from properties of an adequately selected set of parts. This version of reductionism is often associated with methodology of physics expressed, for instance, in modeling biological systems in terms of dynamical systems.

R4. *Causal Reductionism* — a view that upward causation ("the whole is there because of the parts") should be legitimate but downward causation ("the parts are there because of the whole") should be forbidden. Here, a genuine emergence is impossible unless we give up the property of an absolute novelty.

Causal reductionism can be seen as a special case of the epistemological one. It is clear by now that this version of reductionism fails in a most visible way in biology after being extraordinarily successful in physics since Newton. An obvious reason for this failure in the case of complex biological systems is the fact the concept of function, and thereby an idea of a goal of the system's activity, is in fact an important ingredient of a plausible biological explanation. A programmatic rejection of the entire idea of downward causation must upset any methodology, even a reductionistic one, which aims at an explanation of system's behavior in functional and not only structural terms. Living things appear complex because they are in fact convoluted in terms of their function.** An orthodox adherence to causal reductionism could arguably be methodologically sound for studies of structure at best. However all attempts known to this author (including his own attempts) of explaining biological function in terms of structure alone lead to inadequate or even irrelevant (trivial) explanations of their behavior (functioning).

---

* The nonexistence of the largest model and the need for a multitude of complementary models is the actual definition of a complex system given by Robert Rosen [8,34].

** That is why the concept of functional *complexity* in biology needs to be separated from the notion of mere complication due to a very large number of objects and relations between them. We often refer to this latter, combinatorics-related, concept as *algorithmic complexity* [36–38] or *compositional complexity* [9,23,39,40] but in the context of systems biology the word "complication" would be much more accurate. On the other hand, the concept of *convolution* reflects well the functional, causal, or inferential intricacy we are facing while modeling biological systems or interpreting their models. In other words, the property of being convoluted (and not the property of being complicated) is an essence of complex biological systems.

Scientific holism appears to be practiced in at least two following variants [41]:

H1. *Strong holism* (ontological) — a view that putting together the parts will not produce the wholes (such as living systems) or account for their properties and behaviors. This extreme view is probably extinct from science by now because of the programmatic rejection of all mechanistic explanations (and modeling) of the emergence and because the latter is assumed to be a "creative" metaphysical principle [32,41] somehow programmed in the nature of things (the cosmos if you will). As I have mentioned before, this version of holism is almost indistinguishable from reductive idealism.

H2. *Weak holism* (epistemological) — a view that the parts of a complex system (such as a living thing) are involved in a multitude of relations that do not exist for the parts in isolation. In other words, properties of the system cannot be derived by just combining the parts. The natural relations between various configurations of parts need to be taken into account, as well. The existence of relations as legitimate subjects of studies opens the door to the concept of novelty because even if the components of the system (the parts) would not change their properties, the relations represent a new quality. Ergo, novelty is not entirely impossible even if, for pragmatic technical reasons, it may be difficult to make sense of. In this respect, overcoming these technical difficulties that hinder explanation of emergence and/or anticipation would belong to science, mathematics, philosophy, and other fields of creative activity that rely on modeling and reasoning.

In another text devoted to systems biology [10] I wrote:

… Most biologists appear to subscribe to a weak version of holistic worldview according to which properties of biological systems cannot be derived from the properties of their subsystems alone. The main reason why such derivation is questionable, even in principle, is the alleged fact that the observer cannot be naturally separated from the observed living system.* On the other hand a forced (i.e. not natural) separation of the observer from the observed (the epistemic cut [35]) can be done within every model of the system under consideration. The price to pay for this necessary separation is an increased risk of generating inadequate models. In these circumstances a reasonable future for systems biology could be invention of reliable methods of measuring the material adequacy of models and definitions. With such methods at hand biologists would be able to effectively address, identify and perhaps even overcome the real misgivings of epistemological reductionism. In particular an accurate evaluation of material adequacy could help formulating research questions in answerable ways ….

One should perhaps add here that an overwhelming majority of practicing scientists (including this author) could be classified as methodological reductionists in their day-to-day research. However, at the same time, they interpret their findings and ask research-related questions with the weak holistic attitude in mind. (That is particularly true of biologists.) Therefore, it is fair to say that only the methodological variant of reductionism and the weak variant of holism are truly relevant to science because they both make room for explanations of novel (emergent or anticipatory) systemic properties. In biology, the weak holism is likely to become the predominant methodological attitude (complementary to the methodological reductionism) because the process of isolating components of organic systems often destroys (or conceals at best) their pivotal properties that were to be exposed and studied in the first place [4,8].

---

* The analogy with complex number, given by Rosen in [20], comes to mind here. A given complex number can be represented as a pair of real numbers but it cannot be derived within any set of individual real numbers (i.e., a complex number C can be geometrically interpreted as a point in R^2 but not in R^1).

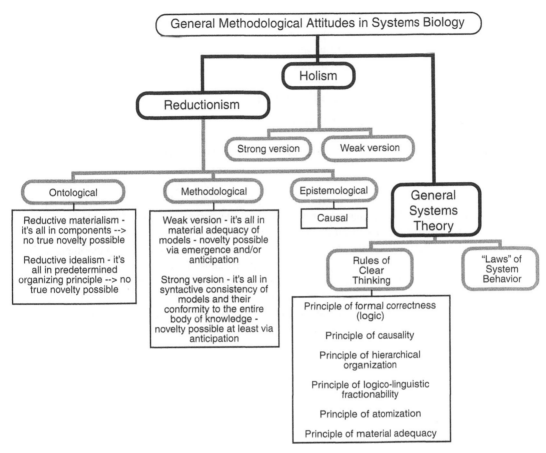

FIGURE 1.1 General methodological attitudes that pertain to systems biology. For simplicity, we assume that rules of clear thinking are a part of general systems theory. This assumption is harmless because the (natural) language and logic reside behind all possible aspects of the reasoning.

Terminological distinctions between main methodological attitudes described thus far in this chapter are schematically summarized in Figure 1.1.

1.5 THE ART OF MODELING

Regardless of the specific mechanism of how the observation is actually sought after, performed, represented, and registered, most individual humans would agree that an observed object, phenomenon, or process should be distinguished from its image (representation) in the observer's mind. In certain circumstances, the representations of observed reality can be included in a (mental) activity that appears to simulate real phenomena or processes. This advanced process of interpretation is often referred to as a model of a specific, named, and observable aspect of reality.

A general research program of how the pairs <observer, observed> behave contains a methodology of modeling *convoluted* (complex) systems. The most general scheme of generating models (the modeling relation due primarily to Rosen [8,31] as well as to Hertz [42], Kant [27], Mach [19,43], and Aristotle [1]) is as technical as it is essential for systems science. Albeit it seems to require a concept of "collective mind" of a class of observers (a *meta-observer* or a *generalized* *observer*), it allows one to discuss convoluted systems (living things would be special cases of those) in a pragmatically and logically clear manner. Needless to say, constraints of the meta-language, in which validity of questions and answers can be judged, constitute a reason for the observer's influence on the reported outcomes of observations.*

1.5.1 MODELS OF CONVOLUTED (COMPLEX) SYSTEMS

1.5.1.1 The Meaning of the Word Model

The word *model* has been used in science in many different meanings. The three most popular general definitions are:

1. A model of a formula of a (formal) language is defined in theory of logic as an interpretation of the language such that the formula comes out true. (Formula is a string of well-formed symbols of an accepted alphabet.)
2. A model of an observable phenomenon or process is another object, phenomenon, or process that usually is easier to control and understand than the original. In this sense, a model is a representation of the observed (subject of observation) designed by the observer in a way that could facilitate further understanding of the subject of observation.
3. A model is a mapping whose domain is an observed real (naturally occurring outside the observer's imagination) object, phenomenon, or process and counterdomain is a formal system with its own alphabet and rules of inference.

It can easily be noted that the metalogical meaning #1 above refers to a concrete realization of abstract things (well-formed strings in a formal language) that can be tested for having certain property (like, for instance, being factually true). The two other notions of a model (#2 and #3) evoke the idea of describing operations that contribute to efficient representing of observed reality by an observer or interpreter of observations. Thereby, the following considerations pertain to the notions 2 and 3 above but to a much lesser extent (if at all) to notion 1.

1.5.2 THE HERTZ–ROSEN MODELING RELATION

The Hertz–Rosen modeling relation [8,31,34,42] is a general protocol of modeling (a meta-model so to speak) based on the assumption that causal entailments within the modeled natural system are "out there" as a property of nature (the facts if you will), while the plausible representations of observable reality can be subject to inferential entailments (inferences) within a suitable formal system. From this perspective, everything the observer can see and analyze is material and symbolic (i.e., not material) at the same time. That is to say the material aspects of the observable reality are complementary to the symbolic aspects thereof and there is no way around this dualism, particularly in biology. One obvious consequence of the coexistence of material and symbolic aspects of observable reality is the distinction between naturally occurring objects (including phenomena and processes) and the laws of nature that "govern" their behavior. (The objects are material all right but the laws are not, even in mechanics). A less obvious scenario can be noted in biology where the actual observed objects are already sophisticated models of reality (such as nucleotide and amino acid sequences of nucleic acids and proteins respectively) and not exactly the generic tidbits of material reality *per se*. As a matter of fact, in order to be able to observe convoluted (complex) systems, the observer usually needs a great deal of preparatory modeling. (That is one aspect of Robert Rosen's [8,31,34] preoccupation with complexity as a generic property of systems that are not simple mechanisms.)

---

* This dependence on language is taken for granted and often forgotten but in the realm of systems biology, in order to discuss the observer's (modeler's) actions, it needs to be reexamined in an entirely new dimension [4–6,10,13,44–47].

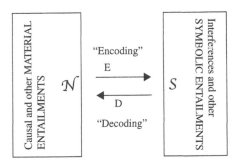

FIGURE 1.2 The meta-model: Hertz–Rosen modeling relation. Selected observations about the natural system N need to be "encoded" into their image in the surrogate formal system S. The image of encoded facts (data) concerning N will be processed within S and then "decoded" back to N. "Encoding" and "decoding" (or representing and counter-representing) are not formally definable within this meta-model. The essence of these operations is to make a semantic and a pragmatic sense to a meta-observer who actually performs the actions of representing and counter-representing. That is to say, the modeling relation is technically a model of all specific acts of modeling (a meta-model) but "encoding" and "decoding" belong to the semantics and pragmatics of the meta-language for describing the meta-models like that (i.e., to a meta-meta-language regarding an object language of the specific models).

The meta-model for modeling itself, the Hertz–Rosen modeling relation, is schematically shown on Figure 1.2. It represents a general protocol that some generalized observer who understands his/her own actions (a meta-observer) would need to follow in order to come out with a materially adequate interpretation of observations of a natural system N. In the most general terms, it can be seen as a schematic description of representing one system by the means of another, surrogate, system in a way that is open to semantic and pragmatic interpretation by a generalized interpreter (observer) who is versed with the object languages of both the actual and the surrogate entity. In principle, several variants of Hertz–Rosen modeling relation are possible but for the purpose of systems biology (and this volume) we will only discuss the meta-model that explores the assumption of partial (local) independence of entailments in the natural system from the entailments in the surrogate system. To further reduce an inconvenient generality of our meta-model, we will also assume (as Newton, Hertz, Mach, Rosen, and many others did before us) that the entailments in the natural system are causal entailments while the entailments within the surrogate system are strictly "mind matters" called *formal entailments* or *inferential entailments* or just *inferences*.

The natural system N and its surrogate S (usually a formal system) need to be brought into a semantic correspondence (congruence) with each other as well as with the operations of encoding N into S and decoding S back into N. This correspondence must be such that the inferences within the formal system will semantically and pragmatically correspond to the entailments in the natural system in the eyes of the generalized observer capable of understanding his or her own activities (a meta-observer). Once the congruence is determined by the meta-observer, we can say that we have found a model of natural system N. Otherwise, a given pseudostate of a modeling relation is a representation of N within S but not a model of N. The congruence is by no means guaranteed and needs to be found by experimenting with adjusting inferences within system S to causal entailments within N.

The foregoing "iterative" process of searching for a model leads to a sequence of representational pseudostates of our meta-model, example of which is shown in Figure 1.3. It can be seen from this Figure 1.3 that all entailments, causal and inferential alike, are ordered in time. Therefore, finding a model of N (i.e., bringing N, S, E, and D in congruence) can only be achieved by a sequence of pseudostates that correspond to moments in time.*

---

* Why is it a case? Well, because both causation and inference are ordered in time, causes by definition precede effects in causal entailments (with the exception of causal loops) while premises precede conclusions in inferential entailments (with the exception of inferential loops).

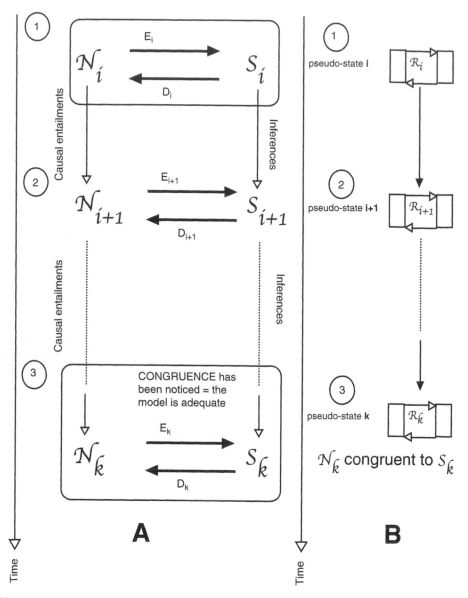

FIGURE 1.3 Modified Hertz–Rosen modeling relation (the meta-model) and its pseudostates. A sequence of pseudostates ends when an adequate representation, a model, of N is found within S (pseudostate marked 3 in figures A and B). That corresponds to a congruence between N_k, S_k, E_k, and D_k. A — subsequent entailment states (the pseudostates). B — simplified view of subsequent representations.

Our discussion thus far indicates that the idea of modeling relation leads us to a language for adequately describing the process of making models. Just like in logic, computer science, and mathematics we need a meta-language to describe an actual (object language) and we need another level of meta-language (a meta-meta-language) to be able to describe the meta-language. From this point of view, the modeling relation corresponds to syntax of a meta-language of making models. Semantic and pragmatic aspects of this meta-language are, however, on a different level that pertains to the modeler. The modeler follows the syntax entailed by modeling relation via representing a natural system in a symbolic form that belongs to a surrogate (formal) system. Then he or she manipulates symbols of the chosen formalism and makes inferences according to his/her interests and expertise (that is a pragmatic aspect of inference). The actual choice of formalism is one clearly nonsyntactitc (i.e., either semantic or pragmatic or both) aspect of this process. The (formal) interpretation of inferences (but not inferences themselves) is clearly a semantic aspect of the process. Another semantic aspect is a decision if the interpretation of formalism should be reinterpreted (decoded) in terms of the natural system that originally induced this formalism. The pragmatic aspect then is the actual decoding back into the natural system. Once again, the sensibility and knowledge of the modeler will determine the outcome of decoding and its meaning (interpretation of the formalism being reinterpreted) in terms of the natural system.

The main lessons that can be learnt from this description are:

1. Syntax of the meta-language for modeling (the modeling relation) will remain the same regardless of specific modelers using it in their specific semantic and pragmatic ways.
2. The specific semantic and pragmatic ways of using the modeling relation will determine which syntactic representations (pseudostates) will bring natural and formal system into congruence (i.e., will be models).
3. The models will typically be independent of each other and not derivable from each other. In rare (nongeneric) situations that one model can be completely derived from another, we can say that they both pertain to a simple (i.e., noncomplex, nonconvoluted) "natural" system such as a mechanical device or a pure formalism.
4. Time matters for modeling! For one thing, the modeling begins with a natural system on the left side of modeling relation and then goes through a series of steps* (entailments and interpretations-encodings or interpretations-decodings), each of which either precedes in time or follows in time some other steps. Another reason for importance of time is the fact (Figure 1.3) that determining congruence of the natural system and the surrogate formal system is the process that begins in certain instance and ends in a later instance (with a model). Finally, the very participation of a natural system in a modeling relation entails the future surrogate systems or future pseudostates of modeling relation (i.e., future pseudostates are likely to be entailed by the modeling relation itself).

1.5.3 Sequences of Representations: The Epistemic Cut

As we have mentioned before, modeling complex systems is intricate because of the very property of complexity and because of the need of the observer (modeler) to communicate (or even think) with a help of the language. Moreover, the specific observers have their favorite ways of interaction with the external world (such as, for instance, through the senses), which, in turn, affects their sensibility regarding potentially observable things, phenomena, and processes. For these various reasons, the actual natural system can be separated from the appropriate (chosen) surrogate system by a sequence of intermediate systems. This situation and some of its consequences are illustrated in Figure 1.4.

The main problem with "encoding" and "decoding" operations in the original Hertz–Rosen modeling relation (Figure 1.2) is to know where the natural system ends and where the surrogate system begins. As we can see from Figure 1.4A–4D, we can almost arbitrarily pick the natural and the surrogate system from the sequence of intermediate representations of the true natural system N. This meta-model reflects the fact that indeed the observer is limited in his or her ability to pick the representations of nature. As we mentioned earlier in this chapter, the limitations come from several sources such as the complexity of natural system itself as well as the observer's dependence

---

* We could call this process a *meta-causation* by analogy to the causation proper, which is always an order of events in time (effects always follow causes.) Indeed, in the meta-language of modeling relation, a given noninitial pseudostate either precedes or follows in time another pseudostate (Figure 1.3). We are also aware of cascades of models (or representations) such that a given system is a natural system for the surrogate system from a moment earlier and a natural system for its own surrogate system. Each system in such sequence of representations is associated with a moment in time (it can be a time interval in an ordered set of nonoverlapping time intervals).

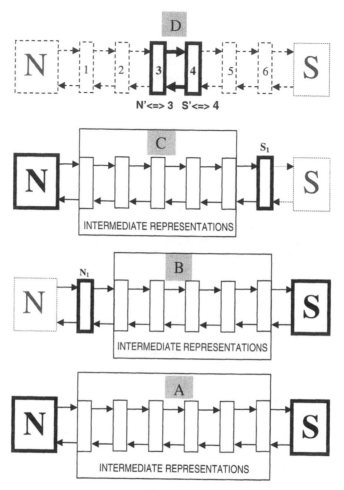

FIGURE 1.4 Sequences of intermediate representations help to explain the concept of *epistemic cut*. A — Natural system N and a surrogate system S are not directly encoded into each other but are connected via a number (6 in this example) of subsequent intermediate surrogate systems. B — What constitutes an observable and modelable natural system is affected by the observer's (experimenter's) sensibility. In this example (Figure 1.4B), the actual natural system N_1 is not the one that should be modeled if the observer was able to handle system N. That is an epistemic cut of type 1: The modeler/observer picks the first modelable surrogate system available (marked as N_1 in our example). C — What constitutes a manageable surrogate (formal in most cases) system is affected by the observer's experience with systems other than the observed natural system N. In this example, the observer, for some pragmatic reasons, picks the surrogate system S1 that is separated from the natural system N by a number of intermediate surrogate systems (4 in our example) but is not the "ultimate" surrogate system S that could perhaps lead to a better formalization or inferences than the chosen S1. That is the epistemic cut type 2: The observer picks the most semantically or pragmatically convenient surrogate system S1. (An example of picking the most elegant notation or formalism comes to mind here). D — Sometimes the observer (modeler) is interested in fidelity of observation(s) and therefore would pick a representation of the modeled system such that there would be no intermediate surrogate systems between the modeled system N′ and its modelable surrogate S′. That is the epistemic cut type 3: The observer picks the first available surrogate of N that he/she can represent by the next surrogate of N in a satisfactorily modelable manner. (In the example on Figure 1.4D N′ and S′ are the third and the fourth intermediate surrogate systems [originally present in Figure 1.4A], respectively.)

on perception via senses, language, knowledge, and experience. To use the modeling relation and eventually find desired models, the observer must choose which of the intermediate systems (Figure 1.4A) will serve as an observable (modelable) natural system and which will be designated as an appropriate (manageable) formal system. The chosen intermediate representations of the natural system, which participate in a given "usable" modeling relation is often referred to as an *epistemic cut** [8,10,35,48].

The epistemic cut is meant to reflect both the complexity of the true natural system and a manifold of observer's limitations. There can be several types of the cut. Three representative examples are shown in Figure 1.4B–D. Finally, it should be noted once again that models resulting from usable modeling relations need not be optimal regarding fidelity of description of "true" natural systems.

The modeling relation and the need for an epistemic cut have consequences in the domain of theory of science. We are aware, for instance, of two major belief systems regarding science in general:

1. The idea of unity of science
2. The idea of separate growth of methodologically distinct specialized fields of research

The belief in the unity of science is a well-known consequence of physicalism as well as epistemological reductionism. It has been particularly disturbing to chemists and biologists until the middle of the 20th century to be instructed by a few neopositivist and physicalist ideologues that the only explanation of complex processes (such as life itself) must come from the lowest levels of organization of matter. This quite poisonous ideology reduced all aspects of natural science to particle physics or a variant thereof. Of course, anyone who was doing research or has just studied science would not take this kind of ideological propaganda seriously because reduction to the lowest levels of organization meant abandoning actual chemical or biological issues at hand and switching to physics instead.

In contrast to the unity of science, the paradigm of independent growth of specialized disciplines has been and continues to be the actual belief of a vast majority of individual scientists. Conformity to the art of modeling and the need for an epistemic cut make it quite obvious that even a single complex system needs to be described by a variety of equivalent complementary models. Because complementary models are not derivable from each other, there is no single largest (or superior) model that would "unify" all explanations of even a single complex process. A single explanatory framework that would unify explanations of all possible systems (complex and simple) is clearly out of question within science. The need for epistemic cut clearly entails a deep, problem-oriented specialization in science while, at the same time, making the idea of unity unrealistic.

I hasten to mention that the foregoing perspectives and views are not meant to contradict opinions of specialized philosophers of science. My views are that of a practicing scientist and I have neither an intention nor expertise to argue a novel philosophical agenda. That said, I would not be surprised if systems biology would inspire serious philosophical thinkers toward developing a systematic methodology of meta-modeling and (epistemic) cutting. Moreover, I would be delighted if this volume (or at least this chapter) could be instrumental in such a task (of inspiring).

1.6 METAPHORS IN SYSTEMS BIOLOGY

It is well known that metaphors play an extraordinarily important (if not pivotal) concept-generating role in all fields of science [6,18,49–52]. Particularly, when we are dealing with processes that are too complex to be easily described or even named, metaphors can help.

---

* Sometimes, the act of choosing the representations is called epistemic cut as well, but it should more appropriately be called "making an epistemic cut" or "epistemic cutting."

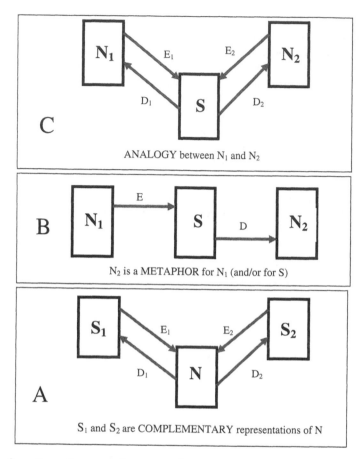

FIGURE 1.5 Configurations of connected representations. A — Two or more different surrogate systems (S1 and S2 in this example) can represent a given natural system. We say that S1 and S2 are complementary representations of N. If there is congruence between N and S1 as well as between N and S2, we can say that S1 and S2 are complementary models of N. B — "Decoding" without "encoding" is characteristic of a metaphor [8]. The figure shows an example of the simplest metaphor in which the participating three systems are indeed subsequent to each other in terms of sequence of encodings. C — If two or more natural systems can be represented by the same surrogate system, we say that the natural systems are analogous to each other. In particular, natural systems represented by the same formalism are analogous.

Figure 1.5 shows the difference between complementary models (Figure 1.5A), metaphors (Figure 1.5B), and analogies (Figure 1.5C). From the perspective of modeling relation, metaphor can be simply defined [8] as "decoding" without the previous "encoding" (Figure 1.5B).

As we can see from the Figure 1.5C, analogies are simply two or more natural systems that are effectively modeled (or at least appear modelable) by the same formal system. An example here can be a pendulum (N1) and an electrostatic condenser (N2); both are usually represented by the same set of differential equations (formal system S). We can notice here that an analogy is also a reversible (bidirectional) metaphor but the reverse does not need to be true (i.e., most metaphors are not analogies).

Figure 1.5A defines the notion of complementarity of representations and models. If two or more surrogate systems (S1 and S2 in our example from Figure 1.5A) can be brought into congruence with the same natural system N, we call S1 and S2 complementary models of N.

I hasten to add here that there are several other definitions of metaphor that are complementary to the one formulated in terms of modeling relation with the help of schemata like the one shown on Figure 1.5B. More precisely, our modeling relation provides a universal syntax of the meta-language that can be used for making semantically specific definitions in appropriate object languages.* Among these specific definitions, the ones adopted from linguistics claim simply a transformation of meaning of symbols (including situations, like in allegories) into a distinctly different domain that may or may not be a member of a larger family of meanings. A more detailed discussion of definitions of metaphors exceeds the scope of this chapter and will need to be delegated to a more specialized future work.

A while ago, in another text [18], I described the relationship between cascades of models and metaphors as follows:

> ... As far as science is concerned metaphors relate to models and their representation in terms of other (new) models. More often than not a given natural system is inaccessible for direct observation (interpretative reasoning included) in which case we need a model to represent it and to talk of it. The model may still be inaccessible for observations in which case we need a model of a model that in turn may require yet another model. The process of modeling consecutive levels of models may continue until we feel comfortable with the sensual and conceptual categories generated by a "final" model in the cascade (series) of models. Of course each time we map a model onto another model we take a risk of missing or misrepresenting some properties of the input model as well as adding new properties which originally were not there. We also take a risk of misrepresenting original relations between properties. That is to say, ideally, our intended mapping should preserve the relational structure of the input model. In practice this invariance of mapping may not always be achieved because of profoundly convoluted (complex) properties of relations that should be preserved or simply because of human errors in perceiving the outcome of the mapping. At this point metaphors enter the methodological picture: instead of modeling a natural system (or its earlier-generation model) we can model its metaphor. The advantage of this deviant but common procedure is the fact that an intuitive understanding of the system provided by a good metaphor helps us to create workable models within which we can ask adequate questions and judge the completeness (or incompleteness) of answers. This in turn is an effective way of deciding if we need further models in the cascade or if we are satisfied with the current model as a "final" one. [Metaphors are particularly useful for representing complex systems or their aspects.] ...

To complete this section, let us consider a few examples of the most prominent metaphors in systems biology.

1.6.1 The Machine Metaphor

A biological object (such as cell, organ, organism, population, and so on) is thought of as a machine (such as a mechanical device, clockwork mechanism, and so on). The simplified configuration of the shades of meanings of this metaphor in biology is shown in Figure 1.6.

1.6.2 The Language Metaphor

Biological objects (such as genomes, proteins, metabolic pathways, and so on) are thought of as texts written in an unknown script over an unknown language. The meanings conveyed by such alleged texts are seen as subjects to "deciphering" by industrious biomedical savants. The simplified configuration of the shades of meanings of language metaphor in biology is shown in Figure 1.7.

---

* Ever since Lakoff and Johnson published their *Metaphors We Live By* [53], the interest of scientists in scientific metaphors has continued to grow. In the time of writing of this text, there should be more than 200 nontrivial publications that deal with metaphors in science alone. The literature on use of metaphors in other fields such as architecture or law (not to mention the most obvious field of fiction writing) is about 20 times larger than that.

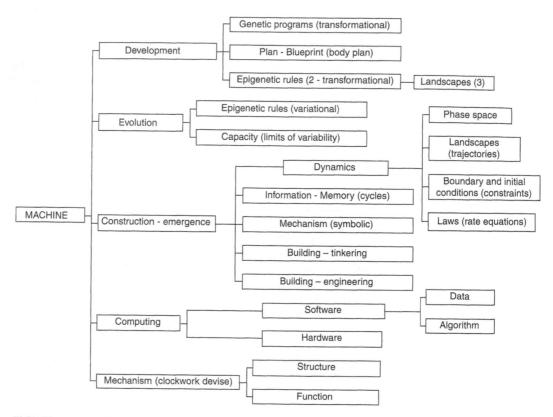

FIGURE 1.6 Machine metaphor in biology. Most characteristic configuration of meanings.

1.6.3 The Organic System Metaphor

Biological objects have genuine organic properties that are responsible for the state of being alive. However, these properties if combined in isolation from living things do not need to lead to the actual living organism. They can simply materialize in a form that, in many respects, is similar to an organism but not an organism itself. Such preorganismal configurations of matter can evolve toward organisms but there exist other, nonorganismal, options for such evolution as well. (Examples from ecology [47] are the most prominent here.)

1.6.4 The Metaphor of Organism

An organism is considered to be a meaningful unit of living matter capable of adaptation and evolution. The organism can be characterized in a similar way as organic system but it also entails its own history and memory thereof. In principle, the organic system plus history would exhaust the main biological meaning of this metaphor. The origin of the term "organism" relates it to machine metaphor.

1.6.5 The Metaphor of Information

The Aristotelian "soul" of things (including living things) refers to their internal organization as a system. Information would be a representable (or communicable) essence of such organization. Specific meanings of this metaphor (the submetaphors) can be found within all other metaphoric concepts mentioned above. For example, Figure 1.8 shows the configuration of meanings within the machine metaphor while Figure 1.7 shows the meanings of "information" within the language metaphor.

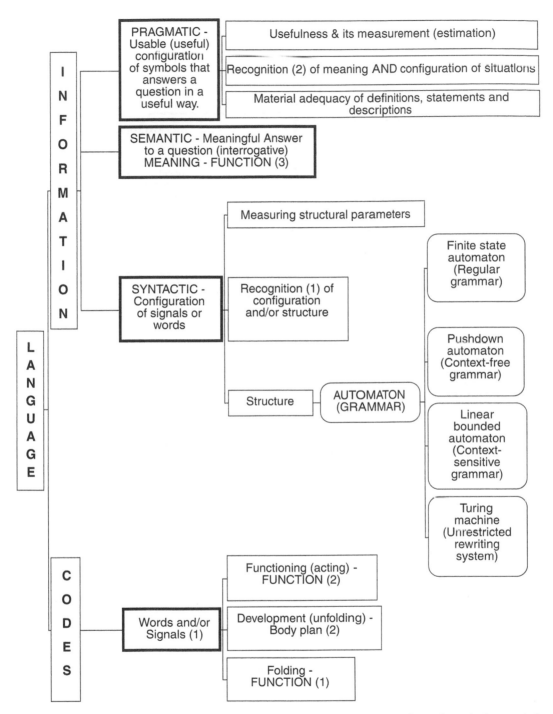

FIGURE 1.7 Language metaphor in biology. Most characteristic configuration of meanings. A characteristic configuration of meanings for the metaphor of information (a linguistic variant thereof) is shown on the top of this tree.

Studies concerning systematic reasoning with the use of metaphors will certainly continue in the future, within systems biology and computer science at large. Ideally, it should be possible to develop a method of selecting (if not designing) metaphors that would pertain to a given factual scenario with the highest adequacy.

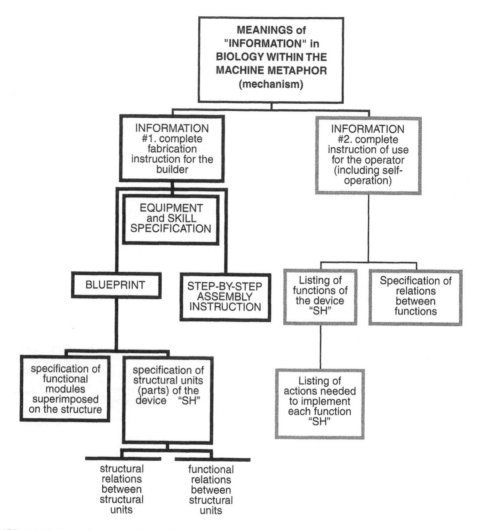

FIGURE 1.8 Information metaphor in biology. Most characteristic configuration of meanings within the machine metaphor. The configuration of meanings within the language metaphor can be found in Figure 1.7. Interestingly, the statistical theory of electromagnetic signal transmission (Shannon theory indicated by symbol "SH") enters this configuration of meanings in only three places. All other shades of meaning of "information" in biology seem to have nothing to do with Shannon theory (that so unwisely has been named by him and others the "information theory"). Readers interested in the concept of "information" in the biological context may wish to consult reference [54] as a reader-friendly but brief expose of the topic.

Our introductory expose of fundamentals of systems biology needs to end here. I believe the readers who have read this chapter will have a firm grasp of basic intuitions and concepts. Reading the sequel to this volume should help in understanding pivotal aspects of today's interdisciplinary research, which are focused on biological questions in a way that evokes diverse styles of systems-level thinking.

REFERENCES

1. Barnes J. Complete Works of Aristotle: The Revised Oxford Translation. Vol. 1 and Vol. 2. Princeton, NJ: Princeton University Press, 1984.
2. Lamarck JB. Philosophie Zoologique. Paris: GF-Flammarion, 1994:1809.

3. Smuts JC. Holism. Encyclopedia Britanica, 1929:640.

4. Mayr E. This is Biology: The Science of The Living Word. Cambridge, MA: The Belknap Press of Harvard University Press, 1997.

5. Konopka AK and Crabbe JC. Practical aspects of practicing interdisciplinary science. Comput Biol Chem 2003; 27:163–164.

6. Lewontin R. The Triple Helix: Gene Organism and Environment. Cambridge, MA: Harvard University Press, 2000.

7. Lewontin R. It Ain't Necessarily So: The Dream of the Human Genome and Other Illusions. 2nd ed. New York: New York Review Books, 2001.

8. Rosen R. Life Itself. New York: Columbia University Press, 1991.

9. Konopka AK. Theoretical molecular biology. In: Meyers RA, ed. Encyclopedia of Molecular Biology and Molecular Medicine. Vol. 6. Weinheim: VCH Publishers, 1997:37–53.

10. Konopka AK. Systems biology: aspects related to genomics. In: Cooper DN, ed. Nature Encyclopedia of the Human Genome, Vol. 5. London: Nature Publishing Group Reference, 2003:459–465.

11. Bertalanffy von L. Modern Theories of Development: An Introduction to Theoretical Biology. Oxford: Oxford University Press, 1933.

12. Bertalanffy von L. General System Theory: Foundations, Development, Applications. New York: George Braziller, 1969.

13. Bertalanffy von L. Perspectives in General System Theory. New York: George Braziller, 1975.

14. Ruelle D. Chance and Chaos. Princeton: Princeton University Press, 1991.

15. Wolfram S. Cellular automata as models of complexity. Nature 1984; 311:419–424.

16. Thom R. Structural Stability and Morphogenesis. Boulder, CO: Westview Press, 1972.

17. Pattee HH. Hierarchy Theory. New York: George Braziller, 1973.

18. Konopka AK. Grand metaphors of biology in the genome era. Comput Chem 2002; 26:397–401.

19. Mach E. Contributions to the Analysis of the Sensations (English edition). La Salle, IL: Open Court, 1897.

20. Rosen R. On the limitations of scientific knowledge. In: Casti J and Karlqvist A, eds. Boundaries and Barriers: On the Limits of Scientific Knowledge. Reading, MA: Addison-Wesley Publishing Company, Inc., 1996.

21. Tarski A. The concept of truth in formalized languages. In: Corcoran J, ed. Logic, Semantics and Metamathematics. Indianapolis: Hackett, 1933:152–278.

22. Konopka AK. All we need is truth. Comput Biol Chem 2004; 28:1–2.

23. Konopka AK. Sequences and codes: fundamentals of biomolecular cryptology. In: Smith D, ed. Biocomputing: Informatics and Genome Projects. San Diego: Academic Press, 1994:119–174.

24. Konopka AK. Selected dreams and nightmares about computational biology. Comput Biol Chem 2003; 27:91–92.

25. Hume D. Enquiries Concerning Human Understanding and Concerning The Principals of Morals. 3rd ed. (With text revised and notes by Nidditch PH.) Oxford: Clarendon Press, 1981.

26. Popper K. The Problem of Induction in Popper Selections. (Miller D, ed.) Princeton, NJ: Princeton University Press, 1985.

27. Kant I. Critique of Pure Reason. 2nd ed. (Norman Kemp Smith transl. Engl. 1929.) London: Macmillan & Co. Ltd, 1929:1787.

28. Peirce CS. (1958) Vol. 2, paragraph 511, 623; Vol. 5, paragraph 270. In: Hartshoren and Weiss, eds. 1958.

29. Whewell W. The Philosophy of the Inductive Sciences. London: Frank Cass & Co. Ltd.

30. Ruse M. The scientific methodology of William Whewell. Centaurus 1976; 20:227–257.

31. Rosen R. Anticipatory Systems. New York: Pergamon Press, 1985.

32. Mayr E. What Makes Biology Unique? New York: Cambridge University Press, 2004.

33. Peacocke AR. Reductionism: A Review of the Epistemological Issues and Their Relevance to Biology and the Problem of Consciousness. Zygon 1976; 11:307–334.

34. Rosen R. Essays on Life Itself. New York: Columbia University Press, 2000.

35. Pattee HH. The physics of symbols: bridging the epistemic cut. Biosystems 2001; 60:5–21.

36. Chaitin GJ. On the length of programs for computing finite binary sequences. J Altern Complement Med (Surrey) 1966; 13:547–569.

37. Kolmogorov AN. Three Approaches to the Definition of the Concept "Quantity of Information." Problemy Peredachi Inform (Russian) 1965; 1:3–11.

38. Solomonoff RJ. A formal theory of inductive inference. Info Contr 1964; 7:224–254.

39. Konopka AK and Owens J. Non-contiguous patterns and compositional complexity of nucleic acid sequences. In: Bell G and Marr T, eds. Computers and DNA. Redwood City, CA: Addison-Wesley Longman, 1990:147–155.

40. Konopka AK and Owens J. Complexity charts can be used to map functional domains in DNA. Gene Anal Technol Appl 1990; 7:35–38.

41. Simon HA. The Sciences of the Artificial. Cambridge, MA: The MIT Press, 1996.

42. Hertz H. The Principles of Mechanics. New York: Dover, 1994:1894.

43. Mach E. The Science of Mechanics (English transl.). 6th ed. La Salle, IL: Open Court Publishing Company, 1960:1883.

44. Konopka AK. "This Is Biology: The Science of the Living World" (Book review by Ernst Mayr). Comput Chem 2002; 26:543–545.

45. Pattee HH. How does a molecule become a message? In: Lang A, ed. 28th Symposium of the Society of Developmental Biology. New York: Academic Press, 1969:1–16.

46. Rosen R. What is Biology? Comput Chem 1994; 18:347–352.

47. Ulanowicz RE. Life after Newton: an ecological metaphysics. Biosystems 1999; 50:127–142.

48. Pattee HH. Causation, control, and the evolution of complexity. In: Anderson P, Emmeche C, Finnemann N, Christiansen P, eds. Downward Causation. Aarhus, Denmark: Aarhus University Press, 2000:63–77.

49. Ogden CK and Richards IA. The Meaning of Meaning. New York: Harcourt-Brace, 1923.

50. Sacs S. On Metaphor. Chicago: The University of Chicago Press, 1979.

51. Torgny O. Metaphor — a Working Concept. Stockholm, Sweden: KTH, Royal Institute of Technology — CID, 1997.

52. Lakoff G. The contemporary theory of metaphor. In: Ortony A, ed. Metaphor and Thought. 2nd ed. Cambridge: Cambridge University Press, 1993:11–52.

53. Lakoff G and Johnson M. Metaphors We Live By. Chicago: The University of Chicago Press, 1980.

54. Konopka AK. Information theories in molecular biology and genomics. In: Cooper DN, ed. Nature Encyclopedia of the Human Genome. Vol. 3. London: Nature Publishing Group Reference, 2003:464–469.

2 Understanding through Modeling

A Historical Perspective and Review of Biochemical Systems Theory as a Powerful Tool for Systems Biology

Eberhard O. Voit and John H. Schwacke

CONTENTS

ABSTRACT

The chapter reviews the history of systems approaches to biology. It places particular emphasis on biochemical systems theory, which has unique features that provide superb tools for modern systems biology.

2.1 INTRODUCTION

Science is the attempt to make the chaotic diversity of our sense experience correspond to a logically uniform system of thought.

Albert Einstein (1879–1955)

Over 5000 yr ago, some clever Egyptians figured out how to make bread and beer. The bread was probably leavened by a mixture of yeast and lactic acid bacteria, both of which were ingredients in the beer mash and foam that were used as a starter culture for bread production. The twofold biotechnological production was quite an ingenious accomplishment of bakers and brewers, who learned, with increasing sophistication, how to affect the processes by varying the amounts of flour, water, and starter, temperature and other conditions, even though they did not even know what the active ingredient in beer foam was until the middle of the 19th century [1].

Currie [2] is to be credited with the first significant steps of optimizing the biotechnological production of citric acid, using strains of the black fungus *Aspergillus niger.* His extensive studies of strains and culture media at the beginning of the 20th century were so successful that they allowed the U.S. to become self-sufficient and even export citric acid to Europe. Since Currie's seminal work, probably a thousand or more scientific studies have been published and several hundred patents issued.

Today, when we discuss the role of systems biology, and of mathematical models in particular, we often begin by emphasizing that we seek an understanding of how biology works [3,4]. Did the Egyptians not understand baking and beer production? Did Currie not understand citric acid production by the mold? Do we understand it now? These are simple yet puzzling questions that force us to define what we actually mean by *understanding*. Just by pondering the early examples above, we see that there are clearly different degrees of understanding that have to do with knowledge of biological details and mechanisms. What may not be immediately evident is that understanding is also subtly but crucially shaped by the philosophical surroundings and the cultural and scientific milieu of every society's time and place in history [5]. To appreciate our current concepts of understanding and the emerging biological thinking in terms of systems, it is useful to traverse the history of science in a flash and to identify turning points and shifts in thinking that changed the ways with which biology and medicine were approached. By reviewing the different eras of scientific pursuit, it will become clearer why a systemic approach to biology is not simply the next fad along the path of the traditional experimentation of the previous century, but a fundamental shift in mind-set.

Following the brief historical review, we will outline challenges that systems biology has to overcome, propose concepts that can address these challenges, show some specific examples, and discuss our expectations for the short-term future. In all these considerations, we will focus almost exclusively on modeling aspects, and even within this limited domain, we will restrict ourselves to just a few concepts and methods that address dynamical systems and models that are scaleable to biological networks of realistic size. This limited scope of our presentation is particularly blatant in our neglect of describing modern high-throughput experimentation, which is an absolute requirement for most of the modeling and mathematical systems analyses. We will mention some experimental techniques that appear especially promising for modeling purposes, but otherwise refer the reader to the growing literature in the area.

2.2 HISTORICAL BACKGROUND

Trying to understand natural phenomena in biology and medicine, has probably been a challenge to the mind as long as humans have had conscious thought. Today, we argue within the logical framework of causality and are concerned with problems of complexity that stem from the enormous numbers of constituents even within a single cell and from the intricacies of mechanisms at all levels of biological organization. In ancient times, these concerns were not of particular interest. Instead, what dominated the thinking about nature from antiquity well through the Middle Ages was, first, the general philosophy of nature and, second, an entirely pragmatic quest for medical and pharmacological answers to problems of disease. The philosophical thinking about nature strongly emphasized the quest for origin and matter. It explored the implications of five fundamental ideas that continued to determine mainstream thinking about biological science for 2000 yr. These ancient ideas were documented in China, India, and Greece, accepted with slight extensions throughout the Arabic world between 600 and 1200 A.D., and remained at the center of scientific thinking until the Renaissance. They are described in splendid detail in the work of Leicester [6], from which much of the following is excerpted.

The first idea held that all consisted of *elements*, or *atoms*. The second idea suggested that *transmutations* transformed the elements into each other. The third idea described a *dynamic equilibrium* of all processes in the universe. The fourth idea postulated a strict correspondence between macrocosm and microcosm, and the fifth idea was the concept of *teleology*, which maintained that every thing in nature had a purpose or was created for a purpose.

Eastern and Western philosophies alike postulated elements, of which all was composed, sometimes even emotions such as love. Chinese philosophers suggested five types — water, fire, wood, metal, and earth — but maintained that for most properties in the physical world, substances were not as important as operations. In Indian philosophy, the world was made of water, fire, and wind. Water was considered a liquid energy of life and, in the form of phlegm, lubricated the body and held it together. Thales (624–546 B.C.) believed that water was the essence of everything. It was the element of which all was composed and to which all returned. Fire created body heat and was found in the body as bile, whose function it was to extract energy from food and process it. Wind and breath were responsible for the voluntary and involuntary movements of the body parts and for the exchange of messages. Empedocles of Sicily (492–432 B.C.) postulated that the four elements ("roots") fire, air, water, and earth moved around according to a divine, mechanistic law. All matter and all organisms were composed of the four elements, and different combinations of their amounts explained the differences in species. For instance, fire dominated in birds, which explained why they were able to fly. Aristotle (384–322 B.C.) based nature on four principles, matter, form, movement, and purpose, and discussed these within this theory of logical causation. For specific arguments, he accepted Empedocles' four elements, but added *pneuma* (breath of life) as the fifth element (*quintessence*) of which stars were made. Aristotle used additional elements, such as bones, flesh, and soul, the latter of which he alleged to be responsible for the distinction between life and death.

With variations, the atomistic doctrines remained in effect for almost 2000 yr. During the great period of Arabic science (800–1200 A.D.), Aristotle's teachings were still considered dogma, but by then the atoms were thought to have not only innate properties, but also attached "accidents" such as rest, motion, life, death, ignorance, and knowledge. Even in the 18th century, the idea of a few elements was not completely abandoned. At that time, the elements were spirit, phlegm, oil, salt, and earth.

In addition to elements, the ancient thinkers postulated various transmutations that provided "mechanistic" explanations of metabolism and life. In fact, the word used by the Greeks was *khēm(e)ia* ("chemistry"), which denoted the art of transmutations practiced by the black (khem) people of Egypt. In Aristotle's teaching, the fundamental biochemical process was *pepsis* (coction) of the elements in the form of heat, cold, moisture, and dryness. A specific transmutation that was taught for many centuries was that of blood into flesh. The alchemists of the Middle Ages still used Aristotle's "equations": Heat + Dryness = Fire, Heat + Humidity = Air, Cold + Dryness = Earth, and Cold + Humidity = Water. Furthermore, water was thought to transmute earth into a liquid, but then this liquid transmuted back into earth. The ultimate goal was to extract, with precisely executed experiments at astrologically perfect times, the quintessence — the fifth eternal element.

While transmutations were continually changing the elementary composition of an organism, the organism and the universe in Eastern as well as Western philosophy remained constant overall. Illustrating this dynamic equilibrium theory, Heraclitus explained: "Change is essential and everywhere: You can never step twice in the same river, for fresh waters are ever flowing in upon you; fire burns continually, but is always different in its sources and products." The elements fire, water, and earth that comprise the human body constantly change into one another, but the body as a whole does not appear to change, even though it was acknowledged that minor oscillations in the interconversion of the basic substances could account for such phenomena as sleep and wakefulness."

Of course, nobody had actually seen fire or air as driving forces of metabolism, but these elements were postulated by the need to fulfill particular functions and by the idea that macroscopic processes also governed the microscopic roles of metabolism. The living human body is warm, it was argued, and the simplest macroscopic way to explain warmth is fire; thus, fire was postulated as the force warming the body. Today, we would have our doubts about this line of argument, but because it was logically defendable, the hypothesis was satisfying to the Greeks and others who followed their doctrines. This is probably the greatest difference between scientific thinking in antiquity and modern times: The Greek thinkers were extremely curious and logical and loved to develop consistent theories explaining why things behaved the way they did. But in contrast to today's scientific thinking, they did not consider it necessary to test theories against experiments, as long as the logic of the explanation was intact. By the same token, they did not see a need to distinguish or decide between competing theories, as long as these could be logically defended. The concept of systematically using controlled experiments to assess the validity of theories did not come to fruition until the late Renaissance [7].

A real change in the way scientists thought about nature came with a period of confusion and transition during the 16th and 17th centuries. Science could not quite let go of the concepts of a few universal elements and the correspondence between macrocosm and microcosm, and even the famous nonconformist physician and pharmacologist Theophrastus Bombastus von Hohenheim (1493–1541), better known as Paracelsus, worked with a triad of "primitive materials," which were sulphur, representing all vapors, mercury, representing combustibility, and salt, representing solidness. Nonetheless, nagging doubts began to emerge about Aristotle's ideas, and key elements of the ancient doctrines were now challenged with arguments of the new scientific thinking. The Belgian scientist Jan Baptista van Helmont (1577–1644) boldly contested Aristotle's suggestion that heat was the mechanism of digestion by arguing that food in warm places just spoiled and that fever did not appear to improve digestion.

Those were convincing arguments, and they slowly caused science to experience a shift from the ancient concepts toward the more modern scientific concepts of controlled experimentation. The idea of general concepts made place for specific hypotheses that were tested in decisive experiments and mathematical theories. It became commonplace to dissect corpses and to execute animal experiments. Instead of viewing the body as an integral entity, individual organs or functions became the focus of investigation. Using Kuhn's [8] terminology, science of this period underwent a paradigm shift in which the old conceptual models were replaced with new ideas and new ways of performing science. This shift was very slow and painful, especially since it conflicted with church teachings, but had to occur since the old paradigm encountered obvious failures and irreconcilable inconsistencies.

The envy of biologists and chemists at the time was astronomy, which viewed the physical universe as an exquisitely designed giant mechanism that obeyed elegant deterministic laws of motion [9] and was characterized by precision and predictability. Aided by new mathematical concepts like differential calculus, astronomy allowed the description of natural phenomena with unprecedented rigor, and scientists began to discard theories — for instance, those pertaining to theology or the concept of the soul — which either could not be tested against observations or failed the test. This scrutiny gained acceptance as the new norm, because it was considered the reason for the great strides in astronomy. Astronomy was being formulated quantitatively with the new tools of science, and its miraculously precise predictions resulted from well-controlled experiments and stringent mathematical analyses, which suggested that chemistry and biology should attempt the same. Textbooks began to characterize chemical compounds according to their measurable physical and chemical properties. Blood pressure measurements became a standard diagnostic tool, and the old-fashioned visual uroscopy was replaced by quantitative urine analysis, since it was recognized that it provided quantitative information about the nature of changes in the body [6].

Quantification was the first significant ingredient of the rapidly accelerating exploration and discovery of mechanistic laws governing biochemistry and physiology. The formulation of kinetic phenomena in terms of enzymatic rate laws became a standard procedure in biochemistry, and the evaluation of parameters such as the maximum velocity and the Michaelis constant became the objective of thousands of biochemical studies. In a similar way, physiological research concentrated on providing mechanistic explanations in mathematical form and pursued physicochemical concepts of oxygen utilization, thermodynamics, mass transfer, and neurotransmission.

The second crucial ingredient of the success of science in the 18th, 19th, and 20th centuries was the concept of *reductionism*. Paralleling the search for mechanistic explanations of physiological and biochemical processes was the idea that, to understand an organism, one would have to understand its organs, and to understand those, one would have to study tissues, cells, and subcellular components. Ultimately, one would arrive at the most fundamental elements and processes, and once all these elementary units were known, one could use this knowledge to reconstruct cells and organisms and, hence, understand life. This approach of reductionism was diametrically opposite to the old holistic thinking that required that the entire organism be considered within its environment if one wanted to explain metabolism and disease, as the father of medicine, Hippocrates, had suggested. Now, the reductionist goal was to find the most basic constituents that made an organism work.

Anatomical and physiological similarities between different species, even between animals and plants, suggested that it might be useful to begin the search for the secrets of life with convenient, simple organisms, such as *Escherichia coli* and *Drosophila melanogaster.* The requirement of experimental controls and reproducibility usually eliminated analyses of whole organisms, and the gold standard of biochemical and physiological experimentation became instead the well-controlled *in vitro* experiment. With the growing sophistication of methods and technologies, research concentrated on increasingly specific details. The analysis of overall organ function shifted toward specific physiological processes and the biochemical characterization of pathways. The regulation of pathways was found to be intimately related to enzymes and factors, and it became an objective to find out how these were regulated, which genes controlled them, and what the sequences of these genes were. The approach of reductionism directly led to the present-day mainstream experimentation in the life sciences and provides the rationale for endeavors like the Human Genome Project.

2.3 BEYOND REDUCTIONISM

While biology became extremely successful in teasing out intricate features at smaller and smaller scales, it turned out that reductionist methods, even if executed to perfection, could not solve all problems. Reductionism was creating a parts catalog, but as Henri Poincaré (1952) said: "The aim of science is not things in themselves but the relations between things." A crucial aspect was missing. Even equipped with knowledge about many of the constituents of cells, it remained impossible to predict with reliability how an organism would respond to untested conditions or stimuli. Thus, even the rapidly growing amount of data and experience with microorganisms and cell lines did not yield the desired true understanding of biological phenomena. Very slowly, it became apparent that another paradigm shift was necessary. New laws had to be postulated — not laws of "life forces" or teleology but laws of integrated systems and of organized complexity — one step beyond the Newtonian view of organized simplicity, and two steps beyond the classical worldviews of divinely ordered or imaginatively envisaged complexity [9]. In Wolkenhauer's [10] estimation, one of the most significant achievements of the systems biologist Robert Rosen in the 1960s may have been his introduction of biological theory sufficient to investigate final causation without implying teleology.

Key concepts of the shift toward objective system thinking were not an invention of the 21st century, but can already be found in the seminal work of isolated biologists, physicists, and mathematicians, most notably Lotka [11] and Ludwig von Bertalanffy [12–14], who laid the foundations of modern systems science and stressed mathematics as the language of choice for describing biological systems and their dynamics. Even though many of von Bertalanffy's insights regarding open and closed systems, perturbations, stability, and dynamic equilibria [13] are still valid and may almost sound commonplace today, they were received all but warmly by his peers and by and large ignored by the community of biologists. Things improved somewhat for systems analysis in biology in the late 1940s, when several fundamentally novel concepts appeared almost simultaneously. Wiener's cybernetics [15], Turing's ideas of automata and computability [16], von Neumann and Morgenstern's game theory [17], and Shannon's information theory [18] enjoyed considerable interest in engineering and various areas of the physical sciences, and eminent mathematicians and engineers pushed for the development of systems theory [10].

Some forward-thinking scientists, such as Rashevsky [19], Ashby [20], Rosen [21,22], and Mesarović [23,24], actually proposed applying systems concepts to biological phenomena. Weaver [25] proposed a science of complexity in biology. Rashevsky even developed a doctoral program in mathematical biology at the University of Chicago and in 1939 created the *Bulletin of Mathematical Biophysics.* Applications often focused on kinetic phenomena in biochemistry, questions of growth, development and pattern formation, and the nervous system. As early as 1940, Rashevsky

[19,26] and Turing [16] proposed abstract systems of two morphogens with autocatalytic features that described features of organismal development. In a similar vein, Kacser [27] demonstrated that even very small networks of biochemical reactions can exhibit switches from one steady state to another, for instance, by a simple change in temperature or by a change in a single influx or efflux. These mechanisms created the conceptual basis for the famous work of Gierer and Meinhardt [28] on the development of hydra and other organisms, which has been dominating the study of biological pattern formation for the past 30 yr [29].

Goodwin [30,31], Heinmets [32], and others began developing kinetic models of gene regulation through induction and repression, describing in mathematical terms the ideas of Jacob and Monod [33] on the functioning and regulation of operons in bacteria. In the same context, Rosen [22] discussed the importance of negative and positive feedbacks in living systems and developed abstract, algorithmic ideas of cellular regulation.

Based on the theory of finite automata and Turing machines, Stahl and Goheen [34] investigated "algorithmic cell models," in which cellular control mechanisms were represented through the logical manipulation of large molecules such as DNA, RNA, or proteins. These models contained dozens of enzymes and products and allowed rather complex biochemical simulations with different initial conditions [35]. Automata were also shown to be useful for the formation of complicated patters, when each "cell" was governed by rules that accounted for the cell's position on a grid [36,37]. These ideas are still popular today, as exemplified in John Conway's "Game of Life" [38]. Apter [39] used growing automata nets to explain spatial differentiation and development, and Sugita [36,37] developed digital models of regulatory gene circuits, based on logical functions such as "and," "or," and "not" and showed their equivalence with McCulloch–Pitts neural nets [40]. In addition to these early systems analyses, several biomathematicians, notably Rashevsky [41] and Hodgin and Huxley [42], worked on issues of signal transduction in nerve cells. In 1961, the *Journal of Theoretical Biology* was born.

Thus, many seminal ideas of today's systems biology emerged during this period. But the time was just not yet ripe for mainstream acceptance of systems thinking in biology. Most biomathematicians continued to be isolated and biology remained to be synonymous with laboratory work. As is typical for every major paradigm shift, the community was not enthusiastic about changing the proven and successful ways of doing science. The tradition of reductionism continued to be the overwhelmingly dominant approach throughout the 20th century.

Given the huge success of reductionism, one must indeed ask why any biologist should do anything but reductionism. The answer is twofold. An almost practical aspect is that reductionist research is generating so many data that it is becoming mandatory not just to collect and store these data in databases, but to develop a functional context within which each experimental result becomes meaningful. Data must become information [43]. Without a functional context, data are isolated factoids, simple descriptions of features, and only their integration within their physical, spatial, and functional surroundings yields insight and, ultimately, knowledge and understanding. As Davenport [44] noted: "Data are not information. Information is data endowed with relevance and purpose. Knowledge is information endowed with application. Wisdom is knowledge endowed with age and experience."

Of course, experimentalists never operate in a vacuum. There exists probably no pure reductionist, and every experimentalist is aware of the importance of scientific context. In fact, every experiment is necessarily guided by a mental model that serves as the basis for the formulation of hypotheses. This model is certainly not always mathematical and may not even be explicit, but it may nevertheless be very sophisticated and complex. It could well be the characterizing feature of superb experimentalists that their mental models are crisp, predictive, testable, comprehensive, and very often accurate. What is emerging with the new scientific paradigm of systems biology is therefore not so much a shift away from current experimentation or reductionism, but a more structured, mathematical formulation of hypothesis-generating, predictive models and a type of large-scale experimentation that targets comprehensiveness, if not completeness.

A deeper aspect of the shortcomings of reductionism is the observation that components of systems often behave differently *in situ* than in isolation. Sometimes, these differences are insignificant, but in other cases they may very drastically affect the role and function of a component. If so, any purely reductionist approach must fail, because its key tenet is complete isolation and perfect purification. To put it more positively, reductionism must be accompanied by a phase of reassembly of components [5,45]. It is mandatory to pay attention to the interactions and relationships among components, wherever possible, in their natural context and environment.

2.4 CHALLENGES

If this step of reassembly makes so much sense, why has there been such reluctance to accept ideas of systems thinking, as von Bertalanffy had proposed it many decades ago? Outside the inertia that delays any paradigm shift, the dominant reason for not studying whole systems is the enormous complexity of living cells and organisms, which may be characterized by three aspects. First, even the simplest cells consist of many more components than our minds can keep track of. A simple yeast cell has over 6000 genes; humans have proteins numbering in the hundred thousands. Even naming such vast quantities and remembering the names is a challenge.

Second, these components interact with each other in multitudinous ways, directly or indirectly. Simple mathematics tells us that the number of possible connections may easily grow quadratically with the number of components, even if there is only one type of interaction between each pair of players. However, it is well known that intra- and intercellular connections are of very many different types and functions, many of which we do not yet understand. Furthermore, the same protein may have entirely different roles, and the same gene may be spliced in different ways, thereby coding for different end products. Lipids are important building blocks of membranes, but they may serve also as signaling molecules. Reductionist investigation readily discovered their important role in membranes, but a full characterization of their signaling function will require more global study *in vivo*.

Not only can the same components have different functions, most of the important cellular control systems also exhibit considerable redundancy, where particular functions can be accomplished by alternative sets of components. As a consequence, experiments knocking out one component at a time may never reveal the underlying control structure. Complicating the situation even further, biological systems have stochastic features, which may, for instance, be caused by small numbers of molecules and their spatial dispersion throughout the cell. This stochasticity is difficult to grasp intuitively and mathematically and seriously confounds predictability. There is no doubt that biological networks are structured and regulated in more complex ways than man-made systems and daunting both in size and complexity.

Third, the interactions within biological systems are seldom linear, with the consequence that increasing an input does not necessarily evoke a similarly increasing output. In linear chains of causes and effects, we easily predict the response to a perturbation, but this is not so in nonlinear networks. Increases in substrate lead to faster growth and higher yield in a microbial population, but only to a point. It is very difficult to predict the threshold where this monotonic direct relationship ceases to hold. Only a small difference in sun exposure changes a tan into sunburn. Unaided by mathematical and computational analysis, we even have difficulties reliably to predict the effects of simple feedback inhibition, which is one of the ubiquitous mechanisms in biological regulation [46]. It is similarly difficult to assess the advantages of inhibition of product formation as opposed to activation of its degradation, or the advantages or disadvantages of different types of gene regulation in a given situation [5].

As a specific example, consider the simple pathway in Figure 2.1, which illustrates an unbranched pathway with regulation [47]. The initial substrate X_0 is converted into X_1 and then into X_2, which is removed from the system. The concentration of X_0 is maintained outside of the

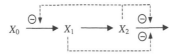

FIGURE 2.1 Simple linear pathway with several inhibitory signals that make predictions difficult.

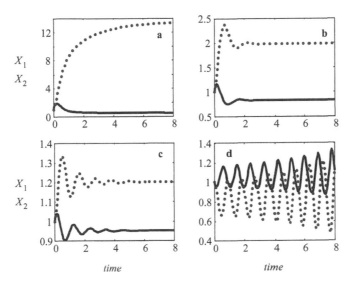

FIGURE 2.2 Time courses of X_1 (solid) and X_2 (dotted) for different values of X_0 (a: 2.0, b: 1.2, c: 1.05, d: 0.95).

illustrated system. The conversion of X_0 to X_1 is inhibited by X_2. The conversion to X_2 is unregulated and depends only on the concentration of X_1. The degradation of X_2 and its removal from the system are inhibited by both X_1 and X_2. While apparently rather simple, it is difficult to predict how the system would respond to slight perturbations, for instance, in the level of X_0. Indeed, this is not merely a matter of purely thinking hard, but depends on the numerical characteristics of the involved processes. In the illustrative numerical set-up of [47], the responses of the system are dramatically affected by the amount of material flowing into the system. For $X_0 = 2$, the system approaches a steady state in a simple, almost monotonic fashion (Figure 2.2a). By decreasing the input (for instance, $X_0 = 1.2$ or $X_0 = 1.05$), the system starts behaving differently, exhibiting damped oscillations (Figure 2.2 [b and c]). If the input value drops below one, the steady state of the system is no longer stable, and the unstable state is surrounded by a stable limit cycle oscillation (Figure 2.2d). Theory shows that even a minute change in X_0 from above one to below one renders the steady state unstable and leads to the emergence of a limit cycle that simply did not exist before.

Following Caesar's motto to divide and conquer, we are trained to apportion big problems into manageable tasks, and this strategy is quite successful if the system is linear. However, for nonlinear processes, subdivision into subsystems becomes problematic when important structural or functional connections are disrupted in the dissection process. Similarly, much can be learned from isolating a cellular structure and studying its features, but as the introduction or extinction of a species in an ecosystem has far-reaching consequences that are difficult to predict, it is hard to make reliable prognoses of what would happen if this cellular structure were changed or missing in the living cell. Thus, a complete understanding of the function of molecules or cellular structures can seldom be achieved by a reductionist analysis alone. In addition to characterizing the properties of cellular features, an integrated analysis that addresses the functioning of these features in their natural contexts is needed. This type of integrated analysis is at the heart of systems biology.

2.5 RECONSTRUCTION

The paradigm shift from reductionism to a systems approach does not mean that well-controlled *in vitro* experiments should be abandoned. Quite the opposite is true. Only if we are equipped with detailed knowledge about the constituents of a cell is it possible to assemble them functionally in an adequate systems description. This step of assembling obtained details through reductionist experiments has been coined *reconstructionism* [48] and is a central component of systems biology.

Having discussed the virtues and shortcomings of reductionism and the emergence of systems thinking as a necessary complement, we should return to the question of what understanding means in the new context of systems biology. Clearly, the Egyptians had some understanding of bread and beer making, yet it did not even include the main player, yeast. The concept of understanding in modern systems biology implies knowledge of all components that contribute to a biological phenomenon, of all processes and interactions that govern the dynamics of this phenomenon, and beyond this knowledge an explanation of why a structure is organized in a particular fashion or why a particular process is operated in a particular manner and not in an alternative manner that could, at first glance, satisfies the same needs.

As a specific example, we need to understand why some genes are regulated through repressors and others through activators, or why in some metabolic systems the production of a metabolite is inhibited and in other systems the degradation is activated. In these alternative situations, the same primary effect is accomplished: a gene is activated when needed, or the metabolite level is decreased if it is in excess. However, systems analysis shows that these alternatives differ in other respects, such as response times or sensitivity to perturbations. This type of understanding addresses the rationale behind the design and operation of organisms and is hoped to lead to deeply rooted, general rules governing life. Savageau pioneered this search for design principles and, among other accomplishments, developed *Demand Theory*, a theory that very generally explains modes of gene regulation, based on an organism's natural environment [49,50]. Similar methods led to the elucidation of metabolic pathway design, both in idealized generic pathway structures [51–55] and in specific, observed pathways [56]. A thorough understanding of such cellular design rules is a prerequisite for any rational engineering of cells [57].

For another example, consider a knockout experiment that eliminates the function of a gene. As the first case, suppose the organism is not viable. Is this gene the "life" gene? In some sense, the answer is yes, even though we may hesitate to assign such an important label, as we would also not claim that a light switch is the cause of electricity to lighten up the room. While the example is crass, this argument has been used to hail genes as "breast cancer genes" or "Alzheimer genes." It is clear that many genes, transcription factors, enzymes, and metabolites need to operate in a well-orchestrated effort to fulfill a physiological need and that any disruption of significance can cause malfunctioning. Expressed differently, the role of each component is embedded in a regulated network of processes, and this role needs to be assessed by studying the entire network with methods of systems analysis. The second case is even more difficult to address with reductionist methods. Imagine that the knockout has no visible effect whatsoever. Does this imply that the gene is not needed? In some cases, the gene may only be expressed in very specific situations that may be rare, but if they occur, lead to lethality without the availability of the gene. In other cases, the gene may code for a redundant process. Is it therefore superfluous? According to purely reductionist argumentation, the answer must be yes, whereas a systems approach would assess its contribution to the overall robustness of the organism under different perturbations. A specific example of such redundancy is the transduction of signals, for instance, through the mitogen-activated protein (MAP) kinase pathway. This pathway consists of dozens of components, and about any single deletion is ineffectual. Nonetheless, evolution and preservation of the pathway structure indicates that it must be advantageous over simpler designs. Only a comparative systems analysis will be able to quantify these advantages.

Redundancy is probably the rule rather than the exception in natural designs. In a recent study [56], we analyzed the regulation of the trehalose pathway in yeast, comparing models that included observed control signals with hypothetical models that were otherwise equivalent but did not possess one or more signals. We found that every regulatory signal contributes to the performance of the pathway under different natural conditions with a slight benefit. Each signal could be deleted, and according to the model, the cells would be able to survive without much problem. However, deletion of all control signals simultaneously led to much inferior performance.

Analyses of these types address the fundamental principles with which cells and organisms are designed and operated. Once we understand these principles, we will not only be able to explain an observed system, but to make predictions about the consequences of alterations. An expert in mechanical engineering can look at a new pulley system and quickly figure out why it is designed in a particular fashion. Similarly, by studying the functional and regulatory modules of natural systems we will one day be able to explain the workings of newly investigated organisms and even to construct new organisms according to desired specifications. Reaching this point is the dream of systems biology.

2.6 GOALS OF SYSTEMS BIOLOGY

The core belief of systems biology is that the biological components at every level of organization we investigate interact with each other and thereby may lead to responses or properties that are not explainable by the study of components in isolation or at only one level. Its premier goal is to capture these "emerging properties" and to understand how they arise, why they are implemented in the cell or organism, and what consequences ensue if they are altered, be it physiologically, pathologically, or for biotechnological purposes. This is a challenging goal, and according to Wolkenhauer [10], today's systems biology is in a situation comparable to half a century ago, when engineers were expected to analyze and control complex dynamical systems for which empirical means were insufficient.

Once we have reached this level of understanding, we will be able to give substantive explanations of observed functional modules and entire biological systems and predict how alterations would affect the function and responses of systems. In the next step, we will use these predictions to manipulate and optimize organisms in a desired fashion, be it for medical treatment, biotechnological production of organics, the creation of pest-resistant crops and live stock, or the preservation of endangered species. Ultimately, the insights may be sufficient to formulate a theory of biology, which would have the same predictive and explanatory power of theories in physics and mathematics. This theory would not simply be a fantastic achievement of the human mind, but have consequences for our daily lives that are unimaginable now. As the famous mathematician David Hilbert allegedly said, there is nothing more practical than a good theory.

To pursue these goals, systems biology uses two strategies that need to be integrated: one is experimental, the other mathematical, utilizing analytical and computational tools [58]. As Ideker et al. [59] summarized it, "systems biology studies biological systems by systematically perturbing them (biologically, genetically, or chemically); monitoring the gene, protein, and informational pathway responses; integrating these data; and ultimately, formulating mathematical models that describe the structure of the system and its response to individual perturbations." A good review of recent trends and successes was recently presented by Yao [60].

While the main purpose of this chapter is the review of a particularly powerful computational framework for systems modeling, it is useful to discuss briefly what kinds of experimental data are being produced right now or are to be expected in the near future. The distinguishing feature between traditional reductionist and systems-oriented experimentation is that the latter aims at high throughput with the ultimate aim of completeness. While traditional investigations might have targeted one particular gene, protein, or metabolite, system biology attempts to assess the expression of most or all genes of an organism simultaneously. It strives at characterizing the existence and quantities of all proteins in a diseased state and comparing this profile with the corresponding normal state. It would like to establish snapshots of all metabolites and their quantities at a given point in time. It would like to discover all connections within a complete transcription factor network and identify all physiological interactions between proteins within a cell.

The types of experimental techniques that can yield comprehensive information are the object of intense investigation. In terms of genome research, the dominant approaches use various types of DNA or RNA arrays, which quantify the expression of many genes under some treatment of interest relative to the expression in a control. If the experiment is performed as one snapshot, the data are analyzed with statistical means, such as cluster analysis or principle component analysis, which tease out significant differences in the expression patterns and show which genes show similar patterns of expression. Because the majority of genes are not yet annotated, strong expression of an unknown gene may trigger the subsequent search for its function. From a true systems point of view, gene arrays become more interesting if taken in time series. We will discuss the analysis of such time series later.

Gene expression is often somewhat removed from the real physiological action, because it is well known that many transcripts are secondarily modified. Furthermore, one cannot expect a one-to-one relationship in quantity between transcribed DNA and protein quantity. Thus, even if gene expression is known, it is only an indirect indication of the physiological response of the cell. Proteomics tries to fill this gap. By measuring which proteins are present and in what quantities, it is hoped that the active state of the cell is much better characterized. Experimentally, this is typically accomplished with two-dimensional (2D) gel electrophoresis, which separates proteins by charge in one dimension and mass in the other. The result is a 2D spot picture, where ideally each spot shows the presence of a particular protein with its relative quantity and modification state. Subsequent mass spectrometry and a comparison with information from protein databases determine which protein corresponds to each spot of interest. Again, a single snapshot does not reveal much about the physiological state of the cell. A comparison between treatment and control indicates which proteins are affected by the treatment, but more information can be deduced if the measurements are executed as time series.

Radioactive labeling has been used in biological research for a long time. Systems biology makes use of a sophisticated variant, called isotopomer analysis [61,62]. In this method, a specifically labeled precursor is made available to the cell; typical examples are carbon-labeled glucose molecules that carry label exclusively in the C1 position, C6 position, or on all carbon atoms. By measuring the labeling state of downstream metabolites, such as glycolytic hexoses and trioses and metabolites of the tricarboxylic cycle, it is possible to determine *in vivo* how hexoses are cleaved and how much flux is diverted into different pathways at branch points. At present, the system is analyzed once it has reached a steady state, from which the flux distribution is computed. However, dynamic measurements and analyses in the future will allow the deduction of much more information, for instance, about the regulation of the pathway.

Metabolic profiling is fast moving onto the stage of systems biology. Currently, several methods are being explored in parallel. Most developed are mass spectrometry and nuclear magnetic resonance, but other methods include tissue array analysis and phosphorylation of protein kinases. These techniques have in common that they allow simultaneous measurements of multiple metabolites. Presently, they are in their infancy and typically limited to snapshots of a large number of metabolites at one time point [63,64] or to short time series measuring a relatively small number of metabolites [65–67]. However, it is almost certain that these methods will soon be extended to relatively dense time series of many concentration or protein expression values [68]. The intriguing aspect of these profiles is that they implicitly contain information about the dynamics and regulation of the pathway or network from which the data were obtained. The challenge for the mathematical modeler is thus to develop methods that extract this information and lead to insights about the underlying pathway or network. We will discuss such analyses later.

2.6.1 MODELING APPROACHES

By now, it should have become clear that mathematics and computational methods are needed for analyzing biological systems of realistic size. The question thus shifts toward the search for the most useful mathematical tools. What is useful in this context is determined by the questions to be asked. To identify the best tools, it is useful to summarize specifically the demands of systems biology. The first is the need for mathematical structures that capture the dynamics of systems with potentially large numbers of components. Since we are interested in dynamic responses, suitable network models will have to be time dependent, which will almost always require differential equations. The second is the complexity and nonlinearity of processes and interactions. There are no general guidelines as to what constitutes good model structures, but the choice of a modeling framework has to ensure that it does not *a priori* exclude relevant phenomena, such as damped oscillations or limit cycles. The third is the ability to scale biomathematical models so that they can address issues of real relevance. Of course, one can always add more equations, but the question becomes whether the analytical and diagnostic tools are able to handle increasingly larger models. The fourth is the allowance for both deterministic and stochastic phenomena. In most cases, deterministic systems are more tractable, for instance, with ordinary differential equations, but there are obvious situations where stochastic effects dominate a phenomenon. An example is the dynamics of systems with very few molecules, where the fundamental laws of kinetics and thermodynamics are no longer applicable. The fifth necessary tool is an efficient handling for spatial processes. Again, if possible, one will try to avoid spatial effects, because they require incomparably more difficult partial, rather than ordinary, differential equations. Nonetheless, if the subject of study is a diffusion process, one will be hard-pressed not to acknowledge explicitly its spatial nature. Mathematics and engineering have been working on these fronts for a long time, and it is the task of systems biology to adapt these methods, extend and refine them, and develop additional tools that are specifically suited for handling the peculiarities of biological phenomena, such as properties that only emerge at the systems level, but are not existent in any of the components.

Because biological systems are so complex, an important task of systems biology is to determine to what degree simplifications and approximations are valid in the development of representations for biological systems. This task is a challenge in itself, because, first, the true functions are not known and can therefore not be evaluated against any approximate representations, and, second, our knowledge of biological details, while growing rapidly, will probably never be sufficiently complete to allow comprehensive parameter estimations of all processes and interactions in a large system. Much has been written about the virtues and problems of abstraction and of omitting detail — the paradigm being the butterfly flapping its wings in Beijing, and thereby directly or indirectly, ever so slightly, affecting everything else in the world.

Simplifications are suggested by nature in several forms [55]. First, many systems are constructed in a modular form. Organisms consist of cells, and while these communicate, they constitute identifiable entities that often allow investigation in the absence of surrounding tissues, vascularization, and means of communication. Often, these modules are associated with spatially identifiable compartments, and it may be valid to isolate these compartments and study their features. The modules may also be functional, such as in the case of hormonal communication. Bacterial genomes are organized in operons, and for some purposes, it may be useful to consider these as building blocks, while ignoring their structural and functional details. It is well acknowledged that metabolism is a multidimensional phenomenon and representations like the Boehringer wall map are merely cartoons. Nonetheless, it may be useful for specific studies to group metabolites in pathways. For instance, in the context of signaling in yeast in response to stress conditions such as heat, it might be sufficient to note that the trehalose pathway is activated and the Ras–cAMP–PKA pathway is downregulated [69] and to ignore details at the level of specific enzymes and reactions.

Second, every process has its time scale, and if a process tangential to the one of interest is very slow, it may validly be considered constant. For instance, aging can likely be ignored in an investigation of pathways of amino acid synthesis. Finally, many metabolites are kept rather constant through a variety of homeostatic mechanisms. This relative constancy may allow a model to consider some of the metabolites as true constants, thereby simplifying the mathematics of its analysis.

Precursors of systems biology have been utilizing these simplifications throughout the short history of mathematical modeling and for systems analysis in engineering. Most notably, engineers have successfully employed linear systems theory for a long time because of its unmatched mathematical and computational tractability. Linear system theory applies to a special class of systems that are homogeneous and additive — two properties that have important consequences, as a simple example may demonstrate. Consider a system with a single input and output where the input is determined external to the system and the response of the system is observed at the output when this input is applied. If one doubles the magnitude of the input and observes a doubling of the output, the system is said to be homogeneous. Now consider two different input signals added together and applied to the system. If the system is also additive, the output for this sum will be equal to the sum of the outputs for each signal applied independently. When systems obey these two properties they are said to be linear and obey the principle of superposition. Often, linear systems are also found to be time invariant (or shift invariant) and so applying the input signal tomorrow yields the same response as it did when applied yesterday. Linear time-invariant systems obeying these properties satisfy the following equation

$$H\left(c_1 \cdot f_1\left(t\right) + c_2 \cdot f_2\left(t\right)\right) = c_1 \cdot H\left(f_1\left(t\right)\right) + c_2 \cdot H\left(f_2\left(t\right)\right) \tag{2.1}$$

for constants c_1 and c_2, input functions $f_1(t)$ and $f_2(t)$, and system response $H(\cdot)$. This property has enormous implications for the applicability of linear systems theory, because if we decompose any input into a linear combination of a set of known functions, we can determine the output simply by knowing the response for each function in the set. Analyzing the response of the system to a complicated input can thus be decomposed into the analysis of its response to a series of simpler functions. If closed-form solutions for the response to these simpler functions exist, it is possible to develop closed-form solutions for arbitrarily complicated inputs decomposed in this way. Clearly, the advantages of having systems that demonstrate these properties are great and a comparable theory for biological systems would clearly be very useful.

Unfortunately, with notable exceptions, as shown below, biological systems rarely meet the criteria needed to exploit the methods of linear system theory due to their genuine nonlinear nature. In engineering, many systems are also nonlinear but have linear regions in which the system can be designed to operate. This is not often the case in biology, and so we are forced to consider approaches that offer at least some of the benefits of linear systems theory within the realities of the nonlinear nature of biological or biochemical systems.

A prominent example of the usefulness and power of a linear approach to a biochemical phenomenon is the distribution of metabolic fluxes throughout a cell. By coding each metabolite as a node and each possible flux as an edge, metabolism can be seen as a directed graph that can be represented as a big adjacency matrix, as will be discussed in more detail later. This matrix forms the core of a dynamical equation that describes the flow of material through the metabolic system

$$\frac{dX}{dt} = S \cdot v \tag{2.2}$$

where X is the vector of all metabolite concentrations, S is called the *stoichiometric matrix,* and v is the vector of all fluxes. If the system is in a steady state, where the influxes and effluxes at every node are in balance, the left-hand side of the stoichiometric equation equals zero, and the analysis reduces to a straightforward task of linear algebra, which is easily scaled up to systems of potentially very large size [70,71]. For instance, the rank of the stoichiometry matrix, which is easily computed with standard methods of linear algebra, is a reflection of conservation relationships between metabolites at steady state [72,73]. In other words, the rank signifies how many reactions are actually linearly independent. Typical analyses of stoichiometric systems address possible pathways leading from a given substrate to a desired product [74,75], the consequences of knockouts, and questions of optimal flux distributions, which are either reached through natural evolution or through biotechnological manipulation.

Adding linear physicochemical constraints to the stoichiometric equation, Palsson and others proposed an approach to analyzing metabolic networks, which is called *flux balance analysis* (FBA) [76,77]. Whereas the stoichiometric equations alone typically allow for a large space of possible solutions, which are mathematically but biologically unrealistic, FBA limits this space to realizable results. Still, many solutions are usually possible, and one strategy of FBA is to find a solution that is optimal, for instance, with respect to biomass production [78]. Another use of such models is to detect gaps in our metabolic understanding in terms of missing fluxes and the identification of genes that might code for enzymes in such fluxes [79].

Stoichiometric approaches focus on the topology of the metabolic systems and record the distribution of fluxes. Even if constrained, they ignore processes such as diffusion and convection, as well as most kinetic details [73] and do not truly account for regulatory signals, which would necessarily lead to nonlinearities and thereby destroy the advantages gained by the stoichiometric matrix representation. Thus, if such kinetic or regulatory details are known, they cannot be used and important information is lost. While this is a genuine problem with stoichiometric models, we will show in the following section that it is possible to design models that account for essential nonlinearities, while still retaining some advantages of linear mathematics.

A first attempt of combining stoichiometry with kinetic features would probably be to replace the simple flux quantities with traditional nonlinear rate functions, such as the famous Michaelis–Menten rate law. The advantages and shortcomings of this rate law have been discussed elsewhere [80–82], and it suffices to mention here that the result of such a strategy leads to mathematical models that are very difficult to analyze. One could argue that modern computing power should be able to solve systems of nonlinear equations in reasonable time and therefore to permit simulation studies. Indeed, over 30 yr ago, Garfinkel [83] and others began constructing such computer models. However, even though the numerical computation of a particular model is usually not an issue, the simulation strategy alone does not provide the types of insight systems biology is interested in. As Heinrich and Rapoport [84] eloquently remarked on this topic, simulations present several problems. First, the output of a simulation does not easily identify which factors and processes are important and which are not. Second, the genesis of responses in a complex system is difficult to extract. Third, simulations require many *ad hoc* assumptions about processes and their parameter values, which are difficult to support by experimental data, and the simulations themselves do not always tell how weakly or strongly these assumptions may affect the computed outcome. As an example, one should consult Schulz's book [85], which details features of kinetic rate processes and shows how dramatically different the mathematical representations are, depending on which mechanism one supposes. In an overall evaluation of simulation models, Heinrich and Rapoport [84], as well as others, therefore strongly recommend striving for mathematical models that permit analytical evaluations, in addition to computational simulations. In contrast to *ad hoc* models, such as those based on traditional kinetic rate laws, the desired brand of models should have a homogeneous structure that is more or less size independent and therefore allows the same types of analyses and diagnostics, no matter how large the model is. We will introduce such *canonical models* in the next section.

42   Systems Biology: Principles, Methods, and Concepts

As an example of the type of problem one may encounter with simulation models, consider a moderate-scale model of the tricarboxylic acid (TCA) cycle in the slime mold *Dictyostelium discoideum* [86]. The authors had collected an enormous amount of kinetic information from the literature and used it to construct a very detailed kinetic model in the tradition of Michaelis–Menten rate laws. The fact that the reactions were expressed as rational functions precluded the authors from executing algebraic analyses and restricted their investigations to computer simulations. The model actually represented many aspects of the TCA cycle well, but some conclusions of the simulation analysis were later shown to be faulty, because an unexpectedly large range in time scales among the governing processes masked some of the model's slow dynamics [87]. Of note is that the root cause of the problem was not the size of the system, but rather the inconvenient mathematical nature of the rational functions, which did not allow application of the mathematical diagnostic tools available for linear systems or the canonical models described in the next section.

2.7 BIOCHEMICAL SYSTEMS THEORY

Faced with the issues discussed in the previous sections, Savageau suggested, in the late 1960s, a theoretical framework that combines some of the convenient features of linear systems with the ability to capture the essence of nonlinear phenomena and was coined *biochemical systems theory* (BST; [88–90]). The following sections describe this theory and its applications, but before we get lost in detail, it is useful to preview in a nutshell why this theory is so particularly useful. The main ingredient of the theory is a power-law approximation of all rate processes in a system of differential equations. This approximation is nonlinear and, even though it is simple as a single individual term, can model essentially any continuous nonlinearity, if it is combined through sums and differences with other power-law terms of the dynamical model [91]. The particularly useful form of an S-system is based on a second crucial ingredient. Instead of individually approximating each reaction that is entering or leaving a metabolite pool with one power-law function, all fluxes entering a pool are first aggregated, and the aggregate is represented by a single power-law term, and the same is done for all fluxes leaving a pool. Thus, whether each pool in the system has one or many production fluxes and one or more consumption fluxes, each equation of the describing S-system contains exactly two terms, which are both power-law functions. As we will see later, this has the unique advantage that the steady state of such a system is characterized by linear algebraic equations, for which there exists an enormous repertoire of analytical methods. Furthermore, this system design and the feature of linearity at steady state are true for systems with any number of components, which is enormously important for the scaling to biological phenomena of realistic size.

As an alternative to the S-system form, the fluxes may not be aggregated. In this case, each reaction is approximated individually with a power-law term, and these terms are added together, with a plus sign for incoming fluxes and a minus sign for outgoing fluxes. This alternative construction may thus have any number of terms and is called a *general mass action* (GMA) system. Because these model representations always have the same structure and therefore allow a streamlined repertoire of methods for design and analysis, they are called canonical models [91,92]. Other canonical models, whose structure is size independent, are linear systems, Lotka–Volterra systems (e.g., Peschel and Mende, 1986), and generalized Lotka–Volterra systems (Hernández-Bermejo and Fairén, 1996). Some historical remarks on the early development of BST, S-systems, and GMA systems can be found in Savageau (1991).

2.7.1 Representation of Reaction Networks

Biochemical systems are formed from pathways and interactions of linked reactions. The enzymes, metabolites, and cofactors appearing in one reaction often appear in other reactions, thereby creating a complex network of mass fluxes and regulatory signals. To build models using BST, we must first understand how to transform these networks into systems of differential equations describing

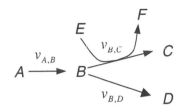

FIGURE 2.3 Example of a simple pathway consisting of six species A, B, C, D, E, and F and three reactions with rates $v_{A,B}$, $v_{B,C}$, and $v_{B,D}$. The branch at B shows two independent paths by which B is degraded. The degradation of B to C requires two molecules of E, which are consumed in the process to produce a molecule of F (e.g., ATP to ADP).

the network's behavior. First, consider a simple reaction that produces a molecule of metabolite P from substrate S. The process is discrete in that one molecule of P is produced for each molecule of S completing the process, but for large numbers of molecules, we can use averages and formulate continuous processes based on "rates." The rate at which P is produced and the rate at which S is degraded are linked and we can formally describe this relationship with ordinary differential equations. If the reaction occurs at a rate of v conversions per second, the rate at which molecules of S are used up is v, and the rate at which molecules of P are produced is also v.

In many reactions, the process is more complicated in that other metabolites or cofactors are required for the reaction to proceed. As a simple example, let molecule C be the product of a reaction that requires one molecule of substance A and one molecule of substance B. Both A and B must be present in order for this reaction to proceed. Again, if the reaction occurs at a rate of v conversions per second, one molecule of A and one molecule of B are used and one molecule of C is produced every $1/v$ sec. The rate of change of A and B is $-v$ and the rate of change of C is v. This model must account for the situation that more than one molecule of each substance is either consumed or produced in the reaction. The specific proportion in which a reactant or product participates in a reaction is called its stoichiometry. For example, let two molecules of A and three molecules of B produce a single molecule of C at a rate of v conversions per second. Under these conditions, the reaction consumes molecules of A at $2 \cdot v$ per second, molecules of B at $3 \cdot v$ per second and produces molecules of C at $1 \cdot v$ per second. Thus, one can easily design equations for these simple reactions describing the time rate of change of the reactants and products from the stoichiometry and the rate of advancement of the reaction. We have yet to discuss how we might describe v and will return to that subject later.

In biological systems, reactants, products, and cofactors seldom participate in only one reaction and so their dynamic behavior is determined by the contributions of many reactions that form a network. Consider the network illustrated in Figure 2.3. It consists of five metabolites, generically referred to as species, and three reactions. Species A is consumed to produce species B in the first reaction and B is converted to either C or D in subsequent reactions resulting in a branch in the pathway. The conversion of B to D also involves two molecules of substance E that are modified in the reaction to form two molecules of substance F. We let the rates for the reactions be $v_{A,B}$ for the reaction producing B from A, $v_{B,D}$ for the reaction producing D from B, and $v_{B,C}$ for the reaction producing C from B. Using the reaction rates and stoichiometry of this network, we can easily construct the set of differential equations that describes the behavior of this system. For each species appearing in the model, we first identify all reactions in which that species participates. If the species participates as a reactant, we include a negative term formed from the multiplication of the species' stoichiometry and the rate of the associated reaction. These terms take a negative sign as they represent reactions where the species is being used up or consumed. If the species participates as a product, the term is formed similarly but takes a positive sign, indicating that the reaction produces that species. Combining these results we write the system of differential equations as follows.

$$dA/dt = -1 \cdot v_{A,B}$$

$$dB/dt = +1 \cdot v_{A,B} - 1 \cdot v_{B,C} - 1 \cdot v_{B,D}$$

$$dC/dt = +1 \cdot v_{B,C}$$

$$dD/dt = +1 \cdot v_{B,D}$$ (2.3)

$$dE/dt = -2 \cdot v_{B,D}$$

$$dF/dt = +2 \cdot v_{B,D}$$

For large networks, it becomes convenient to use a more compact notation based on linear algebra. In the above-mentioned system of equations, for instance, the reaction rates for the three different reactions are used in linear combinations to form expressions for the rate of change for each of the five species. The coefficients of these linear combinations are supplied by the stoichiometry of the reactions. For a tighter representation, it is also necessary to generalize the naming of metabolites and use, instead of intuitive names for the species (e.g., A, B, C, or, Glu, ATP, NADH, PK), the more generic X_i and assign each species to an index in this naming (e.g., $X_1 \leftrightarrow$ Glu, $X_2 \leftrightarrow$ ATP). As suggested in an earlier section, we can then organize this as a matrix-vector product where X is the vector of all species appearing in the model, S is a matrix giving the stoichiometry of the system, and v is a vector of the reaction advancement rates. Thus, the previous system becomes

$$\begin{bmatrix} dA/dt \\ dB/dt \\ dC/dt \\ dD/dt \\ dE/dt \\ dF/dt \end{bmatrix} = \begin{bmatrix} -1 & 0 & 0 \\ +1 & -1 & -1 \\ 0 & +1 & 0 \\ 0 & 0 & +1 \\ 0 & 0 & -2 \\ 0 & 0 & +2 \end{bmatrix} \cdot \begin{bmatrix} v_{A,B} \\ v_{B,C} \\ v_{B,D} \end{bmatrix}$$ (2.4)

This, in turn, may be compactly written as $dX/dt = S \cdot v$.

The stoichiometries of the reactions appear as columns of the S matrix where the $(i,j)^{th}$ entry of the matrix contains the stoichiometry of the ith species in the jth reaction. If the species participates in the reaction as a reactant, the entry is negative and if as a product, the entry is positive. The utility of this more compact form will be evident throughout the following sections.

2.7.2 Rate Laws

Clearly, the reaction rates are key to the behavior of the system and so we now begin to investigate the functional forms used to represent the relationship between these rates and associated products, reactants, and cofactors. The rate at which a reaction proceeds is obviously dependent on the availability of the species participating in the reaction. In the earlier example, the reaction producing P from S can occur only if molecules of S are available and it is reasonably expected to occur more rapidly with larger supplies of S. In biochemical systems, many of the reactions of interest are catalyzed by enzymes, involve more than one metabolite or cofactor, and may be inhibited or activated by additional cofactors, metabolites, or proteins. As a result, the rate of a given reaction may depend not only on the supply of substrate to be converted but also on the current quantity of many other species. In the most general case we write the reaction rate of the ith reaction as $v_i = V_i(X) = V_i(X_1, X_2, X_3, \ldots, X_n)$, which may depend on the current availability of some or every

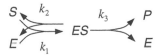

FIGURE 2.4 Reaction scheme for an enzyme-catalyzed reaction converting substrate S to product P with the aid of enzyme E. The first step involves a reversible reaction, with forward and reverse rate constants k_1 and k_2, that forms complex ES. The process is completed in an irreversible reaction where the complex decomposes into enzyme E and product P, with rate constant k_3.

species in the system. Many functional forms of $V_i(.)$ have been proposed, but the most prevalent are formulations based on mass-action kinetics, enzymatic rate laws, and power laws. BST is based purely on the power-law form, but since the other forms are widely applied, an understanding of the relationships between the alternative forms is an essential part of model building using BST.

2.7.2.1 Mass Action Kinetics

Models based on mass-action kinetics are typically used to simulate reaction networks consisting of elementary reactions. Consider the reaction scheme for the simple enzyme-catalyzed reaction shown in Figure 2.4. In this example, substrate S is transformed into product P with the aid of enzyme E through elementary steps. The first step forms, from substrate S and enzyme E, a complex ES, which through a reverse reaction may decompose to S and E. In the second step, the product dissociates from the enzyme to yield product P and the free enzyme E. The reaction rate in each of these steps is determined by the participating reactants and their stoichiometry and is given in the form of a product of power functions. In general, this rate is written as $v = k \cdot X_1^{g_1} \cdot X_2^{g_2} \cdot \ldots \cdot X_n^{g_n}$ where k is a positive constant called the rate constant and g_i are positive integers called kinetic orders that reflect the stoichiometry of the reaction. For example, in the reaction $2A + 3B \rightarrow C$, the rate would be given as $v = k \cdot A^2 \cdot B^3$ where A appears in the second power because two molecules of A are required in the reaction and B appears in the third power because three molecules of B are required. For the enzyme-catalyzed reaction described above, all steps require one molecule of each type involved, and the system of equations is therefore be given as

$$dES/dt = +k_1 \cdot S \cdot E - k_2 \cdot ES - k_3 \cdot ES$$
$$dP/dt = +k_3 \cdot ES$$

(2.5)

where all nonzero kinetic orders are equal to 1. Models based on mass-action kinetics have the advantage of being determined directly from the elemental reactions and their stoichiometry. They suffer, however, from the fact that a large number of rate constants must be determined in order to implement the model. In many cases, these elemental reactions are not experimentally observable and the parameters would thus be difficult to acquire.

2.7.2.2 Michaelis–Menten Rate Law

Enzyme-catalyzed reactions in biochemical systems are often effectively modeled using rate laws representing the net flux through a collection of elementary reactions as modeled in Equation 2.5. The best known of these is the Michaelis–Menten rate law [93], which can be derived from the mass-action model described above under assumptions regarding the steady-state behavior of the complexes and the ratio of substrate and enzyme concentration. The resulting law gives a nonlinear relationship between the quantity of available substrate and the reaction rate and is illustrated in Figure 2.5a. The rate law is defined by two parameters, V_{max} and K_m, that give the saturating rate of the reaction and the substrate quantity at which half of this maximum rate is achieved. Thus,

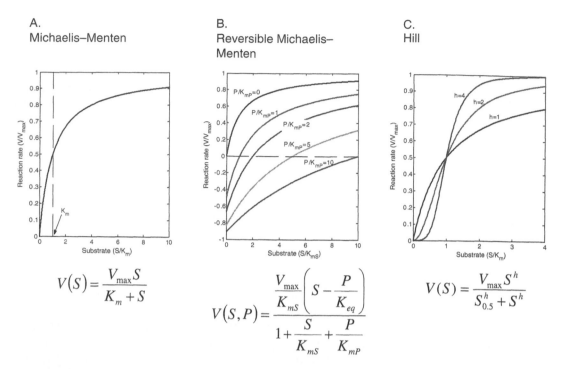

$$V(S) = \frac{V_{max}S}{K_m + S}$$

$$V(S,P) = \frac{\dfrac{V_{max}}{K_{mS}}\left(S - \dfrac{P}{K_{eq}}\right)}{1 + \dfrac{S}{K_{mS}} + \dfrac{P}{K_{mP}}}$$

$$V(S) = \frac{V_{max}S^h}{S_{0.5}^h + S^h}$$

FIGURE 2.5 (**See color insert following page 150.**) Examples of commonly used kinetic rate laws. Panels A, B, and C give the functional form of the Michaelis–Menten, reversible Michaelis–Menten, and Hill rate laws and graphs of reaction rate vs. substrate. The reversible Michaelis–Menten reaction schemes yield negative rates when $P/K_{eq} > S$.

there were originally three rate constants in the mass-action model described above, which have been replaced by two parameters in the rate-law model. While this model is supported by many *in vitro* experiments, it does not account for possible inhibiting effects of available product or reversibility, which constitute realistic situations in biochemical systems *in vivo*. The reversible form of the Michaelis–Menten rate law [94], derived through similar assumptions, accounts for this effect and is illustrated in Figure 2.5b. Examining this equation, we see that as the quantity of product P goes to 0, the equation simplifies to the original Michaelis–Menten rate law, as one would expect. As the product quantity, relative to substrate, approaches the equilibrium constant K_{eq}, the reaction rate approaches 0 and then becomes negative as the quantity of product increases further. As expected, the flux goes to 0 at the substrate–product ratio given by the equilibrium constant and reverses on either side of this point. Another commonly employed rate law, originally proposed by Hill [95], considers enzymes with multiple binding sites that are identical and cooperative. Cooperative binding occurs when a ligand binding to one of several identical binding sites on an enzyme increases (positive cooperativity) or decreases (negative cooperativity) the affinity for ligands at the unbound sites. Under these assumptions, the rate vs. substrate concentration follows a sigmoidal shape (Figure 2.5c) that was originally observed in oxygen-binding curves of hemoglobin. The parameter h is called the Hill coefficient and is a measure of the cooperativity of the binding of enzyme and its ligands. These are just three representatives of a wide range of rate laws that have been proposed, and we shall use these as examples in subsequent discussions.

2.7.2.3 Power-Law Rate Laws

Another form of rate-law, called a power law, forms the basis for BST [89]. Most biological reaction rates, when plotted in a log–log graph against the reaction substrate, appear to have broad linear regions (Figure 2.6). Thus, constructing a linear approximation to these functions in log space gives

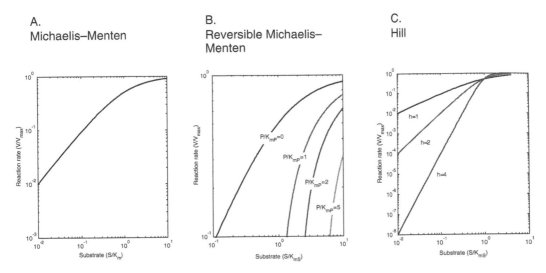

A.
Michaelis–Menten

B.
Reversible Michaelis–
Menten

C.
Hill

FIGURE 2.6 (See color insert following page 150.) Examples of kinetic rate laws plotted in logarithmic coordinates. Nearly linear behavior is often found over wide ranges of substrate concentration. Linear regions in logarithmic coordinates indicate power law behavior in linear coordinates.

a suitable relationship of the form $\log(v) \approx g \cdot \log(X) + a$ where g gives the slope of the line in the linear region and a gives the y-intercept. When transformed back to Cartesian space, the relationship follows a power law given by $v = \alpha \cdot X^g$ where the slope g gives the exponent for the substrate and $\alpha = e^a$ gives a scaling factor for the term. Realistically, the rate is often a function of more than one species, and it is therefore of importance that the approach can be extended directly to two or more variables. In this case, we need the approximation of a surface in log–log coordinates, which is given by a plane (for two variables) or a hyperplane (for more than two variables) and takes the mathematical form $\log(v) \approx g_1 \cdot \log(X_1) + g_2 \cdot \log(X_2) + \cdots + g_n \cdot \log(X_n) + a$. Transforming this back to Cartesian coordinates yields a power law of the form $v = \alpha \cdot X_1^{g_1} \cdot X_2^{g_2} \cdots X_n^{g_n}$. We recognize this as the form of rate laws using mass action kinetics where α is the rate constant and g_i are the kinetic orders. In the power-law form of BST, however, kinetic orders are allowed to take fractional or negative values, and g_i therefore is more strictly referred to as the *apparent kinetic order*. The g_i are, nevertheless, often simply called kinetic orders in the literature when it is clear that the discussion occurs in the context of a power-law model.

The rationale for the power-law form of a rate law is intuitively reasonable, and it can also be backed up with a strictly formal argument, based on a Taylor series approximation upon log–log transformation. Any continuous, differentiable function can be written as an infinite series, called a Taylor series [96], about a selected expansion point, which, for our purposes, is often referred to as the operating point. Let $f(\cdot)$ be the functional representation of the log of the rate vs. the log of the substrate. The Taylor series expansion of this function $f(\cdot) = \log(v)$ about the operating point $y_0 = \log(X_0)$ and for $y = \log(X)$ is defined as

$$f(y) = f(y_0) + f'(y_0)(y - y_0) + \frac{f''(y_0)}{2!}(y - y_0)^2 + \cdots + \frac{f^{(n)}(y_0)}{n!}(y - y_0)^n + \cdots \quad (2.6)$$

If $f(\cdot)$ is approximately linear in some region of interest, the second-order and higher derivatives are nearly 0 and thus may be dropped from the expression, leaving $f(y) = f(y_0) + f'(y_0)(y - y_0)$. One also notes that in the area near the expansion point y_0, the difference $y - y_0$ is small and terms containing $(y - y_0)^n/n!$ become very small as n increases, providing further justification for dropping higher-order terms from the series. Comparing these insights from Taylor expansion with comments made before, we see that the slope g and intercept a of the power-law rate law can be obtained from the function $f(\cdot)$, its first derivative $f'(\cdot)$, and the selected operating point y_0 to give $a = f(y_0) - f'(y_0) \cdot y_0$ and $g = f'(y_0)$. Furthermore, this result is valid also for higher-dimensional systems with any number of variables. A power-law rate expression is therefore equivalent to a first-order approximation to an arbitrary rate equation written in logarithmic coordinates.

2.7.3 SOLUTIONS TO THE SYSTEM OF EQUATIONS

The rate laws discussed are linear if all fluxes are of constant magnitude. In all other cases, the laws are nonlinear in the quantities of the species and, as a consequence, no closed-form solution exists for the system of dynamic equations $dX/dt = S \cdot v$ or even for the steady-state equations $dX/dt = S \cdot v = 0$, except under fortuitous conditions. This restriction may be overcome in several ways including simulation based on numerical integration, linearization in the original coordinate system, and linearization in the logarithmic coordinate system.

2.7.3.1 Numerical Integration

We have seen that it is quite easy to compose equations describing the behavior of a biochemical system, given the biochemical map, selected rate laws, and the associated parameters. The system of equations, coupled with modern computing systems and algorithms for the numerical solution of differential equations, gives us the opportunity to compute numerically the dynamic behavior of the system from a given set of initial conditions. Unless the system is very large, this solution is typically obtained rather quickly. However, while this strategy allows us to explore one trajectory the system might take, it provides little to our understanding of the spectrum of all possible paths. Different starting points may yield qualitatively different behaviors. The system may find a steady state, enter a limit cycle, show damped oscillations, or become unstable. Without further analysis, we would need to exhaust the space of possible initial conditions to determine, with reliability, how the system could potentially behave. While simulation provides a valuable tool for the analysis of the dynamic behavior of the system, by itself it does not provide the complete picture we seek.

2.7.3.2 Linearization

As an alternative, we might consider forming a linear approximation of the system and use the wealth of tools from linear system theory to analyze its behavior [97]. This is a valid strategy in principle. However, as pointed out above, the rates and thus the differential equations describing biological systems are usually nonlinear and thus a linear approximation would be applicable only in a small region about the chosen operating point of the Taylor series expansion. Attempts to explore the dynamic response or infer behavior from this approximation would be difficult unless the system remained in a sufficiently small region about the operating point. Nevertheless, we will see that linearization about a steady-state point actually provides us with valuable information regarding the assessment of the system's stability and robustness at that point. The specifics of this assessment will be presented later.

2.7.3.3 Power-Law Approximation

We have already seen that rate laws based on power functions correspond to first-order approximations to arbitrary rate laws in logarithmic coordinates. These power-law approximations capture more of the nonlinear nature of the reaction rate than the linear approximation discussed previously. Consider, for example, approximations to Michaelis–Menten, reversible Michaelis–Menten, and Hill rate laws as shown in Figure 2.7. In each case, much of the curvature in the region about the operating point is captured in the power-law approximation. Also as noted above, the apparent

A.
Michaelis–Menten

B.
Reversible Michaelis–Menten

C.
Hill

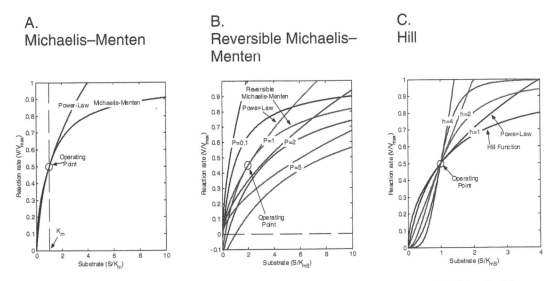

FIGURE 2.7 (See color insert following page 150.) Power law approximations to the Michaelis–Menten, reversible Michaelis–Menten, and Hill rate laws. The power law approximation fits the given rate law exactly at the operating point and typically provides a good approximation to the rate law about that point.

kinetic order for each of these reactions can be determined from the rate law by evaluating the first derivative of the rate law in logarithmic coordinates at the selected operating point. While the power-law approximation is quite close in a range around the operating point, one must caution that this is not necessarily true for very large deviations. For instance, the power-law form does not capture the saturating effect and ultimate approach of the maximal velocity V_{max} found in the Michaelis–Menten, reversible Michaelis–Menten, and Hill rate laws. Similarly, at this level of approximation, the power-law representation does not exhibit the negative rates shown by the reversible Michaelis–Menten rate law for substrate quantities below P/K_{eq}. If these phenomena are of specific interest, it is possible to represent the rate laws at a lower level, as it is given in Equation 2.5, where the power-law representation is exactly equivalent over the entire domain of concentrations. In most situations, however, the substrate concentrations do not vary that widely *in vivo*, and the power-law approximation at the higher level is sufficient in accuracy.

Although it may appear from the description of these methods that they are mutually exclusive, they are in fact all essential tools in the application of BST. Because the dynamic equations are nonlinear, simulation by numerical integration is the only method by which the evolution of the system over time can be determined and thus forms the foundation of dynamic analysis. As we will describe next, linearization provides us with a method for the analytical assessment of the system of equations at steady state, and, of course, power-law approximation is the mathematical connection between BST models and other rate functions.

2.7.4 Nonlinear Canonical Models in BST

As with linear system theory, BST provides canonical forms describing a biological system. The availability of a canonical representation provides significant advantages when the goals are the development of mathematical theory and analytical and computational methods, because it allows us to formulate results applicable to entire classes of problems, rather than *ad hoc* results that only apply to a very specific function under a defined set of conditions. Several canonical forms are possible within BST, but the most frequently used are GMA and S-systems.

2.7.4.1 Generalized Mass Action Systems

The GMA form is the most general of the canonical representations used in BST, and all other forms (linear systems, S-systems, Lotka–Voltera systems, generalized Lotka-Voltera systems, and half-systems or Riccati systems) are direct special cases of a GMA system. Returning to the general system description in vector notation, $dX/dt = S \cdot v$, the GMA representation requires that each rate function be formulated as $v_k = V_k\left(X_1, X_2, \ldots, X_n\right) = \theta_k \cdot X_1^{f_{k,1}} \cdot X_2^{f_{k,2}} \cdots \cdot X_n^{f_{k,n}}$, where θ_k is the rate constant and $f_{k,1}, \ldots, f_{k,n}$ are the apparent kinetic orders for species $X_1, \ldots, X_n$ in the kth reaction. The stoichiometric matrix S contains both positive and negative values corresponding to species participating as either a reactant (negative value) or a product (positive value) in a given reaction.

It is useful to group all of the positive values into one matrix S^+ and the absolute value of the negative values into another matrix S^- such that $S = S^+ - S^-$. The matrix S^+ now gives the stoichiometry for the production of all species and the matrix S^- gives the stoichiometry for the consumption of all species. The differential equation for species X_i in a network of p reactions can then be written as follows:

$$\frac{dX_i}{dt} = \sum_{k=1}^{p} s_{i,k}^+ \cdot v_k - \sum_{k=1}^{p} s_{i,k}^- \cdot v_k$$

$$= \sum_{k=1}^{p} s_{i,k}^+ \cdot \theta_k \prod_{j=1}^{n} X_j^{f_{k,j}} - \sum_{k=1}^{p} s_{i,k}^- \cdot \theta_k \prod_{j=1}^{n} X_j^{f_{k,j}}$$

(2.7)

Combining the constants $s_{i,k}$ and θ_k into single terms $\gamma_{i,k}$ yields the commonly used, general form of a GMA system where the $\gamma_{i,k}$ can take on positive and negative values corresponding to production and consumption of X_i:

$$\frac{dX_i}{dt} = \sum_{k=1}^{p} \gamma_{i,k} \prod_{j=1}^{n} X_j^{f_{k,j}} .$$

(2.8)

This form is structurally very rich and general, because for integer $f_{k,j}$, the system can represent any multinomial and thus a multidimensional Taylor series approximation to any continuous and differentiable rate law. We also see that models based on mass-action kinetics and linear models are simply special cases of this GMA system.

2.7.4.2 S-Systems

As in the development of the GMA form, we split the stoichiometric matrix naturally into a difference of two terms, one associated with the production of a species and one associated with its consumption, $S = S^+ - S^-$. For any species that is produced through only one reaction and consumed through only one reaction, the summations in the GMA form have only one term each and thus the differential equation immediately reduces to the difference of two products of power functions. When a species appears at a branch where it is consumed (or produced) through independent reactions, we can approximate the sum of rates for these parallel consumption (or production) paths as a single product of power functions. As a consequence, the time rate of change of a species can be written as the difference of just two power-laws, one representing the aggregated production, or influx, of this species and one representing the aggregated consumption, or efflux, of the species. In cases where the species is involved in none or only one production or consumption pathway, the result is exactly the same as for the GMA form given above. In other cases, the two forms are exactly equivalent at the chosen operating point, similar when close to this point, and diverge as the species deviate farther from the operating point. This form with one aggregated production term and one aggregated consumption term is referred to as an S-system, where S indicates the synergistic nature of this nonlinear form [89]. In mathematical notation, the result is

$$\frac{dX_i}{dt} = V_i^+ \left(X_1, X_2, \ldots, X_n \right) - V_i^- \left(X_1, X_2, \ldots, X_n \right)$$

$$= \alpha_i \prod_{j=1}^{n} X_j^{g_{i,j}} - \beta_i \prod_{j=1}^{n} X_j^{h_{i,j}}$$

(2.9)

The parameters α_i and β_i are the positive rate constants for the influx and efflux terms and $g_{i,j}$ and $h_{i,j}$ are the (apparent) kinetic orders for the effect of species X_j on the influx (g's) and efflux (h's) of species X_i. If an increase in species X_j tends to increase the rate of production of X_i, the parameter $g_{i,j}$ has a positive value. This might be the case if X_j is a reactant or activator in a process that produces X_i. By contrast, if X_j inhibits the production of X_i then $g_{i,j}$ has a negative value, and if X_j has no effect then $g_{i,j}$ has a value of zero. The relationship between the sign of the kinetic order and the influence of one species on another species' production or consumption is a valuable characteristic of the GMA and S-system representations. It is a direct consequence of the fact that each kinetic order gives the slope of the functional relationship between a production or consumption rate and the quantity of a species when expressed in logarithmic coordinates. By examining the kinetic orders, we can immediately determine the direction and magnitude of the influence of one species on another.

One may ask why it is useful to explore two different forms within BST. Several answers can be given, but maybe the most important is that the GMA form is closer to biochemical intuition, because every process is modeled individually and therefore more easily identified in the equations representing a biochemical map. However, the S-system form has unique and significant mathematical advantages for analyses of the system in the neighborhood of its steady state, and these analyses yield insight into the robustness of the mathematical model, as we shall discuss later. The S-system form is also uniquely suited for optimization tasks and the identification of pathway structure from experimental time course data.

Often, the Michaelis–Menten rate law and its extensions are accepted as "automatic defaults." This is understandable, because these rate laws have had a long and successful history, where many experimental studies supported the mechanisms and functional forms of these rate laws *in vitro*. However, many assumptions that are valid *in vitro* do not necessarily hold *in vivo*. For instance, the rate expression for mass action kinetics, and thus by extension for the Michaelis–Menten mechanism, assumes that the reactions occur in dilute, spatially homogeneous solutions. In living cells, this assumption is not always appropriate, as reactions tend to occur on membranes, in channels, or in complexes and thus result in dimensional restrictions that change the kinetic properties of the reaction. Extensions of the Michaelis–Menten rate law, which, to some degree, account for these effects lead to fractal mass-action kinetics and other functions that are much more complicated than the traditional formulation [80]. Even these more complicated functions are easily converted into the corresponding power-law representations, either in GMA or S-system form, which are mathematically and computationally more efficient, yet still capture the essence of the alleged rate laws, at least in some range about a chosen operating point.

2.8 WORKING WITH MODELS DESCRIBED BY GMA AND S-SYSTEMS

2.8.1 FROM BIOCHEMICAL MAPS TO SYSTEMS OF EQUATIONS

An important consequence of the GMA and S-system forms is the direct mapping of the biochemical network into the system of differential equations that describe the system's behavior. Biochemical maps, or pathway diagrams, are often used to depict the network of reactions found in metabolic systems. These maps, when drawn using a specified set of rules, are a valuable aid in the visualization and analysis of biochemical systems. The rules for constructing maps form a graphical language describing the relationships between the species in the system. This language should use a well-defined vocabulary of symbols with clear and unambiguous meaning and include a procedure with which the dynamic system equations are extracted from the map.

2.8.1.1 Map-Drawing Rules

In the maps used throughout this chapter, we use a consistent format that we will now describe. For our purposes, biochemical maps are graphs consisting of nodes and edges. Nodes indicate species or pools of material, such as metabolites, cofactors, proteins, enzymes, mRNAs, or other molecules of interest. Nodes are labeled with the name or description of the represented pool. Solid edges are used to connect these pools and indicate a flow of material via a reaction or transport mechanism where the direction of flow is indicated by an arrowhead. If more than one species participates as a reactant, the edge is multitailed, and if multiple products result, the edge is multiheaded. In many biologically interesting enzyme-catalyzed reactions, other species within the system activate or inhibit reactions and thus act as catalysts or control signals within the network. We indicate these signals as dashed, directed edges extending from the pool causing the modulation to the edge representing the reaction that is modified. If the effect is to increase the rate of reaction, a plus sign (+) is placed next to the arrowhead; for inhibition a minus sign (−) is displayed. Examples of this symbolic representation are provided in Figure 2.8.

Example	Description
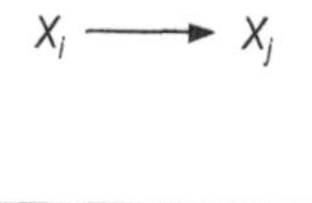 $X_i \longrightarrow X_j$	*Single Substrate, Single Product Reaction.* A solid arrow from one species to another indicates a flow of material. X_j is produced from X_i by this reaction.
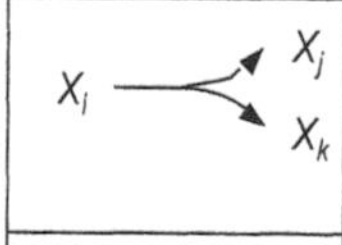 $X_i \prec{\ X_j \atop X_k}$	*Single Substrate, Multiple Product Reaction.* A solid arrow from one species with multiple heads indicates a single reaction in which X_j and X_k are produced from X_i.
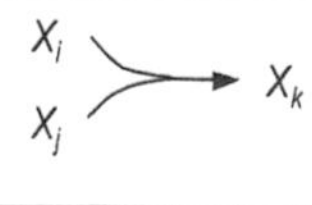 ${X_i \atop X_j}\succ X_k$	*Multiple Substrate, Single Product Reaction.* A solid arrow from more than one species with a single head indicates a single reaction in which X_k is produced from X_i and X_j.
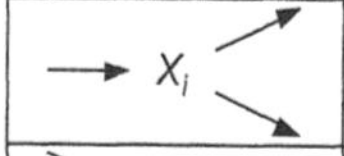 $\longrightarrow X_i \prec$	*Branch.* Species X_i is degraded by way of independent reactions.
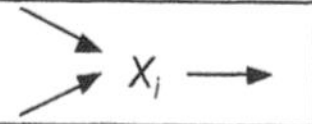 $\succ X_i \longrightarrow$	*Merge.* Species X_i is produced by way of independent reactions.
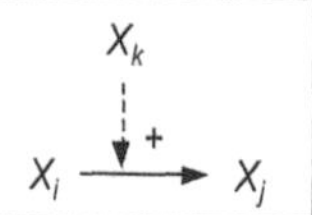 X_k $X_i \xrightarrow{+} X_j$	*Activation.* The rate of production of species X_j from X_i is increased with increasing concentration of X_k.
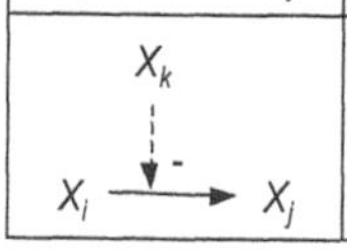 X_k $X_i \xrightarrow{-} X_j$	*Inhibition.* The rate of production of species X_j from X_i is decreased with increasing concentration of X_k.

FIGURE 2.8 Illustrations of the basic components of biochemical maps and map-drawing rules employed throughout this chapter.

Biochemical Map

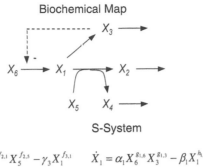

GMA System

$$\dot{X}_1 = \gamma_1 X_6^{f_{1,6}} X_3^{f_{1,3}} - \gamma_2 X_1^{f_{2,1}} X_5^{f_{2,5}} - \gamma_3 X_1^{f_{3,1}}$$

$$\dot{X}_2 = \gamma_2 X_1^{f_{2,1}} X_5^{f_{2,5}} - \gamma_5 X_2^{f_{5,2}}$$

$$\dot{X}_3 = \gamma_3 X_1^{f_{3,1}} - \gamma_6 X_3^{f_{4,6}}$$

$$\dot{X}_4 = \gamma_2 X_1^{f_{2,1}} X_5^{f_{2,5}} - \gamma_4 X_4^{f_{6,4}}$$

S-System

$$\dot{X}_1 = \alpha_1 X_6^{g_{1,6}} X_3^{g_{1,3}} - \beta_1 X_1^{h_{1,1}} X_5^{h_{1,5}}$$

$$\dot{X}_2 = \alpha_2 X_1^{g_{2,1}} X_5^{g_{2,5}} - \beta_2 X_2^{h_{2,2}}$$

$$\dot{X}_3 = \alpha_3 X_1^{g_{3,1}} - \beta_3 X_3^{h_{3,3}}$$

$$\dot{X}_4 = \alpha_4 X_1^{g_{4,1}} X_5^{g_{4,5}} - \beta_4 X_4^{h_{4,4}}$$

FIGURE 2.9 The differential equations for the GMA and S-system representation of the given biochemical map were constructed according to the rules in the text. For example, the production of X_1 is a function of both X_6 (substrate) and X_3 (inhibitor of this reaction). We expect $f_{1,3}$ to be less than 0 because X_3 inhibits this reaction. Terms for the other five reactions are determined similarly. The GMA and S-systems differ only in the degradation of X_1. In the GMA form, two terms are present, one for the path to X_3 and one for the path to X_2. Since both X_1 and X_5 effect the efflux of X_1, the term in the S-system is given by $\beta_1 X_1^{h_{1,1}} X_5^{h_{1,5}}$.

2.8.1.2 Maps to GMA Systems

Systems of differential equations for GMA systems are easily constructed from a biochemical map drawn according to the rules given above. Each reaction (solid edge) is converted into a power law giving the rate of advancement of the reaction. The rate law is constructed by forming a product of power-functions, one for each species that appears as a reactant (solid tail) or as a modifier (dashed edge) and a rate constant. This yields the rate of the reaction in the "by now familiar" form $v_k = \theta_k \cdot X_1^{f_{k,1}} \cdot X_2^{f_{k,2}} \cdots X_n^{f_{k,n}}$. For each time-varying species in the system, we next write the differential equation describing its evolution by constructing a summation consisting of one term for each reaction in which the species participates. Participation can be determined from the map by identifying all of the solid edged tails or heads connected to this species. For each reaction producing the species, we add a term formed from the product of the advancement rate for that reaction and the stoichiometry with which the species participates in that reaction. For reactions consuming the species, we subtract such a term. This yields a system of equations matching the earlier general expression $dX/dt = S \cdot v$. Each edge in this diagram has an associated kinetic order in the model and we can make reasonable assumptions as to the sign of these kinetic orders based on their appearance in the map. Species appearing as reactants or as activating modifiers are expected to have positive kinetic orders and those acting as inhibiting modifiers are expected to have negative kinetic orders. Our ability to infer the sign of the kinetic orders from the map has important implications for the numerical estimation of parameters and for controlled comparisons of biochemical systems, which we shall discuss later. An example of the process of converting biochemical maps to a system of differential equations is illustrated in Figure 2.9.

2.8.1.3 Maps to S-Systems

S-systems are special cases of GMA systems in which only two terms appear in each differential equation, one describing the aggregated influx rate and one describing the aggregated efflux rate. Thus, while GMA systems emphasize reactions, S-systems are more species centric, and this requires a slightly different set of rules for the construction of the system of differential equations. For each species, we identify all reactions in which that species appears as a product and make a list of all other species that participate as reactants or as modifiers in any of those reactions. A single power-law term is constructed as the product of all species in our list, raised to their individual powers, and a positive-valued rate constant. A term representing the efflux is constructed similarly by examining all reactions in which this species participates as a reactant. The difference of the influx and efflux terms gives the differential equation for this species. Differential equations are constructed in this manner for all time-varying species in the system. The process is illustrated in Figure 2.9.

The parameters are so far not numerically specified, but it is important to note that the parameters appearing in these expressions are often constrained by precursor–product relationships. If, for example, a reaction consumes one species and produces another, the efflux term of the reactant and the influx term of the product are constrained.

2.8.1.4 GMA Systems to S-Systems

An S-system model and a GMA model of a biochemical system differ only at points in the system where branches occur. Branches result when two independent processes produce or degrade a species in the system. In a GMA system, these independent fluxes appear as independent terms in the model, whereas in S-systems the fluxes are aggregated. It is often convenient first to construct a GMA system and secondarily to construct an S-system model from this GMA model. In this transition, the S-system terms approximate sums of power laws found in the GMA model as single power laws at each of the branches being aggregated. We recall that the slope, in log–log space, of the rate law that is being approximated determines the kinetic order of the power-law approximation. To approximate the sum of power laws, given in the GMA by

$$V_i^+ = \sum_{k=1}^{p} v_k = \sum_{k=1}^{p} \gamma_{i,k} \prod_{l=1}^{n} X_l^{f_{k,l}} \tag{2.10}$$

We determine, one by one, the kinetic orders of each variable in the S-system by recalling the relationship between slopes and kinetic orders in log–log space. Instead of explicitly transporting the sum into logarithmic coordinates, we take advantage of the relationship $d \ln(x)/d \ln(y) = (dx/x)/(dy/y)$ and write the kinetic order for X_j in the influx term of the S-system equation for X_i directly as

$$g_{i,j} = \frac{\partial V_i^+}{\partial X_j} \frac{X_j}{V_i^+} = \frac{X_j}{V_i^+} \frac{\partial}{\partial X_j}\left[\sum_{k=1}^{p} \gamma_{i,k} \prod_{l=1}^{n} X_l^{f_{k,l}}\right] = \frac{\sum_{k=1}^{p} f_{k,j} \cdot v_k}{V_i^+} \tag{2.11}$$

The kinetic order for X_j in the single power-law approximation of the S-system form is thus given simply as the weighted sum of the kinetic orders for X_j from each term in the GMA form. Each weight is determined by the contribution to the total flux provided by that term. This rather intuitive result is a direct consequence of the power-law representations used in the GMA and S-system forms. The S-system kinetic orders are functions not only of the GMA system kinetic orders but also of the fluxes and, therefore, the quantities of each species affecting X_i.

The GMA and S-system can, in general, have the same influx, efflux, and slope at only one state. The kinetic orders computed as shown above are therefore to be evaluated at this one point, the operating point, which often is selected as the steady state of the system. The rate constant of the power-law approximation is determined by requiring that the flux of the S-system approximation and the original GMA system match at this operating point. Having computed the kinetic orders, the rate constant is therefore determined as

$$\alpha_i \prod_{j=1}^{n} X_j^{g_{i,j}} = V_i^+$$

$$\alpha_i = \frac{V_i^+}{\prod_{j=1}^{n} X_j^{g_{i,j}}} = \frac{\sum_{k=1}^{p} \gamma_{i,k} \prod_{l=1}^{n} X_l^{f_{k,l}}}{\prod_{j=1}^{n} X_j^{g_{i,j}}} \qquad (2.12)$$

The straightforward procedure for constructing an S-system from a GMA system has several advantages. The resulting system is guaranteed to have the same flux and slope at the operating point. For systems that stay near the operating point, the S-system approximation often provides a very good approximation to the GMA system [98]. If the operating point is a steady state for the GMA system, the S-system will also have that steady state. The influx term of the S-system has the same value as the GMA sum of production terms and the efflux term has the same value as the GMA sum of consumption terms. If the GMA system is at steady state at the operating point, the sum of the production and consumption terms is 0. This implies that the same holds for the S-system.

Even though it may seem redundant, it is often useful to construct both the GMA and S-system forms, because they allow slightly different analyses and also indicate how sensitive the model results are to the power-law approximation.

2.8.2 STEADY-STATE SOLUTIONS FOR S-SYSTEMS

A distinct advantage of the S-system form is the computational accessibility of the system's steady state. A system reaches steady state when there is no net change in the quantities of its species. This condition can be reached if either all fluxes are zero or if the net influx and efflux for each species exactly cancel. The first case would occur only if all reactions stopped and would constitute death in a biological system and so is not of interest. The second case can occur anytime the production and consumption of all species are balanced even if those processes are occurring with large magnitude or very rapidly. In S-systems, the steady-state condition $dX/dt = 0$ directly leads to the determination of the system's steady state [89]. For variable X_i this yields

$$\frac{dX_i}{dt} = 0$$

$$0 = \alpha_i \prod_{j=1}^{n} X_j^{g_{i,j}} - \beta_i \prod_{j=1}^{n} X_j^{h_{i,j}}$$

$$\alpha_i \prod_{j=1}^{n} X_j^{g_{i,j}} = \beta_i \prod_{j=1}^{n} X_j^{h_{i,j}}$$

$$\sum_{j=1}^{n} \left(g_{i,j} - h_{i,j}\right) \cdot \log\left(X_j\right) = \log\left(\frac{\beta_i}{\alpha_i}\right)$$

(2.13)

The simultaneous solution to all equations in $dX/dt = 0$ therefore can be written as

$$\begin{bmatrix} \log(X_1) \\ \vdots \\ \log(X_n) \end{bmatrix} = \begin{bmatrix} g_{1,1} - h_{1,1} & \cdots & g_{1,n} - h_{1,n} \\ \vdots & \ddots & \vdots \\ g_{n,1} - h_{n,1} & \cdots & g_{n,n} - h_{n,n} \end{bmatrix}^{-1} \begin{bmatrix} \log(\beta_1/\alpha_1) \\ \vdots \\ \log(\beta_n/\alpha_n) \end{bmatrix} \qquad (2.14)$$

which mathematically has the form of the (linear) matrix equation $Y = A^{-1}b$, when the terms are conveniently renamed. For most nonlinear systems, an analytical solution for the steady state is not generally available and simulation or other numerical methods must be employed. S-systems are distinctive in that this solution is so readily available.

The form of the solution provides insights into the relationship between model parameters and the steady state. Kinetic orders appear in the solution as differences between the order of influence in the influx and efflux terms. If species X_j appears in both the influx and efflux terms for X_i, the kinetic orders $g_{i,j}$ and $h_{i,j}$ are nonzero. If we increase or decrease both by the same amount, the difference remains the same. Since the steady-state solution depends only on differences in kinetic orders, the steady state is unaffected by this change. Rate constants appear in the solution as ratios of the efflux rate to the influx rate. So, increasing or decreasing both α_i and β_i by the same multiplicative factor has no effect on the steady state. The factor cancels out in the ratio and thus the steady state solution is unchanged. This result makes intuitive sense as we are simply increasing the amount of material produced and consumed during each time period by the same amount. The total amount remains the same. In many cases, however, we cannot independently modify the influx and efflux rates for species that are constrained by precursor–product relationships.

2.8.3 STABILITY

Biological systems must be robust with respect to an ever-changing environment. These systems have evolved to store nutrients and energy when available and retrieve it from storage when needed — all in an effort to maintain relatively stable concentrations of key metabolites. This robustness is evident in the ability of biological systems to respond to short-term changes in the environment and quickly to return to the desired steady state. The ability to return to the steady state following small perturbations is an indication that the system has a locally stable steady state. Steady-state solutions can be stable, as expected in a biological system, but may also be marginally stable or unstable. A stable steady state can be described as a marble sitting at the bottom of a bowl. Pushing it up the side of the bowl and releasing it perturbs the marble's state. After some time, it settles to the bottom of the bowl at the original, stable steady-state point. In the marginally stable case, small perturbations cause the system to assume a different steady state. A marble sitting on a flat table characterizes this case. Pushing it to one side or another perturbs the marble. When released, the marble stays at its new position, the new steady state. The marble can assume an infinite number of possible steady states corresponding to the possible positions on the table surface. In the unstable case, small perturbations cause the system to move away from steady state. This case can be characterized as a marble sitting on the top of an inverted bowl. As long as the marble is perfectly balanced on the top of the bowl, it remains in that steady state. A slight perturbation in the marble's position will cause it to move away from the steady state. Perturbations in an unstable steady state may cause the system to move to another steady state or may cause quantities of species to go to zero or infinity. In all of these cases, we are considering only the local stability of the system. Even for a locally stable steady state, the system may be moved out of this state with a sufficiently large perturbation.

The local stability of an S-system can be determined by linearizing the system about its steady state. The behavior of this linear system and that of the approximated S-system are similar in a small neighborhood about the steady state and thus an analysis of the linear system reveals the stability of the S-system [99]. The stability of a linear system $d\mathbf{X}(t)/dt = \Lambda\mathbf{X}(t)$ is determined by the eigenvalues of the coefficient matrix Λ. For an S-system, the elements of Λ are obtained by constructing a linear approximation to the expression for the rate of change of X_i about the steady state:

$$\Lambda_{i,j} = \frac{\alpha_i \prod_{j=1}^{n} X_j^{g_{i,j}}}{X_j}\left(g_{i,j} - h_{i,j}\right) = \frac{\beta_i \prod_{j=1}^{n} X_j^{h_{i,j}}}{X_j}\left(g_{i,j} - h_{i,j}\right) \tag{2.15}$$

The eigenvalues of the matrix Λ are then computed using methods from linear algebra, which are available in most scientific computing packages (MathCAD, Matlab, etc.). Eigenvalues are typically complex numbers with an imaginary and real part. The real parts of the eigenvalues of Λ determine the stability of the system. If any of the eigenvalues have a positive real part, the system will have an unstable steady state and if all of the eigenvalues have negative real parts, the system will have a locally stable steady state. For eigenvalues with zero real parts and nonzero imaginary parts, the system is likely to exhibit oscillatory behavior. These effects are illustrated in Figure 2.10, which shows the loci of eigenvalues for the linearized system along with sample time courses for the associated S-system as a function of $h_{2,2}$, one of the system parameters. As $h_{2,2}$ is varied from -1 to 3, the eigenvalues of the linearized system move along the plotted locus curve, starting with positive real parts (unstable), crossing the imaginary axis when $h_{2,2}$ is near $-\frac{1}{2}$ (oscillatory behavior), and ending with negative real parts (stable) throughout the remainder of the locus. Systems having eigenvalues with positive real parts show rapid growth or oscillations with increasing amplitude, which however may be bounded by a stable limit-cycle oscillation. Those with negative real parts approach the steady state either directly or following damped oscillations. Eigenvalue analysis of the linearized system is therefore a valuable predictor of the qualitative behavior of the nonlinear S-system. We are often interested only in evaluating the stability of the system and thus need only determine if the eigenvalues all have negative real parts. Forming the characteristic polynomial of the matrix Λ and using the Routh–Hurwitz analysis can eliminate the complicated process of explicitly computing the eigenvalues [55].

2.8.4 STEADY-STATE SENSITIVITY ANALYSIS

To this point, we have not differentiated between species whose quantities vary with time and those that, within our model, are fixed or controlled by some external or internal mechanism. We will refer to the variables representing the quantities of these time-varying species, whose behavior depends on the differential equations of this system, as *dependent* variables and those whose values are unaffected by the action of the model as *independent* variables. In cases where the values of these independent variables are fixed, their contribution to the rate law can be absorbed into the rate constant and the steady-state solution is as given above. In many cases, these independent variables represent environmental conditions or external signals whose influence we would like to investigate. To accomplish this, we need to make explicit the contribution of each independent variable on the steady state of the system. Following the results from above, we can partition the vector of variables into n dependent variables and m independent variables where, by convention, the dependent variables appear in the first n indices. We can then write the following solution for the dependent variables, $dX_D/dt = 0$, as

$$Y_D = A_D^{-1}b - A_D^{-1}A_I Y_I \tag{2.16}$$

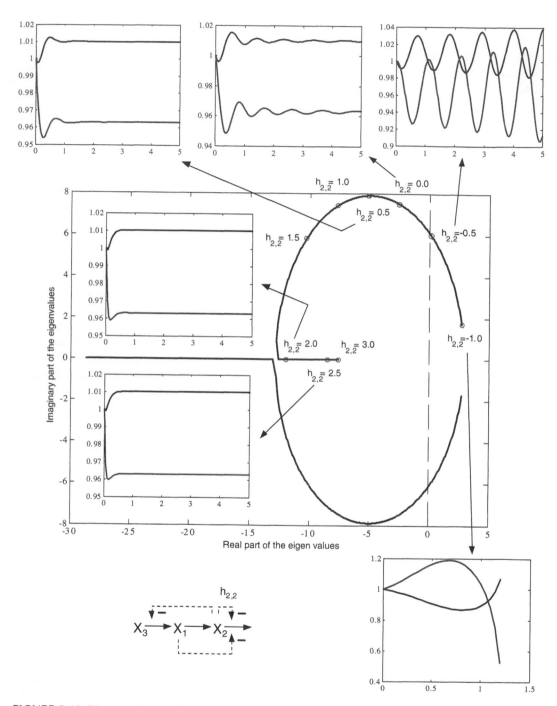

FIGURE 2.10 The central graph plots the locus of the eigenvalues for the given map as a function of parameter $h_{2,2}$, varied from −1 (inhibition) through 0 (no effect) to 3 (activation). For selected points on the locus, we also plot the time course for the system starting at (1,1). For eigenvalues near the imaginary axis, we see strong oscillation in the response with growing amplitudes for cases with positive real parts (unstable cases). The system demonstrates a stable steady state when the eigenvalues have negative real parts.

where

$$
Y = \log(X) = \begin{bmatrix} \log(X_1) \\ \vdots \\ \log(X_n) \\ ----- \\ \log(X_{n+1}) \\ \vdots \\ \log(X_{n+m}) \end{bmatrix} = \begin{bmatrix} Y_D \\ -- \\ Y_I \end{bmatrix} = \begin{bmatrix} \log(X_D) \\ ---- \\ \log(X_I) \end{bmatrix}
$$

$$
A_D = \begin{bmatrix} g_{1,1} - h_{1,1} & \cdots & g_{1,n} - h_{1,n} \\ \vdots & \ddots & \vdots \\ g_{n,1} - h_{n,1} & \cdots & g_{n,n} - h_{n,n} \end{bmatrix} \quad A_I = \begin{bmatrix} g_{1,n+1} - h_{1,n+1} & \cdots & g_{1,n+m} - h_{1,n+m} \\ \vdots & \ddots & \vdots \\ g_{n,n+1} - h_{n,n+1} & \cdots & g_{n,n+m} - h_{n,n+m} \end{bmatrix} \tag{2.17}
$$

$$
b = \begin{bmatrix} \log(\beta_1/\alpha_1) \\ \vdots \\ \log(\beta_n/\alpha_n) \end{bmatrix}
$$

It is again instructive to examine the form of the steady-state solution in logarithmic coordinates. The values of the dependent variables in logarithmic coordinates are determined by an expression linear in the logarithms of the independent variables and, therefore we find that $\partial Y_i/\partial Y_j = -c$ where the constant c is given by the entry (i, j) of the matrix $A_D^{-1}A_I$ for a dependent variable X_i and an independent variable X_j. From the chain rule of calculus we recognize that

$$
\frac{\partial Y_i}{\partial Y_j} = \frac{\partial \log(X_i)}{\partial \log(X_j)} = \frac{\partial X_i/X_i}{\partial X_j/X_j} \tag{2.18}
$$

This quantity can be interpreted as the relative change in the dependent variable $(\partial X_i/X_i)$ due to a relative change in the independent variable $(\partial X_j/X_j)$ and is referred to in BST as a *logarithmic gain*. The logarithmic gain of dependent variable X_i with respect to independent variable X_j is therefore defined as

$$
L(X_i, X_j) = \frac{\partial Y_i}{\partial Y_j} = \frac{\partial \log(X_i)}{\partial \log(X_j)} = \frac{\partial X_i/X_i}{\partial X_j/X_j} = -\left(A_D^{-1}A_I\right)_{i,j} \tag{2.19}
$$

A logarithmic gain of $L_{i,j} = L(X_i, X_j)$ indicates that a 1% change in the quantity X_j will cause an approximately $L_{i,j}\%$ change in the steady-state value of X_i and thus measures the gain of the system from X_j to X_i. Logarithmic gains are useful measures of how sensitive the system's steady state is to changes in the quantity of an independent species.

Biological systems are also robust with respect to changes in rates and affinities associated with the enzymes catalyzing reactions in these systems. In other words, biological systems tend to be insensitive to mutations that effectively change the rate constants and kinetic orders of our models. It has been proposed that robustness, which is related to sensitivity, be used as a measure of plausibility for models of biological systems. The steady-state solution to an S-system facilitates the analysis of sensitivity with respect to these parameter changes. The solution given above indicates that the steady-state solution in logarithmic coordinates is also linear in the logs of the rate constants. We see that $\partial Y_i / \partial \log(\beta_j) = k$ where the constant k is seen to be the entry (i, j) of the matrix $\mathbf{A}_D^{-1}$. As with logarithmic gains, the value k can be interpreted as the relative change in the steady-state value of X_i due to a relative change in the value of β_j. This value is called a parameter sensitivity and, for rate constants, is defined as

$$S\left(X_i, \alpha_j\right) = \frac{\partial Y_i}{\partial \log\left(\alpha_j\right)} = -S\left(X_i, \beta_j\right) = -\frac{\partial Y_i}{\partial \log\left(\beta_j\right)} = -\left(A_D^{-1}\right)_{i,j} \tag{2.20}$$

Again, a sensitivity of S indicates that a change of 1% in the given rate constant will result in an approximate $S\%$ change in the steady-state value of the given dependent variable. We also see that the sensitivities for the rate constants for the influx and efflux terms are related, as any change due to a relative increase in the influx should be equal to the relative change due to a comparable decrease in the efflux.

Similar results can be obtained for sensitivities with respect to changes in the kinetic orders. The sensitivity is defined similar to that given above:

$$S\left(X_i, g_{j,k}\right) = \frac{\partial Y_i}{\partial \log\left(g_{j,k}\right)}. \tag{2.21}$$

While not directly evident in the steady-state solution, kinetic order sensitivities are straightforwardly computed from the rate constant sensitivities, the given kinetic order, and the steady-state solution by way of implicit differentiation [100]. As with logarithmic gains, sensitivities measure the relative change in the steady state due to changes in the parameters of the model. Biological systems and models of those systems should demonstrate low sensitivity, and thus these measures are useful diagnostics in model development [101].

In addition to changes in the steady-state quantities, we are also interested in how the steady-state fluxes change with changing independent variables or parameters. In this case, we wish to determine the relative change in the flux through a species due to relative changes in independent variables or parameters. Flux, as used here, refers to the rate at which material flows into the pool of a given species. Since our analysis proceeds at steady state, this is also the rate at which material leaves the pool. This differs from the use of the term "flux" to describe the rate of flow along a given reaction since in S-systems, all of the reactions that produce (or consume) a species are aggregated into a single influx (or efflux) power-law term. The form and interpretation of logarithmic gains and sensitivities of fluxes with respect to independent variables and parameters are similar to that described above.

2.8.5 Precursor–Product Constraints

In most modeling applications, the modeler must be concerned with the conservation of flux or of moieties that result in constraints on the parameters of a GMA or S-system model. In metabolic pathways, for example, the efflux from species A is often the influx to species B as they participate as reactant and product in the same reaction. This implies that the power-law describing the rate of consumption of A must be the same as the production of B, thus placing constraints on the rate

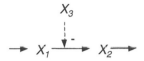

FIGURE 2.11 Simple linear pathway with a regulated step producing X_2 from X_1. The flow of material from X_1 to X_2 creates a precursor–product relationship that introduces constraints into the S-system model.

constants and kinetic orders in those two terms. These so-called precursor–product relationships must be considered in the design and analysis of a model.

The pathway illustrated in Figure 2.11 shows a system with three species X_1, X_2, and X_3 and three reactions. Writing the S-system directly from this map gives

$$\dot{X}_1 = \alpha_1 - \beta_1 X_1^{h_{1,1}} X_3^{h_{1,3}}$$
$$\dot{X}_2 = \alpha_2 X_1^{g_{2,1}} X_3^{g_{2,3}} - \beta_2 \tag{2.22}$$

and it appears at first that the system is defined by eight parameters. From the map, it is clear that the consumption of X_1 to produce X_2 must be balanced and therefore the influx to X_2 must be constrained to be equal to the efflux from X_1. Under these constraints ($\alpha_2 = \beta_1$, $g_{2,1} = h_{1,1}$, and $g_{2,3} = h_{1,3}$), the system has only five parameters and the following system of equations:

$$\dot{X}_1 = \alpha_1 - \beta_1 X_1^{h_{1,1}} X_3^{h_{1,3}}$$
$$\dot{X}_2 = \beta_1 X_1^{h_{1,1}} X_3^{h_{1,3}} - \beta_2 \tag{2.23}$$

As a result, when computing sensitivities with respect to parameters, we can no longer just take the (1,1) entry of the matrix $\mathbf{A}_D^{-1}$ to calculate the sensitivity of X_1 with respect to β_1. The desired sensitivity is given by the difference of the (1,1) entry and the (1,2) entry in $\mathbf{A}_D^{-1}$. If these entries have opposing effects, the difference may cancel what would otherwise appear to be a high sensitivity. Precursor–product relationships must be properly accounted for prior to the analysis of system sensitivities.

S-systems have a single term aggregating all reactions producing (or consuming) a species. In a simple merging pathway as shown in Figure 2.12, the efflux terms from species X_1 and X_2 are written as $\beta_1 X_1^{h_{1,1}}$ and $\beta_2 X_2^{h_{2,2}}$ and the influx term to X_3 is written as $\alpha_3 X_1^{g_{3,1}} X_2^{g_{3,2}}$, and the precursor–product constraint is

$$\alpha_3 X_1^{g_{3,1}} X_2^{g_{3,2}} = \beta_1 X_1^{h_{1,1}} + \beta_2 X_2^{h_{2,2}} \tag{2.24}$$

FIGURE 2.12 A simple branched pathway introduces a precursor–product constraint that is precisely satisfied only at the operating point in an S-system model.

In general, this expression can be satisfied at only one point. When the parameters α_3, $g_{3,1}$, and $g_{3,2}$ are determined by a power-law approximation to the sum of power laws on the right, the equation is satisfied at the operating point. Again, care must be used in interpreting the sensitivities for α_3, $g_{3,1}$, and $g_{3,2}$ which are functions of β_1, β_2, $h_{1,1}$, and $h_{2,2}$.

2.8.6 MOIETY CONSERVATION CONSTRAINTS

In addition to precursor–product constraints, we must often deal with constraints due to the conservation of moieties in the system. Figure 2.13 illustrates a simple phosphorylation–dephosphorylation reaction pair controlled by an external signal, a pattern common in signal transduction cascades such as the MAP kinase cascade. In the illustrated model, the total amount of the enzyme E is conserved but shifted from the inactive (unphosphorylated) to the active (phosphorylated) form as a result of an external signal. The S-system model of this system is given by

$$\dot{X}_1 = \alpha_1 X_2^{g_{1,2}} - \beta_1 X_1^{h_{1,1}} X_3^{h_{1,3}}$$

$$\dot{X}_2 = \beta_1 X_1^{h_{1,1}} X_3^{h_{1,3}} - \alpha_1 X_2^{g_{1,2}}$$

$$(2.25)$$

where the precursor–product relationships are included. We can immediately see that $\dot{X}_1 = -\dot{X}_2$ and, as a result, the system is found to have an infinite number of steady-state solutions. When we constrain the sum X_T of the active and inactive forms to be fixed, the system reduces to a single differential equation of the form

$$\dot{X}_1 = \alpha_1 \left(X_T - X_1 \right)^{g_{1,2}} - \beta_1 X_1^{h_{1,1}} X_3^{h_{1,3}}$$

$$(2.26)$$

A steady-state solution for this equation can be determined using numerical methods. In S-systems, conserved moieties cause the matrix A_D to have less than full rank and thus a unique closed-form solution cannot be obtained using the approach given above. The solution under the constraint $X_1 + X_2 = X_T$ appears at the intersection of the line $X_2 = X_T - X_1$ and a power-law curve giving the space of possible solutions to the underdetermined system represented by the two differential equations mentioned above. The numerical approach to solving the single differential equation above is equivalent to finding this intersection. It is also possible to find the steady state by numerically integrating the pair of equations given above. Since the moiety conservation relationships are inherent in the differential equations, the steady-state solution is guaranteed to adhere to these constraints at least to the accuracy of the numerical solution. This approach has the advantage of eliminating the need to create a reduced set of equations at the expense of a numerical solution to the differential equations.

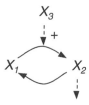

FIGURE 2.13 Conserved moieties create constraints that must be considered in the model development process. In the map shown above the total $X_1 + X_2$ remains constant which leads to loss of a degree of freedom and thus a constraint on the parameters of the model.

2.8.7 System Dynamics

2.8.7.1 Solving the System

The preceding paragraphs focused on the steady-state solution to the S-system and the properties of that solution. The stability of the steady-state solution, sensitivity with respect to parameter variation, and sensitivity with respect to changes in the independent variables were all shown to be easily determined from the steady state. These properties give us limited, yet important insight into the dynamics of the system. Dynamical analyses beyond stability and robustness of GMA or S-system models require numerical methods that integrate the differential equations starting from a selected set of initial conditions and predict the quantities of each species at a future time using the current state and the value of the derivative at the current time. In the simplest of these methods, Euler's method, the future state is projected from the current state with the formula $x(t + h) \approx x(t) + h \cdot dx(t)/dt$, where h is the size of the time step of the integration. More complicated methods avoid the accumulation of numerical error in the solution, which would result from this first-order approximation. Numerical integration of biological systems models is also complicated by the fact that many biological systems include reactions operating on time scales ranging from milliseconds to days. Systems of differential equations with very different time scales often lead to what are called *stiff* systems [102]. These systems are particularly difficult to integrate using numerical methods, and special algorithms have been developed to accommodate these difficulties. The choice of algorithm for the solution of these systems is problem specific and must be addressed for each model. Modern scientific computing packages such as MathCAD and Matlab include a variety of easy-to-use methods that trade off compute time, accuracy, and support for stiff systems.

The power-law terms in GMA and S-systems offer particular advantages in developing numerical solutions. It has been shown that writing the solution as a Taylor series in logarithmic coordinates allows the necessary higher-order derivatives to be computed from a straightforward recursion involving the current state of the system, the kinetic orders, and rate constants [103]. The ability to compute many higher derivatives efficiently allows us to construct high-order Taylor series about the current state with great ease. With these Taylor series, future values of the system state can be quickly computed and the error of prediction estimated. An order and step-size adaptive algorithm has been implemented using this approach [104,105].

2.8.7.2 Visualization of Time Courses

Visualizing the dynamic response of a system is an important part of understanding the system's behavior. Typically, two types of graphs are used to display modeling results: the time course plot and the phase-plane plot. Time course plots are simple graphs of the quantity or concentration of a species, plotted on the y-axis, as a function of the simulation time, plotted on x-axis. The quantities of several species can be plotted on the same graph or as a series of stacked graphs in the form of a strip chart. Stacked strip charts are often useful when each of the species have very different ranges. Time course plots are useful in determining, for example, the steady state of the system, the time required to settle to that steady state, and the behavior of the system while approaching the steady state (e.g., a damped oscillation). For systems with periodic behavior, the time course gives us the period of the oscillation, the minimum and maximum amplitudes, and the relative phase of oscillation in the modeled species. Time course plots are also useful when overlaying experimental data with predicted data. Examples of time courses are given in Figure 2.14.

2.8.7.3 Visualization of Dynamics in the Phase Plane

Phase plane plots provide a convenient method for viewing the behavior of one species as a function of another as a trajectory of points in time. This format is often useful in viewing limit cycles and other oscillatory behaviors. For each time step, one plots a point on the phase-plane whose x-value

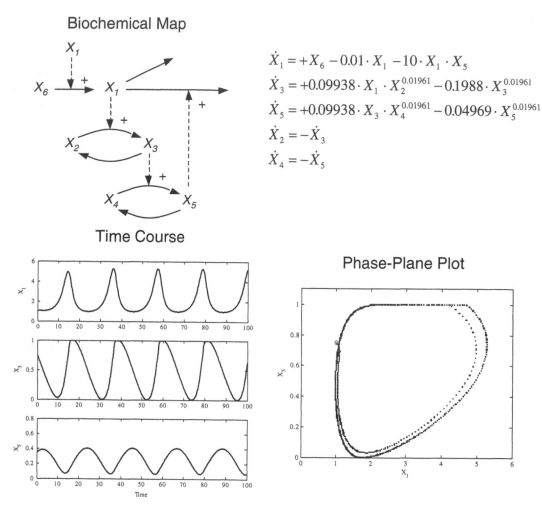

FIGURE 2.14 Example of time course and phase plane plots for a simple negative feedback oscillator. The original model, taken from (Tyson 2003), was converted into a GMA system using the procedure described above. The system was then integrated for 100 time units and variables X_1, X_2, and X_3 were sampled during the integration. Time course (left) and phase plane plots (right) were then constructed. The phase plane plot shows data points sampled uniformly in time and the spacing between successive points is therefore an indication of the rate of change of the system.

is given by the current quantity of species A and whose y-value is given by species B. Successive points in time are connected by lines to create a trajectory in the domain of species A and B. Phase-plane plots provide a view of the behavior of a pair of species in the system. If, for example, the system enters a stable limit cycle, the trajectory will repeatedly trace out a closed curve in the phase-plane showing the trajectory of the limit cycle. If the system exhibits damped oscillations settling to a stable steady-state point, the phase-plane trajectory will spiral toward the stable point. Several starting points can be plotted on the same graph to illustrate the path taken by the system as it evolves under different starting conditions, often approaching the same limit cycle or stable point. Phase-plane plots, however, lose the sense of time found in a time course plot since there is no indication of time along the trajectory *per se*. This can be amended by plotting symbols or points for each pair. If the time samples in the plotted data are equally spaced, the spacing of the points gives an indication of the rate of change of the two species along the trajectory. Points that are spaced closer together indicate a slow rate of change while those far apart indicate a rapid rate of change. Another disadvantage of phase-plane plots is that we can only view the relationship between two species at a time. Three-dimensional phase-plane plots can be constructed but must usually be rotated into a 2D projection or include pair-wise projections to give a feel for the data. Examples of phase-plane plots are given in Figure 2.14. Both time course and phase-plane plots are useful in the visualization of dynamic response data and can be easily constructed from the numerically integrated time course.

2.8.8 PARAMETER ESTIMATION

2.8.8.1 From Rate Laws to Power Laws

BST has been effectively applied to a variety of biological and biochemical systems including anaerobic fermentation [106], citric acid production in *A. niger* [107], the tricarboxylic acid cycle in *D. discoideum* [86], the glyoxylase system [108], trehalose cycle [56] and sphingolipid metabolism in yeast [109], and purine metabolism in man [110–112]. In most BST-based modeling applications to date, the development process began with an initial map of the biochemical network and a set of rate laws and associated parameters based on Michaelis–Menten-type mechanisms. An important step in the development of these models, therefore, was the construction of the corresponding power-law rate laws. We review this step, beginning with the identification of the operating point at which the approximation is to be made. The previous discussion has indicated that, at the operating point, the advancement rate of the reaction and the slope of change in that rate due to any species effecting the reaction match that of the original rate law. Typically, we are interested in the system behavior about an observed steady state and so we often select a steady-state point as operating point. Thus, the Michaelis–Menten system and both the GMA and S-system representations are equivalent at this point. Once the operating point is selected, the computation of the power-law approximations is a matter of straightforward partial differentiation.

Because of its prevalence, the Michaelis–Menten function is a good subject for illustration. Thus, we begin with $V_{MM}(S) = V_{max} \cdot S/(K_m + S)$, where V_{max} and K_m determine the maximum rate and the substrate concentration yielding half of that rate, respectively. There is only one species for which we need to calculate a kinetic order, and we therefore know already the symbolic form of the power-law approximation, namely $V_{PL}(S) = \gamma \cdot S^f$. The kinetic order f is the slope of $V_{MM}(S)$ with respect to S in logarithmic coordinates evaluated at the operating point S_{op}:

$$f = \frac{\partial \log\left(V_{MM}(S)\right)}{\partial \log(S)}\Bigg|_{S=S_{op}} = \frac{\partial V_{MM}(S)}{\partial S} \frac{S}{V_{MM}(S)}\Bigg|_{S=S_{op}} = \frac{K_m}{K_m + S_{op}}. \tag{2.27}$$

Computing kinetic orders in this way guarantees that the rate of change of the rate with respect to each species affecting the rate matches that of the original rate law. Once all kinetic orders have been computed, one determines the rate constant such that the rate of the Michaelis–Menten function matches that of the power-law representation at the operating point. Equating the two yields immediately

$$\gamma = \frac{V_{MM}\left(S_{op}\right)}{S_{op}^f} = \frac{V_{max} S_{op}^{1-f}}{\left(K_m + S_{op}\right)} \tag{2.28}$$

This process is repeated for each reaction in the system. The GMA model of the system is then written directly from these rates and the stoichiometry using the procedures described previously. If the rate laws used to describe biochemical reactions are more complicated, one easily employs modern computer algebra systems such as Maple® or Mathematica® to automate the process of differentiating the entire network of rate laws and evaluating them at the operating point.

2.8.8.2 Parameter Estimation from Time Course Data

Advances in experimental methods are offering the possibility of collecting simultaneous time series data for many of the metabolites in a system. Coupled with nonlinear regression methods or genetic algorithms, one can estimate the parameters of the model directly from these experimental data. At present, much of this work is ongoing and a superior approach has not yet emerged. One of the proposed methods preprocesses the time series data to estimate the derivative of the time course of each measured metabolite using either the slope of the time course or the first derivative of a smooth curve that is computed to run through the time course [113]. Without this step, regression of the parameters on the time series itself requires costly integration of the predicted time courses for each combination of parameter values attempted by the nonlinear regression algorithm [140]. Using slopes as estimates of derivates, the problem of finding the parameters is reduced to a nonlinear algebraic regression problem. Methods using, for example, different means of smoothing and alternative gradient searches and genetic algorithms to accomplish the regression are presently being investigated.

Most directly, the parameters of the power-law model could, in principle, be computed from experimental data measuring the kinetics of a reaction as a function of substrate or modifier concentration. When plotted in logarithmic coordinates, the slope of the line through the data points estimates the kinetic order of the system and the measured rate at a selected operating point.

2.9 APPLICATIONS OF BIOCHEMICAL SYSTEMS THEORY

2.9.1 MODELING AND SYSTEMS ANALYSIS

Systems biology is often associated with the development of models of biological systems, and BST has been used extensively in this regard. Using the methods described previously provides a path from maps of biochemical systems and associated rate laws or time series data to the structure and parameters of a representative model based on the power law formalism. By itself, the model development process often yields insights into the systems as it forces the developer to investigate and question the interactions that have a significant influence on system behavior. For models based on the power-law formalism, this often becomes equivalent to questioning whether a particular kinetic order should be 0 or not. The canonical representations used for GMA and S-system models and the direct association of model parameters and elements of the biochemical map make it possible to translate these questions about system structure into questions about the parameter values in the model. The modeling process also often reveals if variables or processes are missing, because the analysis may show that no parameter combinations could describe some of the observed data. The diagnostic tools of BST are very valuable in this regard.

Completed models offer possibilities for both explaining observed system behavior and predicting new behavior. Consider the model of purine metabolism constructed by Curto and coworkers [110–112]. This model represents a complex network of reactions involving 18 metabolites, 23 enzymes, 37 reactions, and numerous regulatory influences leading from ribose-5-phosphate to uric acid. With this model, the authors were able to examine disease states due to inherited mutations in critical enzymes such as phosphoribosylpyrophosphate synthetase (PRPPS) and hypoxanthine-guanine phosphoribosyltransferase (HGPRT) by appropriately scaling rate constants in steps catalyzed by these enzymes. The new steady state of the system was then determined using the methods described above and compared to clinical observations.

Once validated, these types of models can be used to predict metabolic responses to enzyme deficiencies or persistently increased enzyme activities and thus provide systemic means for exploring a disease condition. In a similar fashion, the models can be used to identify sites that may provide particularly effective therapeutic control points and to evaluate the effects of drugs targeted to affect specific enzymes. Because the models ideally represent the entire relevant pathway, they allow the analysis of desired treatment effects and of potential side effects. In the purine example, the system was primarily assessed close to a healthy or diseased steady state, but since the model also represents the dynamic behavior of the system, it is possible to study transient effects as well. One could, for example, easily simulate the system in response to time-varying conditions, such as the periodic administering of a drug, and study the transient responses of the system to a given dosing regimen.

Numerous examples of model development using BST are available in the literature and the reader is referred to them for additional insights. There is no one solution fitting all modeling needs, and models do not typically emerge in a single pass through the model development process. It is therefore instructional to consult the literature for case studies that show the iterative modeling and refinement process [114].

2.9.2 CONTROLLED COMPARISONS OF BIOCHEMICAL SYSTEMS

A central theme in the application of BST has been the comparative analysis of biochemical systems. As the analysis of biochemical networks has grown, common patterns of design have emerged. Feedback inhibition, for example, is often used as a regulatory mechanism in nature, where the final product of a biosynthetic process inhibits an early step in that process. Since biological systems tend to reuse successful design patterns, one should expect some benefit to the organism from selecting this mechanism of control over some other mechanism. Early in the development of BST, Savageau recognized that a quantitative method was needed to conduct objective comparisons of observed and hypothesized design alternatives, and that these comparisons were instrumental for answering questions of why nature seems to favor particular designs over other apparently reasonable alternatives. As a result, the Method of Mathematically Controlled Comparisons was developed [115]. This method is often described as the computational equivalent of a well-controlled laboratory experiment. The systems being compared are made to be as similar as possible so that the results are not confounded by extraneous differences. In a laboratory experiment, this might correspond to using genetically pure strains of a bacterium, which are maintained at conditions that differ only in the variable of interest. Analogously, the computationally equivalent reference and alternative designs are formulated as BST models and compared using measures of functional effectiveness that are thought to represent evolutionary advantages. Because the two models are constructed to differ only in the one feature of interest, better performance of one design over another can be attributed to this difference. Controlled comparisons have been used to study design alternatives in genetic circuitry, regulation in the immune system, regulation in biosynthetic pathways, and irreversible step positions in biosynthetic pathways. Because of their conceptual importance, it is useful to sketch out the strategy of a controlled comparison in more detail.

The goal is to compare objectively the benefits of one regulatory structure over another with respect to some quantitative measure such as robustness, gain, response time, or flux. The method begins with the development of mathematical models for the reference and alternative structures using the S-system modeling framework. The S-system model is beneficial in this comparison because parameters of the model map directly to regulatory influences in the system. The alternative design is allowed to differ from the reference at one reaction, the subject of the analysis. All other reactions remain the same. In the example shown in Figure 2.15, the two systems differ only in the presence of feedback inhibition from the product back to the first step in the reaction. The parameters for reactions other than at the step of interest are constrained to be equal, thereby forcing the two systems to be *internally equivalent*. In a laboratory experiment, this would correspond to a gene mutation causing a change in the system at only one step without affecting performance of other parts of the system. The parameters of the modified reaction in the alternative are not yet defined and therefore represent degrees of freedom that must be controlled for in the experiment.

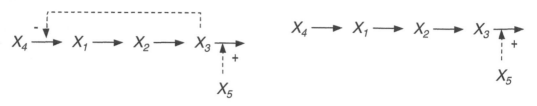

FIGURE 2.15 Two possible designs for a linear biosynthetic pathway. The reference design includes feedback inhibition of the first step by the end product of the pathway.

To this end, one places constraints on the parameters of the alternative system, which make the system as similar to the reference system as possible, except for the difference of interest. Typical constraints assure that external properties, such as steady-state concentrations of certain metabolites, selected logarithmic gains, or selected sensitivities, are the same in both systems. Since the systems are defined as S-systems, closed-form solutions often exist for these systemic properties, and expressions for the property taken from the reference and alternative are set equal and solved for one of the unconstrained parameters in the alternative system. Additional constraints are applied until all of the previously unconstrained parameters in the alternative are fixed. An alternative constructed in this manner is considered to be *externally equivalent* to the reference because, to an external observer, the reference and the alternative exhibit the same behavior with respect to the constrained properties. The reference and its internally and externally equivalent alternative can now be compared with respect to the predefined set of functional performance measures. These functional performance measures can again often be expressed in closed form, which permits a comparison of the reference and alternative in terms of the ratio of these measures. Since the sign of the model parameters is known, the comparison often leads to a very general conclusion of preference that is independent of the numerical parameter values in the system. In cases where the preference cannot be determined independent of the parameter values, statistical approaches can be applied [53]. In this case, sample parameter sets are drawn at random from distributions representing prior beliefs about the range of the parameter values. Equivalent alternatives are constructed for each drawn reference and the functional effectiveness measures are computed for each pair. The ratios of these measures are then used in graphical and statistical methods to assess preference for one design over the other. Studying preferences for one design over another leads to an understanding of the general benefits of the design and helps us find and understand applications of that design elsewhere in biological systems.

As an illustration, both biosynthetic pathways shown in Figure 2.15 implement the synthesis of X_3 from X_4 through intermediates X_1 and X_2, with the demand determined by X_5. In the reference design, the initial step is inhibited by X_3, a regulatory signal not found in the alternative. The S-system for the reference can be directly determined from the biochemical map in the figure and is given by

$$\dot{X}_1 = \alpha_1 X_3^{g_{1,3}} X_4^{g_{1,4}} - \beta_1 X_1^{h_{1,1}}$$

$$\dot{X}_2 = \beta_1 X_1^{h_{1,1}} - \beta_2 X_2^{h_{2,2}} \qquad (2.29)$$

$$\dot{X}_3 = \beta_2 X_2^{h_{2,2}} - \beta_3 X_3^{h_{3,3}} X_5^{h_{3,5}}$$

From the map, we know immediately that the parameter $g_{1,3}$ must be negative as it inhibits the production of X_1. All other parameters are positive. The one difference between the reference and alternative is the presence of feedback inhibition from X_3 to the production of X_1. This inhibition is represented by the kinetic order $g_{1,3}$ in the system of equations given above, and the system of equations for the alternative can be obtained by simply setting $g_{1,3} = 0$. We see that the S-system representation allows us to associate a structural change in the system design (loss of inhibition) with a specific parameter in the model. For the purposes of this comparison, we write the equivalent alternative, without inhibition, as

$$\dot{X}_1 = \alpha_1' X_4^{g_{1,4}'} - \beta_1 X_1^{h_{1,1}}$$

$$\dot{X}_2 = \beta_1 X_1^{h_{1,1}} - \beta_2 X_2^{h_{2,2}} \qquad (2.30)$$

$$\dot{X}_3 = \beta_2 X_2^{h_{2,2}} - \beta_3 X_3^{h_{3,3}} X_5^{h_{3,5}}$$

We have indicated that the parameters α_1' and $g_{1,4}'$ differ in the reference and alternative whereas all other parameters are fixed by internal equivalence. To make these systems as similar as possible, we must determine values for α_1' and $g_{1,4}'$ that preserve selected external properties. For this example, we choose to force the alternative and reference to have the same steady-state value of X_3 and the same logarithmic gain in X_3 with respect to X_4. Closed-form solutions for the steady state of X_3 and for $L(X_3, X_4)$ are determined for the reference and equivalent alternative, set equal, and solved for α_1' and $g_{1,4}'$, giving

$$g_{1,4}' = \frac{g_{1,4} h_{3,3}}{h_{3,3} - g_{1,3}}$$

$$\alpha_1' = \exp\left(\frac{h_{3,3} \log\left(\alpha_1\right) - g_{1,3} \log\left(\beta_3\right) - g_{1,3} h_{3,5} \log\left(X_5\right)}{h_{3,3} - g_{1,3}} \right) \qquad (2.31)$$

Now that we have defined a reference and its equivalent alternative, we can compare selected functional performance measures. We consider how the steady-state value and flux of X_3 are affected by changes in demand given by X_5. The properties of interest are $L(X_3, X_5)$ and $L(V_3, X_5)$, which are easily determined in closed form for the steady-state systems. Taking the ratio of $L(X_3, X_5)$ in the reference to that in the alternative, we find

$$\frac{L_{\mathrm{Ref}}\left(X_3, X_5\right)}{L_{\mathrm{Alt}}\left(X_3, X_5\right)} = \frac{h_{3,3}}{h_{3,3} - g_{1,3}} \qquad (2.32)$$

Since $g_{1,3}$ is negative and $h_{3,3}$ is positive, this ratio is always less than one, which implies that the steady state of X_3 in the design with feedback inhibition is always less sensitive to demand than in the alternative design. Low levels of intermediates and low sensitivity to changing demand and supply are expected to be desirable properties in biosynthetic processes and so we would conclude that the system with feedback is preferred to the alternative. The logarithmic gains of the flux through X_3 with respect to demand for the reference and alternative are given by

$$L_{\mathrm{Ref}}\left(V_3, X_5\right) = -\frac{g_{1,3} h_{3,5}}{h_{3,3} - g_{1,3}} \qquad (2.33)$$

$$L_{\mathrm{Alt}}\left(V_3, X_5\right) = 0$$

We see that in the alternative design the flux through X_3 is unaffected by X_5, unlike in the reference system where X_5 controls this flux. The ability of the demand signal to control the rate of production of end product (efflux of X_3) is expected to be a desirable characteristic of a biosynthetic pathway. Again, the reference design with feedback inhibition demonstrates this behavior, whereas the alternative does not. For both of these effectiveness measures, we were able to compare the designs while only knowing the signs of the parameters. If, however, we had been interested in the settling time of the system in response to a step increase in demand, the statistical methods of Alves [53] and Savageau [55] would have been required.

Controlled comparisons provide insights into deep principles that cause the selection of one design over another. The key to performing these comparisons is the direct mapping of regulatory influences in biochemical systems to parameters in S-system models of those systems. The accessibility of a closed-form solution for the steady state and for properties local to the steady state enables the direct comparison of designs and, in many cases, the identification of preference for one design over another, independent of the system parameters. The comparative analysis of biochemical systems will be an important tool in attempts to understand why biological systems evolved as they did.

2.9.3 System Optimization

Early efforts to optimize production in biotechnological systems often failed as a result of a limited understanding of the distribution of control in biological systems. Increases in the enzyme thought to catalyze the "limiting step" in these systems failed to produce the desired result. The introduction of BST and metabolic control analysis (MCA) provided the quantitative framework needed to understand the relationships in complex systems more fully.

In many biotechnological applications, the primary objective is to modify a biological process by over- or underexpressing one or more enzymes with the goal of maximizing the production of a useful metabolite. The question of how to modify the process is often difficult to answer because the regulatory influences in the biological system tend to compensate for changes that one might want to introduce. To understand how to optimize production of the desired metabolite, we need a faithful model of the system that can be used to predict the results of such changes and that has a mathematical form suitable for solving the optimization problem.

The availability of an algebraic solution to the steady state makes the S-system modeling framework a good choice for solving such optimization problems [116,117]. Additionally, the fluxes in the system, given in power-law form, can also be written as linear expression in logarithmic coordinates. Hence, since the steady-state solution, metabolite and enzyme concentrations, and fluxes are all linear in logarithmic coordinates, the optimization of fluxes or metabolite concentrations is easily performed using standard linear programming techniques. Reported applications of these techniques include the optimization of citric acid fermentation by *A. niger* [107], ethanol, glycerol, and carbohydrate production in *Saccharomyces cerevisiae* [116,117], the catalytic efficiency of triosephosphate isomerase [118], and tryptophan [119] and L-(-)-carnitine production [120] in *Escherichia coli*.

2.10 METABOLIC CONTROL ANALYSIS

Parallel to BST, Kacser and Burns [121] and Heinrich and Rapoport [84] independently developed an approach on the basis of sensitivity analysis, which later coalesced into what is now called MCA. While the methods of BST and MCA came from different lines of thinking, they arrive at mathematical solutions that are quite similar. We include this short discussion because MCA has been widely applied, is occasionally compared to BST, and is in fact closely related to BST.

Early efforts to characterize regulatory enzymes in metabolic pathways focused on the identification of a so-called *rate-limiting step*. If such a step existed, it was argued, the overexpression of the enzyme catalyzing the limiting step should increase flux through the pathway. This, however, was not found to be the case in many instances, and researchers realized that control had to be distributed throughout the enzymes of the pathway. Kacser and Burns and Heinrich and Rapoport gave an explanation for this shared control by exploiting methods of sensitivity analysis and developing a quantitative approach for attributing control to steps throughout the pathway. They focused on assessing how small perturbations in individual independent variables of the system (e.g., enzyme activities) modified the steady-state behavior of the system as a whole. The small perturbations and changes to the steady state were measured relative to the unperturbed conditions and expressed as a ratio. This ratio was termed a "control coefficient" and is defined as

$$C_{e_i}^{J_k} = \frac{\partial J_k / J_k}{\partial e_i / e_i} \tag{2.34}$$

where J_k is the flux at steady state through the reaction catalyzed by enzyme e_k and e_i is the concentration or activity of the ith enzyme at steady state. Mathematically, this expression is precisely the logarithmic gain in BST of flux k with respect to the independent variable associated with the activity of enzyme i ($L(V_k, X_i)$). Thus, these control coefficients give an indication of one enzyme's effect on each flux in the system and are therefore more specifically called *flux control coefficients*. One can also measure the effect of an enzyme on the steady-state concentration of any one of the dependent variables in the system through the use of concentration control coefficients, which is analogously defined as

$$C_{e_i}^{x_k} = \frac{\partial X_k / X_k}{\partial e_i / e_i}. \tag{2.35}$$

Again, the mathematical form is precisely that of a logarithmic gain in BST. Kacser and Burns noted that the sum of the control coefficients, over all enzymes having an effect on the system equals 1:

$$\sum_{i=1}^{n} C_{e_i}^{J_k} = 1 \tag{2.36}$$

This result, known as the *summation theorem*, has important consequences in MCA. It shows that control is distributed throughout the system and that control coefficients are systemic measures giving each enzyme's contribution to this sum. If a new steady state is chosen or if some modification to the system changes one enzyme's activity, some or all of the control coefficients must change so as to preserve the sum. These results reinforce the concept of distributed control promoted by the developers of MCA.

Control coefficients are systemic properties as they relate changes in an independent quantity (e.g., enzyme concentration) to changes in the steady-state fluxes or concentrations of the entire system. MCA also provides methods for connecting the behavior of individual components of the system to the behavior of the whole. These local properties, called *elasticities*, measure the relative change in the rate of an individual reaction, isolated from the system, as a function of a relative change in a species in the model,

$$\varepsilon_{X_i}^{V_k} = \frac{\partial V_k / V_k}{\partial X_i / X_i} \tag{2.37}$$

Elasticities are mathematically equivalent to the definition of kinetic orders in BST. MCA emphasizes that elasticities, like control coefficients, vary with changes in the system steady state. Elasticities, as local properties, could be measured in experiments isolating an enzyme from the system and measuring the reaction rate under changing concentrations of a substrate. The linkage between these local properties and the systemic properties represented by the control coefficients is given by the *connectivity theorem*. This theorem states that the sum over all enzymes of the product of the flux control coefficient for that enzyme and the elasticity for that enzyme for a given species is 0:

$$\sum_{i=1}^{n} C_{e_i}^{J_k} \cdot \varepsilon_X^{V_i} = 0 \qquad (2.38)$$

This theorem provides the basis for expressing the control coefficients in terms of data measured for each enzyme in isolation.

We have touched on just a few of the important properties and theorems of MCA and recommend that readers interested in a more complete treatment refer to the well-written full-length texts dealing with the subject [122].

2.10.1 Relationship between BST and MCA

MCA emerged from the insufficiency of rate-limiting steps as an explanation for the effects of enzymes on the state of intact systems observed in experimental systems. There was a clear need to understand and quantitatively characterize the distribution of control in biochemical systems imposed by the enzymes of those systems. Control coefficients and elasticities were defined as properties that can be measured in laboratory experiments and, through the theorems of MCA, form an intuitive bridge between theory and application. Thus, this form of biochemical analysis facilitates the experimentalist's understanding of the distribution of control throughout the system by connecting global properties of a system to the local properties of its constituents. BST, by contrast, arose from the need to develop quantitative methods for the exploration of the behaviors of biological systems and their underlying design principles. So, BST has a more extensive foundation in system theory and targets model building, static and dynamic analysis, comparative analysis of alternative designs, and optimization of biotechnological processes. Thus, even though BST and MCA have common mathematical underpinnings, they have evolved along different lines and have often been applied to different types of scientific questions.

2.11 FUTURE

Some trends in the development of BST and the mathematical side of systems biology are quite evident; others, of course, are speculative. Clearly, models will continue to grow in size and increase in complexity. This trend has been and will be tied closely to the availability of more and better data, which allow the estimation of specific model parameters, as well as to increased computer power, wider access to faster computers and the development of higher-level programming languages. In addition to this growth, it will be necessary to develop new modeling techniques or adapt existing techniques to the peculiarities of biological systems. Ultimately, the community of systems biologists must develop theoretical foundations, first for tightly defined phenomena and then for growing domains of biology.

2.11.1 Model Extensions and Needs

Ten years ago, typical models in biology may have consisted of five to ten variables, whereas it is not uncommon now to see models containing dozens of differential equations. Without doubt, this trend will continue, and it is the declared goal of many investigators around the world to construct whole-cell models, first presumably of *E. coli* [123,124], even though other cells, such as hepatocytes or pancreas cells, as well as other microbial organisms are already being considered as representative targets [125–128]. Although several consortia have been formed around this goal [129] and substantial progress has been made, all involved in these efforts agree that it will take many years before a reliable model will be completed.

The first step toward any whole-cell model is almost certainly a complete stoichiometric flux model that contains all metabolites and fluxes. Draft models of *E. coli*, *S. cerevisiae*, and some other well-studied organisms already exist with at least preliminary quantification [130–132]. However, modeling the metabolic flux distribution is only the first, and arguably simplest, step and must be followed up with regulated metabolic models, which will include control mechanisms at different levels, including direct feedback and feedforward signals, signal transduction pathways, and pathway regulation at the genomic level. To understand metabolism at this level, we must also know the dynamics of proteins and, in particular, the concentration and activation state of enzymes and transcription factors. Clearly, these dynamic processes may occur at different time scales and are often subtler than enzyme-catalyzed conversions of metabolites. Furthermore, because of their great importance, the control structures within cells are frequently redundant, which complicates data collection and mathematical analysis.

A significant component of this type of exploration and analysis will entail extracting information from data obtained *in situ*. Especially valuable will be time series measurements, which implicitly reflect not only the mass flow structure of a system but also its regulatory control. As mentioned before, many groups have recently begun to develop methods for identifying systems from time series data; representative trends are reviewed in Veflingstad et al. [133]. These methods differ greatly in approach and complexity, spanning a range from systematic inspection [134] to various means of linearization [133,135–138] and attempts of identifying complete nonlinear systems at once [113,139,140].

At a higher, yet initially coarser scale, it will be necessary to understand the dynamics and regulation of multicellular organisms, which requires data and models for cell communication between like and different cell types. Again, this is not uncharted territory, and beautiful examples, such as bacterium-phage interactions [141] and the regulation of development in sea urchin embryo [142] indicate that analyses at this level will be very exciting and require specific mathematical methods that characterize systems biology.

With the increasing size of models, the question becomes more pressing of whether it is valid and feasible to dissect systems into subsystems or modules and to analyze the entire system by considering the modules as units, which could be studied more or less in isolation. In some sense, this approach runs counter to the idea of systems biology and the integrity of cells and organisms. At the same time, we are learning more about the modularity of actual biological phenomena, and it does not appear to be far fetched to map this modularity onto mathematical descriptions, which still preserve some of the connections between modules and among hierarchical layers of organization and control. Some of these ideas go back at least to Rosen [143] and are now experiencing a revival [58,144,145]. As a specific example, it is possible to decompose complex metabolic networks into simpler units performing a coherent function by means of elementary-mode analysis [146]. This type of analysis permits reconstruction of network topology, analyses of robustness, redundancy, efficiency and flexibility, and the prediction of functional aspects of genetic regulation.

A different way of thinking about hierarchy and modularity is to associate modules with what we learn from the exploration of design and operating principles. The argument would be that an organism has to execute particular tasks, such as the conversion of substrate into easily interchangeable energy or an effective protection against some environmental stress. Having studied and truly understood different designs in other organisms, for instance, as a result of controlled comparisons, along with optimal strategies for accomplishing given tasks under specific conditions, it might be possible to predict quite reliably the best-suited designs and operating strategies and import them into models of the organism of interest, without studying the details of this module further, either experimentally or through modeling. As an architect designs the blueprint for a house without initially worrying about every detail of amounts and types of materials to be used in the actual construction, we will be able to develop biological models at different organizational levels without being impeded by minute details at other levels. At an abstract level of network analysis, simple structural design motifs have been found in transcription factor networks, neurons, food webs, and technological systems [147].

If the modular structure of a biological phenomenon extends throughout different layers, it will be an important task of systems biology to map the hierarchical control onto effective mathematical and computational representations that permit analysis and insight [148,149]. The telescopic property of S-systems [150,151], which allows the structure-invariant inclusion of modules from lower levels into higher-level models via power-law approximation, may render BST a particularly powerful framework for hierarchical systems.

The vast majority of dynamic biological models rely on ordinary differential equations, which describe rates of production, conversions, and consumption. This representation implicitly assumes some degree of continuity, homogeneity, and averaging within the investigated systems. For some systems and processes, these assumptions may be satisfied, but it is becoming increasingly clear that they are often violated to a point where one has to question the applicability of continuous, deterministic models. Three classes of problems arise. One has to do with the fact that the cell is not a bag holding some homogeneous soup of life, but is highly structured and quite densely packed. The second issue is that some interactions may involve only a small number of a particular system component and therefore become highly stochastic and require corresponding methods. Examples are biochemical reactions, where enzyme or substrate molecules are present in a cell in only very few copies, and interactions between cells and viruses [141]. The third issue is the spatial organization of many biology processes, which ultimately will require partial rather than ordinary differential equations.

2.11.2 COMPUTATIONAL SUPPORT

With the widened focus of systems biology and the growth of models, both in size and complexity, comes the need to exchange models among researchers with greater efficiency than is presently possible. While there is an abundance of specific software packages for particular purposes, only a small percentage of these packages allow the user to add new features or to include modules from different programs. As a consequence, many model analyses are never independently validated outside the author's laboratory and many phenomena are modeled time and again, because it seems easier to compose new code than to understand the models of others in sufficient detail to incorporate them in one's work. To counteract these problems, higher-level modeling languages like Mathematica and MatLab have been developed, and there is a strong push to standardize both data storage and models across platforms. Data and model representation seems to be converging to structures in so-called *mark-up languages*. Originally created for website design, these languages have become more general, and specifically the "eXtensible Markup Language" XML [152,153] may become a universal basis for structuring documents and for high-level modeling systems such as CellML [154] and the systems biology mark-up language SBML [155]. A particular benefit of computational standards is that they facilitate the exchange of topological flow diagrams, such as biochemical maps, as well as the computational and mathematical descriptions of the corresponding models and, therefore, make both the model structures and mathematical representations more easily accessible to independent testing within the community of systems biologists.

As biological systems models grow in size and complexity and require large-scale simulation, it is becoming gradually more difficult to interpret and understand their output and reliability. This renders it necessary to complement simulation efforts with a different type of systems analysis that attempts to discover general — or even universal — design and operating principles. As discussed before, such principles govern the functioning of entire classes of phenomena, and their discovery truly explains why a particular organizational or regulatory structure or procedure is superior to alternative designs. Since BST has been the dominant modeling framework for such analyses, a specific software, BSTLab [156], was developed to facilitate semiautomated controlled comparisons for investigations of design and operation. BSTLab is SBML compatible, which makes it automatically accessible to a growing community of systems biologists.

At present, computational tools supporting systems biology target the extreme ends of sophistication within the user spectrum, which, however, is not only growing in size, but also in diversity of backgrounds. As it stands, nonexperts have no choice but to use the fixed-function tools developed by experts and must rely on those experts to make enhancements to the tools. Experts cannot go beyond existing packages either, because these are typically closed source. They therefore typically resort to general-purpose computation environments and create from scratch the particular code they need, thereby presumably redoing what others composed before. Evolving scientific software for systems biology needs to bridge the gap between experts and nonexperts and thus support a "continuum" of users. Current desktop business applications have accomplished this by providing menu and forms-based interfaces to preprogrammed functions for nonexperts, macros for users needing to automate repetitive tasks, scripting (e.g., Visual Basic for applications [157]) for experts needing to implement customized behavior, and automation (e.g., the component object model [158]) for programmers who need to take control of the software. These features vastly broaden the spectrum of users supported by a given tool and must be made available to systems biologists of different educational backgrounds.

2.11.3 APPLICATIONS

The premier goal of systems biology will of course continue to be a true understanding of how biology works. This goal may sound purely academic, but it is evident that it will be the basis for uncounted applications, most of which are unforeseeable at this time. These applications will revolutionize medicine, biotechnology, agriculture, use of the environment, and about every aspect of dealing with our living surroundings. Leroy Hood and Hiroaki Kitano, two of the strong proponents of systems biology, agree in their prognoses that, based on knowledge of biological systems, predictive, preventive, and personalized medicine and the design of multiple drug therapies and therapeutic gene circuits will dominate medicine of the 21st century [58,159]. Already pharmaceutical companies are investing large amounts of money into high-throughput data generation [160,161], and it is clear that this experimental side of systems biology will soon be accompanied by better computational and systems-based analyses [60].

Combined with the miniaturization of biotechnology and the development of "labs on chips" [162], it may be possible in the future to obtain personalized *in vivo* proteomic and metabolic profiles with minimal invasion and even to perform microperturbations that show the individual's responsiveness to disease agents and treatments. As we now hook up a car to diagnostic computers that show, within seconds, what is wrong with its complex electronics, we may soon be able to obtain a comprehensive picture of the genomic, proteomic, and metabolic state of a patient's target tissue. Whole-cell bacterial and virus interaction models will allow us to predict what types of interventions will be most efficacious in interrupting infectious disease processes without burdening the human body with side effects. We may even truly understand how microorganisms become resistant and how our drugs can stay ahead of them. Controlled cell lines will allow us to grow tissues as per specification.

Biotechnology will be as prominent as computer science and engineering are now. Following the present and ongoing phases of studying and reverse engineering biological systems [144], we will be able to start optimizing not just individual pathways [82] or stoichiometric flow patterns

[130], but whole organisms with all their means of regulation and control mechanisms, their sensing and signal transduction, and their adaptability and mutability. We will begin constructing living systems from modules, "nature-prefabricated" parts, or even from scratch [163–165], and thereby enter a phase of synthetic biology [57], with all its obvious and hidden promises and threats.

2.12 CONCLUSION

For over 5000 yr, mankind has been enjoying bread and beer. The Egyptians had enough of an understanding to be successful with the required fermentation processes, even though they did not know the main ingredient, yeast. Today, we have quite intricate knowledge of this organism, with *Medline* producing about 80,000 hits for keyword "yeast." Yet, do we know exactly whether a yeast cell would survive a so far untested knockout mutation or why bleach kills the organism? What is it that truly distinguishes a living cell from a dead one? In comparison to the Egyptians of antiquity, we understand the biochemistry and physiology of fermentation extremely well, but we are not yet at a point where we could take a yeast cell apart and successfully put it back together, or where we would be able to design from scratch a new fermentation organism. Systems biology strives to achieve such capability with large-scale experimentation and global mathematical and computational analyses.

Systems biology is a young science and its goals are lofty. We cannot expect to complete comprehensive whole-cell or whole-organism models any time soon, and a theory of biology is not even on the horizon. One must therefore ask whether there are achievable shorter-term targets or milestones against which we could measure progress in systems biology. Several decades ago, Mesarović [23] formulated such a milestone, when he wrote: "The real advance in the application of systems theory to biology will come about only when the biologists start asking questions which are based on the system-theoretic concepts rather than using these concepts to represent in still another way the phenomena which are already explained in terms of biophysical or biochemical principles." There are early indications that we are beginning to head in this direction. Biologists are starting to discuss research problems with modelers and systems scientists and some are showing guarded willingness to perform experiments that may not be of great intrinsic interest in a reductionist sense but have value for the construction and numerical implementation of mathematical models. After overcoming their initial, deep-rooted skepticism about the relative simplicity of mathematical models in comparison to biological complexity, some biologists are finding out that systems biology is a mind-set that can help formulate hypotheses and decide which critical variables may support or invalidate a mental model. They also begin to appreciate that it is frequently the *process* of modeling rather than the finished model that is the most valuable outcome [10] and that wrong models often tell more than models that fit some data well. While some biologists are developing a taste for modeling, growing numbers of mathematicians and computer scientists are beginning to study biology in its fascinating detail and thereby subject themselves to the "bilingual" education [45] that is a necessary prerequisite of systems biology.

ACKNOWLEDGMENTS

This work was supported in part by a Quantitative Systems Biotechnology grant (BES-0120288; E.O. Voit, PI) from the National Science Foundation, a Cancer Center grant from the Department of Energy (C.E. Reed, PI), and a National Heart, Lung and Blood Institute grant (N01-HV-28181; D. Knapp, PI). JHS was supported by a training grant from the National Library of Medicine (T15 LM07438-01). Any opinions, findings, and conclusions or recommendations expressed in this material are those of the authors and do not necessarily reflect the views of the sponsoring institutions.

REFERENCES

1. Enfors, S.O., Baker's yeast, in *Basic biotechnology*, B. Kristiansen, ed. 2001, Cambridge, U.K.; New York, NY: Cambridge University Press.
2. Currie, J.N., *On the citric acid production of Aspergillus niger.* Science, 1916. 44: 215–216.
3. Kitano, H., *Systems biology: a brief overview.* Science, 2002. 295(5560): 1662–1664.
4. Kitano, H., *Looking beyond the details: a rise in system-oriented approaches in genetics and molecular biology.* Curr Genet, 2002. 41(1): 1–10.
5. Savageau, M.A., *Reconstructionist molecular biology.* New Biol, 1991. 3(2): 190–197.
6. Leicester, H.M., *Development of biochemical concepts from ancient to modern times.* Harvard monographs in the history of science. 1974, Cambridge, MA: Harvard University Press. 286.
7. Garraty, J.A., P. Gay, and Columbia University, *The Columbia History of the World.* 1st ed. 1972, New York: Harper & Row. xx, 1237.
8. Kuhn, T.S., *The Structure of Scientific Revolutions.* 1962, Chicago: University of Chicago Press. xv, 172.
9. Laszlo, E., *The Systems View of the World; the Natural Philosophy of the New Developments in the Sciences.* 1972, New York: G. Braziller. viii, 131.
10. Wolkenhauer, O., *Systems biology: the reincarnation of systems theory applied in biology?* Brief Bioinform, 2001. 2(3): 258–270.
11. Lotka, A.J., *Elements of Mathematical Biology.* 1956, New York: Dover Publications. 465.
12. von Bertalanffy, L., *Das Gefüge des Lebens.* 1937, Leipzig und Berlin: Verlag und druck von B.G. Teubner. iv[2], 197.
13. von Bertalanffy, L., *Der Organismus als physikalisches System betrachtet.* Die Naturwissenschaften, 1940. 33: 521–531.
14. von Bertalanffy, L., *General System Theory; Foundations, Development, Applications.* 1969, New York: G. Braziller. xv, 289.
15. Wiener, N., *Cybernetics.* 1948, New York: J. Wiley. 194.
16. Turing, A.M., *The chemical basis of morphogenesis. 1953.* Bull Math Biol, 1990. 52(1–2): 153–197; discussion 119–152.
17. Von Neumann, J. and O. Morgenstern, *Theory of Games and Economic Behavior.* 3rd ed. 1953, Princeton: Princeton University Press. 641.
18. Shannon, C.E., *A mathematical theory of communication.* Bell Syst Tech J, 1948. 27: 379–423.
19. Rashevsky, N., *Mathematical Biophysics; Physico-Mathematical Foundations of Biology.* 3rd rev. ed. 1960, New York: Dover Publications. 2 v.
20. Ashby, W.R., *An Introduction to Cybernetics.* 1956, New York: J. Wiley. 295.
21. Rosen, R., *Abstract biological systems as sequential machines. 3. Some algebraic aspects.* Bull Math Biophys, 1966. 28(2): 141–148.
22. Rosen, R., *Recent developments in the theory of control and regulation of cellular processes. 3.* Int Rev Cytol, 1968. 23: 25–88.
23. Mesarović, M.D., Systems theory and biology — view of a theoretician, in *Systems Theory and Biology*, M.D. Mesarović, ed. 1968, Berlin, New York, etc.: Springer. 59–87.
24. Mesarović, M.D. and Y. Takahara, General systems theory: mathematical foundations. *Mathematics in science and engineering*, v. 113. 1975, New York: Academic Press. xii, 268.
25. Weaver, W., *Science and complexity.* Am Sci, 1948. 36(4): 536–544.
26. Rashevsky, N., *An approach to the mathematical biophysics of biological self-regulation and of cell polarity.* Bull Math Biophys, 1940. 2: 15–26.
27. Kacser, H., Some physio-chemical aspects of biological organization, in *The strategy of the genes; a discussion of some aspects of theoretical biology*, C.H. Waddington, ed. 1957, Allen & Unwin: London, UK. 191–249.
28. Gierer, A. and H. Meinhardt, *A theory of biological pattern formation.* Kybernetik, 1972. 12(1): 30–39.
29. Gierer, A., *Theoretical approaches to holistic biological features: pattern formation, neural networks and the brain-mind relation.* J Biosci, 2002. 27(3): 195–205.
30. Goodwin, B.C., *Temporal organization in cells; a dynamic theory of cellular control processes.* 1963, London; New York: Academic Press. ix, 163.
31. Goodwin, B.C., *A statistical mechanics of temporal organization in cells.* Symp Soc Exp Biol, 1964. 18: 301–326.

32. Heinmets, F., *Analysis of normal and abnormal cell growth; model-system formulations and analog computer studies*. 1966, New York: Plenum Press. xiii, 288.

33. Jacob, F. and J. Monod, *Genetic regulatory mechanisms in the synthesis of proteins*. J Mol Biol, 1961. 3: 318–356.

34. Stahl, W.R. and H.E. Goheen, *Molecular algorithms*. J Theor Biol, 1963. 5(2): 266–287.

35. Stahl, W.R., R.W. Coffin, and H.E. Goheen, *Simulation of biological cells by systems composed of string-processing finite automata*. AFIPS Joint Computer Conf, 1964. 25: 89–102.

36. Sugita, M., *Functional analysis of chemical systems in vivo using a logical circuit equivalent*. J Theor Biol, 1961. 1: 415–430.

37. Sugita, M., *Functional analysis of chemical systems in vivo using a logical circuit equivalent. II. The idea of a molecular automation*. J Theor Biol, 1963. 4(2): 179–192.

38. Stuart, S., *John Conway's Game of Life*. www.tech.org/~stuart/life/life.html 2003.

39. Apter, M.J., *Cybernetics and development*. 1st ed. 1966, Oxford, New York: Pergamon Press. xi, 188.

40. McCulloch, W.S. and W. Pitts, *A logical calculus of the ideas immanent in nervous activity*. Bull Math Biophys, 1943. 5: 115–132.

41. Rashevsky, N., *Outline of a physico-mathematical theory of excitation and inhibition*. Protoplasma, 1933. 20: 42–56.

42. Hodgkin, A.L. and A.F. Huxley, *A quantitative description of membrane current and its application to conduction and excitation in nerve*. J Physiol, 1952. 117(4): 500–544.

43. Clarke, R., *Fundamentals of 'information systems'*. www.anu.edu.au/people/Roger.Clarke/SOS/ISFundas.html 1999.

44. Davenport, K., *Letter*. Libr J, 2002. 127(8): 10.

45. Savageau, M.A., *The challenge of reconstruction*. New Biol, 1991. 3(2): 101–102.

46. Savageau, M.A., Critique of the enzymologist's test tube, in *Fundamentals of medical cell biology: a multi-volume work*, E.E. Bittar, ed. 1991, Greenwich, CT: JAI Press. 45–108.

47. Voit, E.O., *S-system modeling of complex systems with chaotic input*. Environmetrics, 1993. 4(2): 153–186.

48. Levine, A.S., *A new reductionism and a new journal*. New Biol, 1989. 1(1): 1–2.

49. Savageau, M.A., *Demand theory of gene regulation. II. Quantitative application to the lactose and maltose operons of Escherichia coli*. Genetics, 1998. 149(4): 1677–1691.

50. Savageau, M.A., *Demand theory of gene regulation. I. Quantitative development of the theory*. Genetics, 1998. 149(4): 1665–1676.

51. Voit, E.O., *Design principles and operating principles: the yin and yang of optimal functioning*. Math Biosci, 2003. 182(1): 81–92.

52. Schwacke, J.H. and E.O. Voit, *Improved methods for the mathematically controlled comparison of biochemical systems*. (submitted), BMC Theoretical Biology and Medical Modelling **1**:1, 2004.

53. Alves, R. and M.A. Savageau, *Comparing systemic properties of ensembles of biological networks by graphical and statistical methods*. Bioinformatics, 2000. 16(6): 527–533.

54. Irvine, D.H. and M.A. Savageau, *Network regulation of the immune response: alternative control points for suppressor modulation of effector lymphocytes*. J Immunol, 1985. 134(4): 2100–2116.

55. Savageau, M.A., *Biochemical systems analysis: a study of function and design in molecular biology*. 1976, Reading, MA: Addison-Wesley Pub. Co. Advanced Book Program. xvii, 379.

56. Voit, E.O., *Biochemical and genomic regulation of the trehalose cycle in yeast: review of observations and canonical model analysis*. J Theor Biol, 2003. 223(1): 55–78.

57. Arkin, A.P., *Synthetic cell biology*. Curr Opin Biotechnol, 2001. 12(6): 638–644.

58. Kitano, H., *Computational systems biology*. Nature, 2002. 420(6912): 206–210.

59. Ideker, T., T. Galitski, and L. Hood, *A new approach to decoding life: systems biology*. Annu Rev Genomics Hum Genet, 2001. 2: 343–372.

60. Yao, T., *Bioinformatics for the genomic sciences and towards systems biology. Japanese activities in the post-genome era*. Prog Biophys Mol Biol, 2002. 80(1–2): 23–42.

61. Wiechert, W., *13C metabolic flux analysis*. Metab Eng, 2001. 3(3): 195–206.

62. Christensen, B. and J. Nielsen, *Isotopomer analysis using GC-MS*. Metab Eng, 1999. 1(4): 282–290.

63. Goodenowe, D., Metabolomic analysis with fourier transform ion cyclotron resonance mass spectrometry, in *Metabolic profiling: its role in biomarker discovery and gene function analysis*, R. Goodacre, ed. 2003, Boston, MA: Kluwer Academic. 25–139.

64. Goodenowe, D. Metabolic network analysis: integrating comprehensive genomic and metabolomic data to understand development and disease. in *Cambridge Healthtech Institute Conference on Metabolic Profiling: Pathways in Discovery*. 2001. Chapel Hill, NC.

65. Neves, A.R., et al., *Is the glycolytic flux in Lactococcus lactis primarily controlled by the redox charge? Kinetics of NAD(+) and NADH pools determined in vivo by 13C NMR*. J Biol Chem, 2002. 277(31): 28088–28098.

66. Gerner, C., et al., *Concomitant determination of absolute values of cellular protein amounts, synthesis rates, and turnover rates by quantitative proteome profiling*. Mol Cell Proteomics, 2002. 1(7): 528–537.

67. McKenzie, J.A. and P.R. Strauss, *A quantitative method for measuring protein phosphorylation*. Anal Biochem, 2003. 313(1): 9–16.

68. Weckwerth, W., *Metabolomics in systems biology*. Annu Rev Plant Biol, 2003. 54: 669–689.

69. Estruch, F., *Stress-controlled transcription factors, stress-induced genes and stress tolerance in budding yeast*. FEMS Microbiol Rev, 2000. 24(4): 469–486.

70. Schilling, C.H. and B.O. Palsson, *Assessment of the metabolic capabilities of Haemophilus influenzae Rd through a genome-scale pathway analysis*. J Theor Biol, 2000. 203(3): 249–283.

71. Jorgensen, H., J. Nielsen, and J. Villadsen, *Metabolic flux distribution in Penicillium chrysogenum during fed-batch cultivations*. Biotechn Bioeng, 1995. 46: 117–131.

72. Gavalas, G.R., *Nonlinear differential equations of chemically reacting systems*. 1968, Berlin, New York etc.: Springer-Verlag. viii, 106.

73. Heinrich, R. and S. Schuster, *The regulation of cellular systems*. 1996, New York: Chapman & Hall. xix, 372.

74. Seressiotis, A. and J.E. Bailey, *MPS: an artificially inteligent software system for the analysis and synthesis of metabolic pathways*. Biotechn Bioeng, 1988. 31: 587–602.

75. Mavrovouniotis, M.L. and G. Stephanopoulos, *Computer-aided synthesis of biochemical pathways*. Biotechn Bioeng, 1990. 36: 1119–1132.

76. Varma, A., B.W. Boesch, and B.O. Palsson, *Metabolic flux balancing: basic concepts, scientific and practical use*. Biotechnology, 1994. 12: 994–998.

77. Edwards, J.S. and B.O. Palsson, *The Escherichia coli MG1655 in silico metabolic genotype: its definition, characteristics, and capabilities*. Proc Natl Acad Sci USA, 2000. 97(10): 5528–5533.

78. Ibarra, R.U., J.S. Edwards, and B.O. Palsson, *Escherichia coli K-12 undergoes adaptive evolution to achieve in silico predicted optimal growth*. Nature, 2002. 420(6912): 186–189.

79. Reed, J.L., et al., *An expanded genome-scale model of Escherichia coli K-12 (iJR904 GSM/GPR)*. Genome Biol, 2003. 4(9): R54.

80. Savageau, M.A., *Michaelis-Menten mechanism reconsidered: implications of fractal kinetics*. J Theor Biol, 1995. 176(1): 115–124.

81. Hill, C.M., R.D. Waight, and W.G. Bardsley, *Does any enzyme follow the Michaelis-Menten equation?* Mol Cell Biochem, 1977. 15(3): 173–178.

82. Torres, N.V. and E.O. Voit, *Pathway analysis and optimization in metabolic engineering*. 2002, New York: Cambridge University Press. xiv, 305.

83. Garfinkel, D., *The role of computer simulation in biochemistry*. Comput Biomed Res, 1968. 2(1): i–ii.

84. Heinrich, R. and T.A. Rapoport, *A linear steady-state treatment of enzymatic chains. General properties, control and effector strength*. Eur J Biochem, 1974. 42(1): 89–95.

85. Schulz, A.R., *Enzyme kinetics: from diastase to multi-enzyme systems*. 1994, Cambridge; NY: Cambridge University Press. x, 246.

86. Wright, B.E., M.H. Butler, and K.R. Albe, *Systems analysis of the tricarboxylic acid cycle in Dictyostelium discoideum. I. The basis for model construction*. J Biol Chem, 1992. 267(5): 3101–3105.

87. Shiraishi, F. and M.A. Savageau, *The tricarboxylic acid cycle in Dictyostelium discoideum. IV. Resolution of discrepancies between alternative methods of analysis*. J Biol Chem, 1992. 267(32): 22934–22943.

88. Savageau, M.A., *Biochemical systems analysis. 3. Dynamic solutions using a power-law approximation*. J Theor Biol, 1970. 26(2): 215–226.

89. Savageau, M.A., *Biochemical systems analysis. II. The steady-state solutions for an n-pool system using a power-law approximation*. J Theor Biol, 1969. 25(3): 370–379.

90. Savageau, M.A., *Biochemical systems analysis. I. Some mathematical properties of the rate law for the component enzymatic reactions*. J Theor Biol, 1969. 25(3): 365–369.

91. Savageau, M.A. and E.O. Voit, *Recasting nonlinear differential equations as S-systems: a canonical nonlinear form*. Math Biosci, 1987. 87: 83–115.

92. Voit, E.O., *Canonical nonlinear modeling: S-system approach to understanding complexity*. 1991, New York: Van Nostrand Reinhold. xii, 365.

93. Michaelis, L. and M.L. Menten, *Die kinetic der invertinwirkung*. Biochem Zeitschrift, 1913. 49: 333–369.

94. Haldane, J.B.S., *Enzymes*. 1965, Cambridge: M.I.T. Press. vi, 235.

95. Hill, A.V., *Possible effects of the aggregation of the molecules of haemoglobin on its dissociation curves*. J Physiol, 1910. 40: iv–viii.

96. Johnson, R.E., F.L. Kiokemeister, and E.S. Wolk, *Johnson and Kiokemeister's Calculus with analytic geometry*. 5th ed. 1974, Boston: Allyn and Bacon. x, 839.

97. Kuo, B.C., *Automatic control systems*. 5th ed. 1987, Englewood Cliffs, NJ: Prentice-Hall. xiv, 720.

98. Voit, E.O. and M.A. Savageau, *Accuracy of alternative representations for integrated biochemical systems*. Biochemistry, 1987. 26(21): 6869–6880.

99. Guckenheimer, J. and P. Holmes, *Nonlinear oscillations, dynamical systems, and bifurcations of vector fields*. 1983, New York: Springer-Verlag. xvi, 453.

100. Sorribas, A., Sensitivity analysis in GMA models, in *Computational analysis of biochemical systems: a practical guide for biochemists and molecular biologists*, E.O. Voit, ed. 2000, New York: Cambridge University Press. 251–259.

101. Morohashi, M., et al., *Robustness as a measure of plausibility in models of biochemical networks*. J Theor Biol, 2002. 216(1): 19–30.

102. Gear, C.W., *Numerical initial value problems in ordinary differential equations*. 1971, Englewood Cliffs, NJ: Prentice-Hall. xvii, 253.

103. Irvine, D.H. and M.A. Savageau, *Efficient solution of nonlinear ordinary differential equations expressed in S-system canonical form*. SIAM J Numer Anal, 1990. 27: 704–735.

104. Voit, E.O., D.H. Irvine, and M.A. Savageau, *The User's Guide to ESSYNS*. 1989, Charleston, South Carolina: Medical University of South Carolina Press. 148.

105. Ferreira, A.E.N., *Power Law Analysis and Simulation*. 1996–2006, www.dqb.fc.ul.pt/docentes/aferreira/plas.html.

106. Curto, R., A. Sorribas, and M. Cascante, *Comparative characterization of the fermentation pathway of Saccharomyces cerevisiae using biochemical systems theory and metabolic control analysis: model definition and nomenclature*. Math Biosci, 1995. 130(1): 25–50.

107. Alvarez-Vasquez, F., C. Gonzalez-Alcon, and N.V. Torres, *Metabolism of citric acid production by Aspergillus niger: model definition, steady-state analysis and constrained optimization of citric acid production rate*. Biotechnol Bioeng, 2000. 70(1): 82–108.

108. Ferreira, A.E.N., A.M. Ponces Freire, and E.O. Voit, *A quantitative model of the generation of N(epsilon)-(carboxymethyl)lysine in the Maillard reaction between collagen and glucose*. Biochem J, 2003. 376(Pt 1): 109–121.

109. Alvarez-Vasquez, F., et al., *Integration of kinetic information on yeast sphingolipid metabolism in dynamical pathway models*. J Theor Biol, 2004. 226(3): 265–291.

110. Curto, R., E.O. Voit, and M. Cascante, *Analysis of abnormalities in purine metabolism leading to gout and to neurological dysfunctions in man*. Biochem J, 1998. 329(Pt 3): 477–487.

111. Curto, R., et al., *Validation and steady-state analysis of a power-law model of purine metabolism in man*. Biochem J, 1997. 324(Pt 3): 761–775.

112. Curto, R., et al., *Mathematical models of purine metabolism in man*. Math Biosci, 1998. 151(1): 1–49.

113. Almeida, J.S. and E.O. Voit, *Neural network-based parameter estimation in complex biochemical systems*. Genome Informatics 14, 14–123, 2003.

114. Voit, E.O., *Computational analysis of biochemical systems: a practical guide for biochemists and molecular biologists*. 2000, New York: Cambridge University Press.

115. Savageau, M.A., *A theory of alternative designs for biochemical control systems*. Biomed Biochim Acta, 1985. 44(6): 875–880.

116. Vera, J., et al., *Multicriteria optimization of biochemical systems by linear programming: application to production of ethanol by Saccharomyces cerevisiae*. Biotechnol Bioeng, 2003. 83(3): 335–343.

117. Rodriguez-Acosta, F., C.M. Regalado, and N.V. Torres, *Non-linear optimization of biotechnological processes by stochastic algorithms: application to the maximization of the production rate of ethanol, glycerol and carbohydrates by Saccharomyces cerevisiae*. J Biotechnol, 1999. 68(1): 15–28.

118. Marin-Sanguino, A. and N.V. Torres, *Modelling, steady state analysis and optimization of the catalytic efficiency of the triosephosphate isomerase.* Bull Math Biol, 2002. 64(2): 301–326.

119. Marin-Sanguino, A. and N.V. Torres, *Optimization of tryptophan production in bacteria. Design of a strategy for genetic manipulation of the tryptophan operon for tryptophan flux maximization.* Biotechnol Prog, 2000. 16(2): 133–145.

120. Alvarez-Vasquez, F., et al., *Modeling, optimization and experimental assessment of continuous L-(-)-carnitine production by Escherichia coli cultures.* Biotechnol Bioeng, 2002. 80(7): 794–805.

121. Kacser, H. and J.A. Burns, *The control of flux.* Symp Soc Exp Biol, 1973. 27: 65–104.

122. Fell, D., *Understanding the control of metabolism.* 1997, London, U.K.: Portland Press.

123. Tomita, M., *Whole-cell simulation: a grand challenge of the 21st century.* Trends Biotechnol, 2001. 19(6): 205–210.

124. Tomita, M., et al., *E-CELL: software environment for whole-cell simulation.* Bioinformatics, 1999. 15(1): 72–84.

125. EASL, *The European Association for the Study of the Liver,* www.easl.ch.

126. Juty, N.S., et al., *Simultaneous modelling of metabolic, genetic and product-interaction networks.* Brief Bioinform, 2001. 2(3): 223–232.

127. Magnus, G. and J. Keizer, *Model of beta-cell mitochondrial calcium handling and electrical activity. I. Cytoplasmic variables.* Am J Physiol, 1998. 274(4 Pt 1): C1158–C1173.

128. Post, E.L., *A variant of a recursively unsolvable problem.* Bull Am Math Soc, 1946. 52: 264–268.

129. E. coli Consortium, *E. coli Community,* www.ecolicommunity.org.

130. Reed, J.L. and B.O. Palsson, *Thirteen years of building constraint-based in silico models of Escherichia coli.* J Bacteriol, 2003. 185(9): 2692–2699.

131. Schilling, C.H., et al., *Genome-scale metabolic model of Helicobacter pylori 26695.* J Bacteriol, 2002. 184(16): 4582–4593.

132. Forster, J., et al., *Genome-scale reconstruction of the Saccharomyces cerevisiae metabolic network.* Genome Res, 2003. 13(2): 244–253.

133. Veflingstad, S.R., J.S. Almeida, and E.O. Voit, *Priming non-linear searches for pathway identification.* BMC Theoretical Biology and Medical Modelling **1**:8, 2004.

134. Vance, W., A. Arkin, and J. Ross, *Determination of causal connectivities of species in reaction networks.* Proc Natl Acad Sci USA, 2002. 99(9): 5816–5821.

135. Chevalier, T., I. Schreibe, and J. Ross, *Toward a systematic determination of complex reaction mechanisms.* J Phys Chem, 1993. 97: 6776–6787.

136. Diaz-Sierra, R., J.B. Lozano, and V. Fairen, *Deduction of chemical mechanisms from the linear response around the steady state.* J Phys Chem, 1999. 103: 337–343.

137. D'Haeseleer, P., et al., *Linear modeling of mRNA expression levels during CNS development and injury.* Pac Symp Biocomput, 1999: 41–52.

138. Godfrey, K.R., M.J. Chapman, and S. Vajda, *Identifiability and indistinguishability of nonlinear pharmacokinetic models.* J Pharmacokinet Biopharm, 1994. 22(3): 229–251.

139. Voit, E.O. and J.S. Almeida, *Decoupling dynamical systems for pathway identification from metabolic profiles. Bioinformatics* **20(11)**, 1670–1681: 2004.

140. Kikuchi, S., et al., *Dynamic modeling of genetic networks using genetic algorithm and S-system.* Bioinformatics, 2003. 19(5): 643–650.

141. Arkin, A., J. Ross, and H.H. McAdams, *Stochastic kinetic analysis of developmental pathway bifurcation in phage lambda-infected Escherichia coli cells.* Genetics, 1998. 149(4): 1633–1648.

142. Davidson, E.H., et al., *A genomic regulatory network for development.* Science, 2002. 295(5560): 1669–1678.

143. Rosen, R., *Subunit and subassembly processes.* J Theor Biol, 1970. 28(3): 415–422.

144. Csete, M.E. and J.C. Doyle, *Reverse engineering of biological complexity.* Science, 2002. 295(5560): 1664–1669.

145. Hofmeyr, J.H. and H.V. Westerhoff, *Building the cellular puzzle: control in multi-level reaction networks.* J Theor Biol, 2001. 208(3): 261–285.

146. Stelling, J., et al., *Metabolic network structure determines key aspects of functionality and regulation.* Nature, 2002. 420(6912): 190–193.

147. Milo, R., et al., *Network motifs: simple building blocks of complex networks.* Science, 2002. 298(5594): 824–827.

148. de la Fuente, A., et al., *Metabolic control in integrated biochemical systems.* Eur J Biochem, 2002. 269(18): 4399–4408.

149. ter Kuile, B.H. and H.V. Westerhoff, *Transcriptome meets metabolome: hierarchical and metabolic regulation of the glycolytic pathway.* FEBS Lett, 2001. 500(3): 169–171.

150. Savageau, M.A., *Growth of complex systems can be related to the properties of their underlying determinants.* Proc Natl Acad Sci, 1979. 76: 5413–5417.

151. Savageau, M.A., *Mathematics of organizationally complex systems.* Biomed Biochim Acta, 1985. 44(6): 839–844.

152. Achard, F., G. Vaysseix, and E. Barillot, *XML, bioinformatics and data integration.* Bioinformatics, 2001. 17(2): 115–125.

153. Barillot, E. and F. Achard, *XML: a lingua franca for science?* Trends Biotechnol, 2000. 18(8): 331–333.

154. Hedley, W. and M. Nelson, *CellML Specification.* www.cellml.org/specifications/ 2001.

155. Hucka, M., et al., *The systems biology markup language (SBML): a medium for representation and exchange of biochemical network models.* Bioinformatics, 2003. 19(4): 524–531.

156. Schwacke, J.H. and E.O. Voit. BSTLab: a Matlab toolbox for biochemical systems theory, in *Eleventh International Conference on Intelligent Systems for Molecular Biology.* 2003. Brisbane, Australia.

157. Lomax, P., *VB & VBA in a nutshell: the language.* 1st ed. 1998, Beijing; Sebastopol, CA.: O'Reilly & Associates. xiv, 633.

158. Rogerson, D., *Inside COM.* 1997, Microsoft Press: Redmond, WA.

159. Hood, L., *Systems biology: integrating technology, biology, and computation.* Mech Ageing Dev, 2003. 124(1): 9–16.

160. Davidov, E., et al., *Advancing drug discovery through systems biology.* Drug Discov Today, 2003. 8(4): 175–183.

161. Werner, E., *Systems biology: the new darling of drug discovery?* Drug Discov Today, 2002. 7(18): 947–949.

162. The Royal Society of Chemistry, *Lab on a chip,* rsc.org/is/journals/current/loc/locpub/htm.

163. McAdams, H.H. and A. Arkin, *Towards a circuit engineering discipline.* Curr Biol, 2000. 10(8): R318–R320.

164. Kholodenko, B.N., et al., *Engineering a living cell to desired metabolite concentrations and fluxes: pathways with multifunctional enzymes.* Metab Eng, 2000. 2(1): 1–13.

165. Carlson, R., D. Fell, and F. Srienc, *Metabolic pathway analysis of a recombinant yeast for rational strain development.* Biotechnol Bioeng, 2002. 79(2): 121–134.

166. Peschel, M. and W. Mende, *The Predator-Prey Model: Do We live in a Volterra World?* Akademie-Verlag, Berlin, 1986.

167. Hernández-Bermejo, B. and V. Fairén, Lotka-Volterra representation of general nonlinear systems. *Math. Biosci.* 1997. **140**, 1–32.

168. Savageau, M.A., 20 years of S-systems, Chapter 1, Voit, E.O. (Ed.). *Canonical Nonlinear Modeling. S-System Approach to Understanding Complexity,* xi+365 pp., Van Nostrand Reinhold, New York, 1991.

3 Thermostatics: A Poster Child of Systems Thinking

Peter Salamon, Anna Salamon, and Andrzej K. Konopka

CONTENTS

This chapter aims to provide a comprehensive introduction to macroscopic thermodynamics (thermostatics) presented so as to highlight its role as a paradigm for systems thinking. It is not our intention to replace the great many excellent textbooks of thermodynamics. Instead, we present thermostatics as a conglomerate of worldviews, methods, and interpretations, which are methodologically and conceptually different from chemistry, physics, or engineering. To benefit from the uniqueness of this methodology, we focus on the modeling aspects of the subject with particular attention to the *construction* of the two state functions that carry information about a thermostatic system: the internal energy and the entropy. This is easiest to do by following the subject's historical development.

Most scientists are aware that there are three laws of thermodynamics. The first law is the conservation of energy and is relatively straightforward. The second law states that energy can only be degraded. It specifies the "arrow of time." This law is subtler and merits much more discussion. The third law concerns the behavior of thermodynamic systems near the absolute zero of temperature and will not concern us in the present treatment. Less well known is the zeroth law, which is part definition, part postulate, and asserts the well definedness of thermal equilibrium and of the ordinal scale of temperature that determines which of two systems is hotter.

3.1 BASIC CONCEPTS

Underlying all of thermostatics is the definition of the sort of system to which thermostatic arguments truly apply. Understanding this definition in detail is important for applying the subject to systems biology where the notion of a "thermostatic system*" will be stretched to (and perhaps beyond) its traditional limits. What can constitute a thermostatic system has proved to be a remarkably robust notion; it will surely be able to accommodate further generalizations appropriate for systems biology. The key to recognizing what can or cannot constitute a thermostatic system is to understand the time scales that separate processes into fast and slow. The reason for this is that the "dynamics" in "thermodynamics" is a misnomer. The theory describes what can happen between *equilibrium* states of thermodynamic systems, i.e., the theory is really comparative statics insofar as it concerns net effects of moving between an initial static situation and a final one [1] (where the chosen time scale affects what is and is not a "static situation" by affecting which processes are considered negligibly slow). That is why we use the term "thermostatics" in this article for what is often called "thermodynamics" in the literature. Thermostatics asserts whether certain net effects are or are not possible without consideration of mechanisms by which these effects could be attained.

One requirement of the above considerations is that we need to recognize what *equilibrium* means. This is not always easy for at least two reasons:

1. Equilibrium is approached asymptotically and thus thermostatic systems never quite reach it.
2. Equilibrium means something very different on "ordinary" vs. on astronomical time scales.

The fact that equilibrium is approached asymptotically can be seen as follows. We consider the *state* of a thermostatic system to be characterized by the values of some set of macroscopically observable parameters (e.g., temperature, pressure, and volume). We use the term *equilibrium state* to denote any state from which the system will not undergo any further spontaneous changes over the time-scale under consideration. If we make the reasonable additional assumption that the rate of change of a system's parameter values is locally proportional to the system's distance from equilibrium, it follows that the system will approach but never reach its equilibrium state.

The fact that equilibrium means something different on different time scales is due to the very large waiting times for relaxation of degrees of freedom corresponding to large activation energies. Examples of such degrees of freedom include nuclear transitions. In fact, the "true" (infinite time) equilibrium of any chunk of matter would have to be ^{56}Fe [2]. Thus equilibrium is a notion that makes sense only once a time scale has been specified.** We can say that a thermostatic system is in equilibrium once all its degrees of freedom change so slowly that they can be regarded as constant (i.e., not changing at all). In other words, the necessary condition for thermostatic arguments is what is referred to as the separability of time scales.

Before considering how the subtle concept of equilibrium should be used in the life sciences, we present the laws of thermostatics in the context of the simplest thermostatic systems where the classical ideas apply rigorously. We thus take our thermostatic system to be *homogeneous*** and occupying a well-defined region in space known as the *control volume*. The paradigm example of such a thermostatic system is a gas in a container. Historically, the ideal gas and the ideal gas law have contributed a great deal to thermostatics. Without this example of a simple thermostatic system, whose simple equation of state was known explicitly and allowed all properties to be calculated analytically, the subject would have taken much longer to develop than it in fact took.

---

* The word "system" is unfortunately used in both the thermodynamics and the biology communities, often with a different meaning in each. To avoid confusion, we introduced the term "thermostatic system" in this chapter to denote what thermodynamicists refer to simply as "system" or sometimes (in chemistry) as "macroscopic system."

** Connected to such a time scale Δt at any temperature T is a closely associated energy scale ΔE, where $\Delta t \approx \exp(-\Delta E/kT)$ gives the mean first passage time out of a valley with excitation energy ΔE. Thus we could alternatively say that equilibrium is a notion that does not make sense until an energy scale has been specified.

*** A thermostatic system is said to be homogeneous when all measurable properties are the same at every point inside the control volume.

3.2 THE ZEROTH LAW

The zeroth law postulates the well-definedness of thermal equilibrium and a phenomenological (ordinal) temperature scale. The postulate describes one possible result of establishing "thermal contact" between two systems: when establishing such contact brings no sensible change in either system, we say that the two systems are *at the same temperature*. Furthermore, we postulate that if thermostatic systems A and B have the same temperature and systems A and C have the same temperature then it must be the case that B and C have the same temperature.

3.3 THE FIRST LAW

The first law expresses the conservation of energy. The law was constructed from separate older conservation laws for mechanical energy in mechanics and for caloric in the theory of heat. It retrospect, the ability to rub two sticks together to produce heat flagrantly contradicts any conservation of caloric, and it was clear to the scientists even before the first law that their laws had only context-limited validity [3]. However, the one-way nature of the conversion — mechanical energy to thermal energy — kept things mysterious and no doubt delayed the formulation of the combined version. With the advent of the steam engine (and its immense economic importance), it became possible to investigate the production of work from heat and the first law soon followed.

This conservation of energy was then used to *construct* a function of state known as internal energy, U. This function is defined for one system by how it changes:

$$\Delta U = Q + W$$

where Q is the heat entering the thermostatic system and W is work done on the system by its surroundings. This formulation uses the interconvertibility of mechanical work and heat to combine the separate conservation laws in mechanics and in the theory of heat into one larger law of conservation of energy. This combined conservation law leads to a profound conceptual generalization: heat and mechanical work are different forms of energy. The extent to which different forms of energy can be converted into each other has been an issue that has initiated the field of thermostatics and dominated it until this day.

The first law can be extended into the chemical domain, i.e., it can include the energetics of chemical reactions. We delay their discussion until after the presentation of the second law.

3.4 THE SECOND LAW

The second law is the only known physical principle that expresses the unidirectionality of time. All other laws of physics allow for each possible process to occur either forwards or backwards. Because the world as we experience it is unidirectional regarding time, the second law should play a pivotal role in understanding the world.

There are many equivalent formulations of the second law [4]. Perhaps the most intuitive is in terms of the second possibility concerning what can happen when two thermostatic systems are brought into thermal contact. The first possibility was contact between two systems at equal temperatures, in which case the zeroth law asserts that nothing happens. The second possibility is thermal contact between two systems at different temperatures, in which case the second law asserts that heat always flows from the hotter to the colder, i.e., from the thermostatic system at a higher temperature to that at a lower temperature. In other words, it is impossible to have a process* whose only net effect is to transfer heat from a cold reservoir to a hot one. Equivalently, it is impossible to have a heat engine whose only net effect is to take heat from a single reservoir and

---

* A thermostatic *process* is defined as a succession of states through which a thermostatic system passes.

convert it entirely into work. In fact, recent studies have adopted the useful convention of viewing work as heat at infinite temperature, in which case the second formulation becomes a specific instance of the first one.

The second law is often expressed by saying that "energy is always degraded." What this means is that it spontaneously moves to lower and lower temperatures. Note that this encompasses both friction in which work (infinite temperature heat) is degraded to heat at finite temperature, and spontaneous heat flow, in which thermal energy at one temperature becomes thermal energy at a lower temperature. In fact, it is remarkable that this one statement encompasses all irreversible processes, i.e., provided no heat moves to a lower temperature, the process is reversible! For mechanical processes, reversibility means only that no friction is involved. For thermal processes, reversibility requires that all heat moves between systems at equal temperatures. For reversible processes, we can *construct* an additional conservation law as shown below.

Based on the second law, we can construct a second state function similar to internal energy. This state function is *designed* to quantify the unidirectionality of time and is known as entropy. While reversible processes cannot in fact occur, they can be approached to arbitrary accuracy, at least as gedanken processes. In such processes, we define a conserved quantity, the entropy, again by how it changes. For one system, the differential change in its entropy, S, is defined in terms of the differential amount of heat δQ_{rev} entering or leaving the system, where the *rev* subscript reminds us that we can count this way only for reversible processes:

$$dS = \frac{\delta Q_{rev}}{T}.$$

By the foregoing discussion, a reversible process will always keep the total entropy S (added up over subsystems) constant because heat moves only between systems at the same temperature. Irreversible processes on the other hand increase entropy. Note that for processes with friction, entropy is produced because the appearance of heat corresponds to the appearance of entropy. Similarly, when heat Q moves from a higher temperature T_1 to a lower temperature T_2, the total amount of entropy increases by

$$\left(\frac{Q}{T_2} - \frac{Q}{T_1} \right).$$

To calculate how the entropy of a certain subsystem changes, we have to calculate what would happen in a reversible process that goes from the system's initial to its final state. It is a fact that all reversible processes would end up transferring the same amount of entropy to/from our subsystem. We usually state this by saying that entropy is a function of state.

It is worth pausing here to discuss these ideas in light of the more modern statistical theories of thermal physics. From this perspective, we think of heat as energy transfer that is completely "thermalized," i.e., as randomized as the temperature allows. Thermal energy at a high temperature is more ordered than thermal energy at a low temperature. In this connection, work can be viewed as heat from an infinite temperature system, i.e., not randomized at all but rather directible entirely into lifting a weight. This randomization is at the heart of the statistical view of thermodynamics. From this perspective, the second law describes how energy gets mixed up as it spreads out to the many molecular degrees of freedom. A quantitative handle on this aspect of the subject comes from the microscopic interpretation of entropy as the logarithm of the number of states consistent with a given macroscopic state. In this context, the increase of entropy corresponds to things getting mixed up and it is *entropy production* that gives us a quantitative measure of irreversibility.

In this regard, we mention a cautionary note: heat and work are *not* state functions but rather transfers of energy. Just as work is energy *flow* that shows up as mechanical energy, heat is energy *flow* that shows up in a purely randomized form. Thus, although the environment does work on our thermostatic system, all or part of this work could enter our system as heat. From a mathematical perspective, heat and work are represented by differential forms that are not exact, i.e., they are not in general the differentials of a function of state. As differential forms, they assign numbers to processes (paths in the set of states) but not to states. (In contrast, functions of state, such as temperature, assign numbers to states.)

Entropy has another interesting property. Recall that if we consider the amount of work extractible from a mechanical process such as a water wheel, one pound of water moving one foot lower can produce at most one foot-pound of work. Our definitions allow an analogous statement for how much work can be produced by a certain amount of entropy moving (reversibly) from T_1 to T_2: the maximum work is the product of the entropy that moved and the temperature difference. To see this, consider a heat engine in which a thermostatic system acts in an intermediary role but starts and ends in the same state. Such intermediary systems are called working fluids. Assume that the working fluid picks up heat Q_1 from system 1 at temperature T_1 and deposits heat Q_2 in system 2 at temperature T_2. By the first law, the work produced by the process will be the difference $Q_1 - Q_2$. Because we are interested in the *maximum* work, we want the process to go reversibly because any irreversibility will merely degrade some exergy (available work). Reversibly means that $Q_1 = T_1 \Delta S_1$ and $Q_2 = T_2 \Delta S_2$, with $\Delta S_1 = -\Delta S_2$. The above fact then becomes

$$W_{\max} = \left(T_1 - T_2\right) \Delta S$$

$$= \left(T_1 - T_2\right) \frac{Q_1}{T_1}$$

$$= \left(1 - \frac{T_2}{T_1}\right) Q_1$$

i.e., the maximum work is the famous Carnot efficiency $\left(1 - \dfrac{T_2}{T_1}\right)$ times the heat.

3.5 STANDARD STATES AND TABLES

For both the internal energy and the entropy, the foregoing discussion talks only about changes. Thus it follows that each of these functions is defined only up to an additive constant, which is usually specified by the selection of a particular state for which the value of the function is to be zero. The choice of such *reference states* with respect to which energies and entropies are calculated is an important freedom that can be used to simplify a particular calculation. The older texts choose standard temperature and pressure to define the standard states for all the elements, for which the internal energy and the entropy are chosen to be zero. More modern sources use zero temperature as the reference state for the zero of entropy, a choice particularly convenient for the third law. Because ultimately our calculations deal with comparative statics, the choice of reference states drops out and as long as we are careful and consistent, all choices are equally correct. Nonetheless, switching between references that differ in their choice of standard states can be confusing.

3.6 STATES VS. PROCESSES

Historically, thermodynamics was born as a theory of heat engines. Its growth was spurred primarily by the success of the steam engine. A full generation later, in the hands of J.W. Gibbs, the subject underwent its most profound transformation — it changed from a theory of processes to a theory of states. Before this, it had only dealt with the transfer of heat and work between systems. After the transformation, it became theory of equilibrium states for one system. The first law became

$$dU = TdS - pdV = \delta Q + \delta W$$

where U, S, T, p, and V all refer to the properties of a single system. The main impetus for the transformation was the need to include chemical processes in the first and second laws.

Gibbs noted that chemical phenomena were associated with heat and work and thus their analysis should be combined with the thermodynamic laws. He achieved the required unification by introducing the notion of chemical potential. The chemical potential, μ_i, of the ith molecular species plays the role for molecules of type i that temperature plays for heat. When two systems are brought into contact of a sort that allows transfer of species i, mass of species i moves from the system with the higher chemical potential to that with the lower one. Furthermore, the maximum work that could be obtained from a spontaneous process of mass flow is the product of the mass dM and the chemical potential difference between the two systems

$$\delta W_{max} = \left(\mu_2 - \mu_1\right)dM.$$

This forces

$$\mu_i = \frac{\partial U}{\partial M_i}$$

and adds new terms to the differential form of the first law

$$dU = TdS - pdV + \sum_{\text{species } i} \mu_i dM_i.$$

The notion of chemical potential extended the definition of a reversible process to mass transfer, with reversibility holding in exactly those processes in which mass is transferred across zero chemical potential differences. More interestingly, it can now be used to define reversible chemical reactions to be those where the stoichiometric combination of the reactants' chemical potentials equals the stoichiometric combination of the products' chemical potentials. For example, for the reaction

$$\alpha\,A + \beta\,B \rightarrow \gamma\,C + \delta\,D$$

between reactant species A and B forming product species C and D with stoichiometric coefficients α, β, γ, and δ, the reversibility condition is

$$\alpha\,\mu_\alpha + \beta\,\mu_\beta = \gamma\,\mu_\gamma + \delta\,\mu_\delta.$$

In fact this is also the equilibrium condition, with the result that a reaction is reversible if and only if it takes place at equilibrium.

Once chemical interconversion is allowed into the set of processes we consider, the reference states with zero energy and entropy are no longer independently selectable. While such reference states remain arbitrarily selectable for the *elements* in their natural states, they imply definite nonzero values for chemical compounds. For example, we can take the energy and the entropy of gaseous hydrogen and gaseous oxygen at STP* to be zero but then are forced to take the energy and entropy for liquid water at STP to be the energy and entropy change associated with the reversible reaction forming water from H_2 and O_2.

---

* Standard Temperature and Pressure: zero degrees Centigrade and one atmosphere.

3.7 REFORMULATIONS

Most physical theories admit reformulations, and such reformulations are important for making the theory usable in a wide variety of contexts. The thermodynamic theory of states developed above gives a complete characterization of equilibrium states for an isolated* system: its total internal energy remains constant and any spontaneous process inside the system, i.e., interaction among its subsystems, will increase the total entropy of the system. It follows that when entropy is as large as it can get, no further spontaneous processes are possible and the system is at equilibrium. Thus the final characterization is that entropy is maximized for a constrained energy.

While energy has become a familiar concept from mechanics, entropy has not fared nearly as well. Gibbs made use of a general duality between objective and constraint functions in an optimization problem to transform the condition of equilibrium into one that seems more palatable for intuitions steeped in mechanics. Instead of saying that entropy is maximized subject to a fixed energy, we can say equivalently that energy is minimized subject to a fixed entropy.

A similar reformulation is possible at the process level. There, the second law's assertion that entropy increases in any spontaneous process changes to an assertion that available work is lost in any spontaneous process. Such available work was introduced by Gibbs and has enjoyed a recent resurgence of popularity among engineering thermodynamicists under the name of *exergy*. The fact that exergy can only be lost is a common way to state the second law.

Closely related to exergy is the notion of free energy, also introduced by Gibbs. In fact, given certain constraints on the allowable processes, free energy is just exergy. For example, for processes at constant temperature and pressure, the change in Gibbs free energy ΔG of some system is exactly the change in exergy of that system.

Free energies are important in another context because isolated systems are not nearly as common as systems that are coupled to their surroundings. If these surroundings maintain the pressure and the temperature of the system at a constant value, then it is possible to consider a larger system made up of our system and its surroundings. If we can consider this larger system isolated, then the laws of thermodynamics as stated above characterize the equilibrium of the composite system. When the surrounding system is large enough so that its temperature and pressure are effectively unchangeable by interaction with our small system, then it turns out that the equilibrium of the small system can be characterized by minimizing the free energy

$$G = U + pV - TS.$$

Such reformulations of the principles are known as Legendre transforms. They have been generalized beyond the usual free energies [5,6] to determine the equilibrium conditions for systems in contact with surroundings that fix any function of state. While useful applications have yet to be found, it is our belief that such generalizations will prove of interest for systems biology.

3.8 IMPLICATIONS FOR LIVING SYSTEMS

As noted above, thermostatics' change from a theory of processes to a theory of states required describing interactions between subsystems, each of which may be viewed as a thermostatic system in its own right. In the process, our notion of thermostatic system lost some of its simple character as a homogeneous region in space with well-defined boundaries. In fact, for chemical reactions, the systems can consist of one type of molecule with a volume that is completely superimposed on the volume of a second system consisting of different molecules. Recent analyses have gone

---

* An *isolated system* is defined as any system that does not exchange matter or energy with its surroundings. A system's *surroundings* are defined as everything external to the system.

further and treated only the (say) vibrational degrees of freedom of certain molecules as a thermodynamic system. These analyses are responsible for our modern notion that what is needed for thermodynamic arguments is a separability of time scales. If equilibration within a certain degree of freedom is sufficiently fast while its interaction with other degrees of freedom is significantly slower, then this interaction can be thought of as taking place between different thermostatic systems, each with its own temperature. In living things, the appropriate time scales for many interactions have yet to be measured, but analyses of the thermostatics of cellular processes should extend our current notions of thermostatic systems to new and biologically important examples.

The implications of the two thermostatic laws for living organisms have been addressed by Erwin Schroedinger in his influential book from the mid-1940s entitled *What is Life?* [7]. He posed and answered the question of how living things manage to maintain their organization despite making extensive use of irreversible processes, which produce entropy. In order to maintain their organization, living things have to dispose of this entropy. They manage to dispose of the entropy produced, because they are not isolated systems but are in fact coupled to their surroundings through the flows of ingested food and excreted waste. As long as the excreted waste carries significantly more entropy out of the organism than the ingested food brings in, the entropy change inside the organism can be maintained close to zero. One example of the molecular details for this mechanism proceeds by the production of molecules such as adenosine triphosphate (ATP) that are rich in free energy and can therefore be used as an exergy currency for cellular processes. Because spontaneous processes are exactly those which degrade exergy, cells can accomplish any needed cellular processes, including those which require input of exergy, by coupling the processes to ATP degradation in such a way that the coupled processes create a net degradation of exergy in conformity with the second law.

3.9 THE ANALOGY BETWEEN SHANNON "INFORMATION" AND THERMOSTATIC ENTROPY

The above-mentioned connection between entropy and randomness has a flip side — it can be used to create a formal analogy between entropy and Shannon "information." In fact, as defined in the classic work by Claude Shannon [8–10], the "entropy" of a random variable* is exactly the average amount of "information" per elementary signal in a very long sequence of signals: the log of the number of possibilities. This metaphor has led to thermostatics-like analyses of coding algorithms and has resulted in thermostatics-like theorems regarding the extent to which noisy communication channels can be made immune to the noise levels. Despite these attempts, there is little agreement regarding how exactly to merge the modern theories of Shannon "information" and thermostatics. This failure will likely be remedied within some fields of systems biology where the chemical degrees of freedom carry large amounts of Shannon-like "information" that must be transmitted nearly free of error in communication-like processes. Determining the exact details of how the foregoing connections between different theory-laden concepts can be done remains an important open problem.

3.10 FINITE-TIME THERMODYNAMICS

In the foregoing discussion, we mentioned reversible processes and the possibility of approaching them with real processes. In point of fact, heat transfer can only approach reversibility as we slow

---

* The Shannon entropy has reportedly been named "entropy" because of a sarcastic practical joke played by John von Neumann on the thermodynamically illiterate Claude Shannon. The story says that von Neumann, asked by Shannon how to call "this strange H-function," answered: "call it entropy"; no one understands what it is anyway ... " This joke, if the story is true, has created confusion ever since.

the process down. This is also true for chemical reactions and for all transport processes whose rates are proportional to generalized forces (differences in intensive variables) that must be made to approach zero as we approach reversible operation.

This fact has spurred a body of results known collectively as finite-time thermodynamics [11–14]. The central organizing fact is that once we require a process to proceed at a certain rate, we are not able to make it approach reversibly. Bounding its rate away from zero can give positive bounds on minimum entropy production (minimum exergy loss) that must accompany the process. Finite-time thermodynamics has focused on maximizing power and on minimizing the dissipation that must accompany a process taking place at a certain rate. This line of investigation is likely to be useful for systems biology, where higher efficiency at the cost of lower power is not usually of interest. While the efficiency of many of the processes so crucial for the bioenergetics of the cell are well understood as measured in comparison to reversible processes, how they compare to optimal processes subject to the constraint of operating at a given rate is not known. In an attempt to pique the reader's interest in more specialized monographs [11–14], we close our discussion by mentioning a couple of the more intriguing results that have come out of this line of investigation.

In finite time, there are essentially two extremes of optimal operation corresponding to maximum power and minimum exergy loss. These extremes can be loosely identified with the points of view of an industrialist who wants as much product as possible and of an environmentalist who wants to minimize the expenditure of resources.

The maximum power operation of a process has no exact analog in traditional thermodynamics, although it appears that simple and general principles may surround such operation. For example, the efficiency of a heat engine operating at maximum power between a heat source at temperature T_H and a heat sink at temperature T_C is quite generally given by

$$\eta_{MaxP} = 1 - \sqrt{\frac{T_C}{T_H}}$$

a formula whose similarity to Carnot's efficiency is striking.

Minimum exergy loss operation also points to a general principle. It turns out that there is a distance on the set of states of a physical system and the minimum production of entropy is just the square of this distance divided by twice the number of relaxations for the process. The distance is defined using the second derivative of the system's entropy as a metric matrix.* The result can be generalized to quantify the losses that are incurred during the coevolution of two systems when there is a cost to adjustment and one of the systems follows some prescribed path. This result, known as the horse-carrot theorem, has been applied in thermodynamics, economics, and coding theory [15–19]. Its implications for coevolution in biological systems are likely to be interesting.

Our final example is more speculative and concerns the question: Is it possible to construct additional functions of state on par with energy and entropy? One hope for extending the constructions afforded by thermostatics is to focus on a class of processes satisfying some optimality condition such as minimum entropy production subject to some constraints. In this case, the optimality condition represents an additional equation, and we should be able to use the additional equation to construct a quantity that will be conserved for optimal processes of this type. There will of course be a corresponding inequality for processes that are not optimal in the sense considered. We believe that this could be carried out for appropriate models of biological systems; the development presented above could serve as a steppingstone toward achieving this goal.

---

* This metric can be interpreted as an equivalent of Fisher's statistical distance known from genetics.

REFERENCES

1. Tribus M, *Thermostatics and Thermodynamics,* Princeton, NJ: D. Van Nostrand, 1961.
2. Tolman RC, *Thermodynamics and Cosmology,* Oxford, UK: Oxford University Press, 1934.
3. Tisza L, *Generalized Thermodynamics,* Cambridge, MA: M.I.T. Press, 1966.
4. Pippard AB, *Elements of Classical Thermodynamics for Advanced Students of Physics,* Cambridge, UK: Cambridge University Press, 1957.
5. Salamon P, *The Thermodynamic Legendre Transformation or How to Observe the Inside of a Black Box,* PhD Thesis, Department of Chemistry, University of Chicago, 1978.
6. Salamon P, Andresen B, Nulton JD, Konopka AK, *The mathematical structure of thermodynamics.* In Konopka, AK, ed. Handbook of Systems Biology, Boca Raton: CRC Press/Dekker, 2006.
7. Schroedinger E, *What is Life?* Cambridge: Cambridge University Press, 1944.
8. Shannon CE, *A mathematical theory of communication.* Bell. Syst. Tech. J. 1948; 27:379–423 (623–656).
9. Shannon CE, *Communication theory of secrecy systems.* Bell. Syst. Tech. J. 1949; 28:657–715.
10. Shannon CE, *Prediction and entropy of printed English.* Bell. Syst. Tech. J. 1951; 30:50–64.
11. Salamon P, Nulton JD, Siragusa G, Andersen TR, Limon A, *Principles of control thermodynamics.* Energy 2001; 26: 307–319.
12. Sieniutycz S, Salamon P, *Advances in Thermodynamics 4: Finite-Time Thermodynamics and Thermoeconomics.* New York: Taylor and Francis, 1990.
13. Berry RS, Kazakov VA, Sieniutycz S, Szwast Z, Tsirlin AM, *Thermodynamic Optimization of Finite Time Processes.* New York: John Wiley & Sons, 2000.
14. Bejan A, *Entropy Generation Minimization: The Method of Thermodynamic Optimization of Finite-Size Systems and Finite-Time Processes.* Boca Raton: CRC Press, 1996.
15. Salamon P, Berry RS, *Thermodynamic length and dissipated availability.* Phys. Rev. Lett. 1983; 51: 1127–1130.
16. Nulton J, Salamon P, Andresen B, Anmin Q, *Quasistatic processes as step equilibrations.* J. Chem. Phys. 1985; 83: 334–338.
17. Salamon P, Komlos J, Andresen B, Nulton JD, *A geometric view of consumer surplus with non-instantaneous adjustment.* Math. Soc. Sci. 1986; 13: 2.
18. Flick JD, Salamon P, Andresen B, *Metric bounds on losses in adaptive coding.* Inf. Sci. 1987; 42: 239.
19. Salamon P, Nulton JD, *The geometry of separation processes: the horse-carrot theorem for steady flow systems.* Europhys. Lett. 1998; 42: 571–576.

4 Friesian Epistemology

Kelley L. Ross

Systematic epistemology begins with Aristotle because *logic* begins with Aristotle. If we ask for reasons for our beliefs, we must be able to say why the reasons support or justify the beliefs. This is what logic does. In a valid deductive argument, if the premises (the reasons) are true, then the conclusion (the belief) must be true. In the analysis offered by logic, we can see why this is so.

Aristotle realized, however, there was another question. How do we know if our reasons, our premises, are true? We can provide more premises, reasons for the reasons, but this only delays, not resolves the issue, since we can ask all over again for the justification of the new premises. Providing reasons for reasons is the "regress of reasons," and Aristotle realized that an *infinite* regress, being impossible, would settle nothing. So how can the regress of reasons end? Only if there are propositions that, for whatever reason, do not need to be *logically* proven, i.e., do not need other reasons. These would be "first principles," *archai* in Greek, *principia* in Latin. Why first principles would not need to be proven is the "Problem of First Principles." Aristotle's answer to the problem became the dominant one: First principles do not need to be proven, because they are *self-evident*, i.e., one knows them to be true just by *understanding* them.

Aristotle did not believe that we come to understand first principles just by thinking about them. He was at heart an empiricist, believing that knowledge comes from experience. He realized, however, that first principles are going to be general propositions, about universals, while descriptions of experience are particulars, about individual things. To get from the individuals to the universals is then the process of logical *induction*. Experience is examined, and a multitude of particular truths accumulated, until a generalization, to universals, is made. Aristotle realized that there is a difficulty here. How do we know *how many* individuals need to be examined before a generalization can be made? This is the "Problem of Induction." A great many white swans can be examined before one happens on Australia and suddenly discovers black swans. Aristotle's answer connects up with his answer to the Problem of First Principles. The idea is that when we get close enough to a first principle, we are able to make an intuitive leap, an act of "mind" (*noûs* in Greek), by which we achieve an understanding of the principle and its self-evidence. This is possible for a metaphysical reason: the mind absorbs (abstracts) the *essence* (*ousía* in Greek) of the matter from the objects. An essence is what makes something what it is, and in Aristotle's metaphysics, the form or essence of a thing is its *actuality*.

It is important to remember what induction is supposed to be in this theory. It is a process of discovery, not of justification. That is how the Problem of Induction is resolved. As discovery, in turn, induction is a process in which truth is brought to the mind. Individuals are examined until the essence stands out. The mind is thus relatively passive, lying open to experience, until the moment when it intuitively grasps the truth. The early modern theorist of induction, *Francis Bacon* (1561–1626), thus required that investigation takes place without prejudice or preconception. For centuries, this all seemed reasonable.

While Aristotle expected that every discipline of knowledge would take the finished form of a system of deductions from first principles, there was only one that looked anything like that in his day. That was geometry. Euclid's geometry identified five axioms and five postulates (and some definitions) from which a satisfyingly large system of theorems could be derived. An axiom ("worthy") was regarded by all as self-evident, while postulates were somewhat less so. A proposition "postulated" would be one that we discover we need in order to prove theorems that we expect to be true, or that we are just experimenting with. Many mathematicians doubted that Euclid's Fifth Postulate, the Parallel Postulate, was self-evident, and centuries followed in which proofs were sought for it. With a proof from the other axioms and postulates, the Fifth Postulate would be a theorem, not a first principle. No proof was ever found. Logically, postulates work in no different a way than axioms, which is why the first principles of a formal deductive system today tend to just be called "axioms," and the system itself an "axiomatic" system, even if the thought is that such axioms are only "postulated." It is noteworthy, however, that the original terminology reflected some uncertainty about the self-evidence of five of the propositions required for Euclid's axiomatic system. There should be no uncertainty about propositions that *ought to be* self-evident.

Curiously, as the centuries were to pass, nothing besides geometry would be developed as an axiomatic system. Nevertheless, Aristotle's view of knowledge was generally accepted, outside of the Skeptics, by all. Perhaps the first serious crack in the dike came with the Nominalists, like *William of Ockham* (1295–1349), who rejected the existence of Aristotelian essences. This would remove the metaphysical basis for self-evidence, damaging the solutions to the Problems of First Principles and of Induction. But the Aristotelian consensus only really broke down with the advent of modern thought. First of all, there was science. *Copernicus* (1473–1543) turned the universe upside down, but his theory had little to do with observation or induction — at the time, it had no real evidence in its favor. *Galileo* (1564–1642) vindicated Copernicus with actual observations, but he also did something unanticipated by Aristotle, applying mathematics to physics. Mathematics never stood in much need of experience, and mathematicians like *Johannes Kepler* (1571–1630), *René Descartes* (1596–1650), and *Gottfried Leibniz* (1646–1716) were drawn to rationalism, claiming sources of rational knowledge independent of experience. Kepler, although responsible for one of the greatest empirical discoveries in the history of astronomy that the orbits of the planets are ellipses, not circles, nevertheless tended toward an outright Platonism, the likes of which usually draws only sneers from recent historians and philosophers of science, and dismissal as no better than mysticism. Although the *nova scientia*, the "new knowledge," was analyzed in terms of induction by the likes of Bacon, its practice seemed anything but. Mathematics can be belittled as a human invention, as by Aristotle, but this was not the inspiration of its greatest practitioners. Nor could practicing induction alone explain how mathematics applied to the models of the real world so powerfully, opening secrets of nature undreamt by the ancients.

While these developments tended to discredit the empiricist and inductivist side of Aristotle, worse was to follow for the other side. The rationalists, although content with self-evidence and constantly appealing to it (Descartes' "clear and distinct ideas"), discredited it with their practice, since they developed philosophical systems radically different from each other. If these were all based on self-evidence, how could they be different? Self-evidence, indeed, leaves no obvious way to settle disagreements. If someone denies the self-evidence of a first principle, they must just not understand it. It cannot be vindicated by logical proof from more ultimate premises, since by definition, there cannot be any. This painted rationalism into a corner with no escape.

The Aristotelian edifice of knowledge was finally taken down in a combined attack of empiricism, nominalism, and skepticism by *David Hume* (1711–1776). Few propositions were left as self-evident by Hume, since they had to pass the test that they could not be denied without contradiction. This even undermined something that had never been doubted in the Middle Ages — the Principle of Causality — that every event has a cause. Hume could easily imagine the contrary. Things could just happen spontaneously and inexplicably. This returned the Problems of First Principles and of Induction in their full force. Hume, however, is commonly misunderstood. Hume *believed* in causality, he just said that it would not be rationally *justified*. This would make it in Euclid's terms, not an axiom, but a postulate. Yet Hume is commonly taken to have expected that causality *could* be violated. No. He was best understood in this by *Immanuel Kant* (1724–1804), who said that Hume's critics "were ever taking for granted that which he doubted, and demonstrating with zeal and often with impudence that which he never thought of doubting…" Kant's distinction was between the *quid facti*, the content of propositions (the matter of fact), and the *quid juris*, their justification (the matter of law, or right), i.e., where the truth comes from. Hume had no problem with the *quid facti*, just with the justification. Kant also understood that Hume's skepticism actually needed to go further, since he realized that even the axioms of geometry, which Hume regarded as self-evident, could be denied without contradiction. When systems of non-Euclidean geometry were subsequently constructed, this demonstrated in the most tangible way that this was true. Kant even denied that *arithmetic* was self-evident, a notion that remained bizarre until it was demonstrated in the 20th century that an axiomatic system of arithmetic required, indeed, axioms, just like geometry. Few regard these axioms as self-evident: in practical terms, they are postulates.

Kant himself developed an argument that principles like causality were justified because they were "conditions of a possible experience," i.e., that as postulates necessary for the world of ordinary experience, their contrary could not be conceived except in terms of a chaotic, dissociated, and, especially, unconscious reality. This, however, whatever its merits as an argument, limited knowledge to "possible experience" and could not resolve metaphysical or moral questions that did not rest on the factual character of perception. A general reconsideration of justification did not come until a successor of Kant, *Jakob Fries* (1773–1843).

Fries distinguished three modes of justification, which he called "proof," "demonstration," and "deduction." All of these were terms that originally had simply meant proof, i.e., logical derivation. That meaning is confined to "proof" itself in Fries. "Demonstration" now means the justification of a proposition by referring to its ground in perception. This kind of thing had pretty much been taken for granted by everyone for centuries, until called into question by Descartes, who did not see how perception could provide knowledge of external objects when perceptions could be unreliable, even hallucinatory. Empiricism did not help. Indeed, *George Berkeley* (1685–1753) would argue that *matter* did not even exist, since we do not perceive it existing independently of our perceptions. Hume, as a skeptic, could take it for granted that perception, indeed, could prove no such thing. Fries, however, as a Kantian, will see matters rather differently. Kant saw the objects of experience, phenomenal objects, as existing *in* experience, not outside of it. Kant called this "empirical realism." This provided for Fries a content of "immediate knowledge," or "intuition," so that as truth is the correspondence of knowledge to its objects, perceptual knowledge is the correspondence of mediate knowledge, i.e., discursive propositions (in some language), to immediate knowledge, i.e., phenomenal objects. Most modern epistemology still struggles hopelessly with this, and "externalist" theories still think of knowledge as justified by, for instance, a causal connection between objects and knowledge. But an effect does not need to resemble its cause, and such a causal connection suffers precisely from the Cartesian objection that our perceptions might be caused by something other than what we think.

"Immediate knowledge" and "intuition" in Fries are not what we might think. Usually these terms are used to mean beliefs we have with some kind of subjective certainty, or as infallible items of discursive knowledge. But Friesian immediate knowledge does not consist of beliefs at all (just of the phenomenal objects themselves), and the possible errors in matching up immediate knowledge with actual beliefs, i.e., with the discursive statements that express them, means that articulated knowledge, mediate knowledge, will always be *fallible* and *corrigible*. In other words, we can make mistakes in describing what our perceptions reveal, but we can also correct them. The important feature of both procedures is that our perceptions must be consulted. If we wonder whether the bedroom window is open, and cannot remember, we have to go look, or get someone else to look. There is no other way to do it, and no *a priori* reasoning, apart from experience, that will settle the case.

Friesian "demonstration" will not help in the problem of Aristotelian first principles, since these are abstract and universal, and not about particular perceptual objects. Justification for them Fries calls "deduction," borrowing the term from Kant, as Kant had borrowed it, along with *quid juris*, from law, since, he said, the deduction shows the *quid juris*, i.e., the justification. Kant's "Transcendental Deduction" contained his argument for the truth of principles like that of causality. Fries did not think that such arguments could avoid circularity. Instead, the nature of the ground of first principles is found by a process of elimination: It is not mediate, since a mediate justification is going to mean either (1) a proof, which is ruled out for first principles, or (2) self-evidence, which in Kantian philosophy only means *analytic* truths, i.e., propositions that cannot be denied without contradiction (Hume's criterion). This means no more than truths of logic and definitions, where definitions themselves can be "real," i.e., based on some natural connection or merely conventional. A definition, then, can function as a postulate. The ground of a first principle, consequently, or a "real" definition, can only be immediate. But, for a principle like causality, the ground cannot be intuitive since, agreeing with Hume, both Kant and Fries rule out that kind of self-evidence. Therefore, the "deduction" is that first principles are grounded in "nonintuitive immediate knowledge," which is the signature doctrine of Friesian epistemology.

What is "nonintuitive immediate knowledge" going to mean? It means that we know things that we are *unaware of* but that we use constantly anyway. This is paradoxical but now obvious in instances such as where we use the rules of the grammar of a language, without being aware of what those rules are. When we identify such things, there is no logical or intuitive property that secures their truth. Similarly, we act as though every event has a cause, even when we do not articulate such a principle. So what do we do? The precedent, indeed, antedates Aristotle. The idea that there are items of knowledge that we are unaware of but use anyway goes back to *Plato*, who thought that knowledge is recollection, i.e., items in memory but forgotten. As it happened, this feature of Plato's theory was superseded by Aristotle's analysis, not only in the later Middle Ages when Aristotle was accepted root and branch but even in the Neoplatonic revival of Late Antiquity, which attempted to reconcile Plato and Aristotle. Since now the Aristotelian approach of self-evidence is no longer viable, we must consider how Plato handled verification in his theory of recollection. Unfortunately, there is no easy answer to this, since Plato was nowhere near as systematic as Aristotle in this respect, and nothing like formal logic existed yet. What we have are examples, and Plato's examples always involve Socrates. The classic example is in Plato's dialogue, the *Meno*, where Socrates elicits a geometry construction simply by asking questions. This is Plato's notion of "Socratic Method." The logic of this goes back to Socrates himself. Socrates, as it happens, did not ask for "reasons for reasons." Socrates solicited beliefs and then examined their logical consequences, exposing inconsistencies. The logic of this only now is apparent. Aristotle was asking for *verification*, i.e., what it is that makes something *true*, while Socrates was effecting *falsification*, since inconsistencies — contradictions — show that something is *false*. Today, when the grammar of a language is described and rules formulated, they are tested by comparison with relevant examples of usage. If usage violates a rule but "sounds right" to a native speaker, the rule may be wrong.

The significance of Socratic Method for modern epistemology was only properly appreciated by the 20th century philosopher who revived Friesian philosophy, *Leonard Nelson* (1882–1927). Its role, indeed, was not in identifying the *quid juris*, which is handled by the "deduction" of nonintuitive immediate knowledge, but is specifying the *quid facti*, the actual content of the nonintuitive truths. Nelson, however, expected Socratic Method to be easier than it is. Socratic Method is limited by its data, i.e., by the input of beliefs, and by the fact that a contradiction only shows that *some* belief is false, without it always being clear *which one*. Socratic Method thus simply imposes a coherence test of truth, whose greatest demand is on reflection and the imagination. The use of nonintuitive immediate knowledge does not spontaneously produce the corresponding belief. This must be *generated* in mediate knowledge by reflection, and the ability to do so, to come up with new ideas in the light of observation, is a function of imagination. This is then the grist for a coherence test. Coherence, however, does not guarantee truth, since the proper belief may not yet figure in the discussion. Thus, the principle of causality that every event has a cause was hardly even clearly stated until Hume. A Medieval critique of causality, in many ways like Hume's, that of *al-Ghazâlî* (1059–1111), did not question the causal principle itself, only the certainty of laws of nature and the identity of *particular* causes.

The importance of falsification in epistemology was only fully appreciated by the philosopher of science *Karl Popper* (1902–1994), who observed that the practice of science was not Baconian induction, but by falsification. Thus, the process of *discovery* of scientific theories is of no final *logical* significance. Induction, remember, required observation of empirical events with an open mind, waiting for generalizations to suggest themselves, i.e., for the essence to stand out. According to Popper, we can make up any theories we like, about things that are even unobservable, a process where the proper view may even come to us in a *dream* (as has happened in the history of science), and this will all be proper science as long as we can make *observable predictions* from the theory. If the predictions are correct, this may increase our confidence in a theory, but it does not prove the theory true. All the success of Newton's physics did not prevent its replacement by Einstein's. If the predictions are *not* correct, however, then logically the theory or something about it, is falsified. If there is no alternative available to the theory, then the falsifications can be regarded as "anomalies," hopefully to be accommodated; but if a new theory can make predictions that the old theory cannot, the new one will tend to replace the old. Like Socratic Method, however, the final truth may simply not have been thought of yet. This, indeed, is what people in science now tend to think.

Popper saw his own theory as a version of Friesian epistemology, which he had learned about from his cousin *Julius Kraft* (1898–1960), who had been a student of Nelson and an advocate of his work. Popper thought that falsification could be used in all areas of philosophy, but he was not familiar with the practice of Socratic Method, as Nelson was, nor was he any more aware of the limitations of the method. At least in science, if a startling prediction can be made, as in Einstein's prediction that the apparent positions of stars would be displaced around the sun, something observable, though not yet observed, during a total eclipse of the sun, then this places a very hard limit on theories that are or are not consistent with it. In metaphysics or ethics, however, the limits of coherence are more fluid and it is easier to argue about what are the relevant facts of the matter. For instance, those so minded can still disparage capitalism, which has produced unprecedented wealth and freedom, as a system which perpetuates poverty and oppression, even while preferring systems, as in the Soviet Union, which manifestly *did* produce only poverty and oppression. Since Socratic Method requires reflection and imagination, its success will be the most limited with those who most lack these facilities.

If the Kantian *quid facti* and *quid juris* are secured by Socratic Method and the "deduction" of nonintuitive immediate knowledge, then, as in Aristotle, the next question to ask is the metaphysical one. What is the nature of the *existence* of the nonintuitive immediate ground of knowledge? In this respect, the Friesians, both Fries and Nelson, did not advance beyond Kant, who would simply have said that the ground lies in Reason. Kant, however, did not believe that Reason was necessarily just a subjective faculty in our minds. He realized that we have an urge to speak beyond the limits of a "possible experience," to describe objects that do not exist phenomenally, but only among "things-in-themselves," i.e., objects as they exist apart from experience — such as God, free will, and the immortal soul. Kant, with his principle of a possible experience, did not believe we could have *knowledge* of such objects, but he did think they could emerge with some different justification in morality and religion. Fries was less agnostic, since he thought that the principle of a possible experience was unnecessary. If there was nonintuitive immediate knowledge about transcendent objects, it could be handled like anything else.

This does not help much with the metaphysical question. "Reason" still does not answer it. If there are objects of nonintuitive immediate knowledge, our only precedent for describing them is in Plato or Aristotle. Indeed, we need something like Aristotle's essences, but without some of their features. They do not become intuitively evident, through induction or anything else. Just what their content is will always suffer from the uncertainties of the coherence test of evaluation. Thus, we are not *directly acquainted* with essences as Aristotle would have thought. This lack of direct acquaintance is the defining feature, however, of Plato's *Forms*, which are only "remembered" from our sojourns between lives. Aristotle's main objection to the Forms was that they were in a different reality, a transcendent World of Forms, and not in the objects themselves. This point can be well taken. But "transcendence" after Descartes or Kant does not mean a different reality. If the "Form" is in the thing-in-itself, then it is indeed transcendent, i.e., outside of phenomena and possible experience, but still in the present object and the present world. So, as often happens in philosophy, the ancient opponents are both right, just in different ways.

And we must remember that "essences" or "Forms" include things like the Laws of Nature. Scientists tend to break down into realists, who expect that the Laws of Nature are "in" the objects or "out there" in the world, or the positivists, who do not care, profess agnosticism, or like arguments that metaphysical questions are meaningless. A contemporary mathematician, like *Roger Penrose*, does not think that alternatives to realism even make sense — though, of course, as a mathematician, he tends toward the traditional mathematician's Platonism. Now, however, if Kantian and Friesian epistemology can reconcile Platonic and Aristotelian metaphysics, the "out there" view is perfectly reasonable.

Kant had more to go on to limit knowledge to the world than his principle of a possible experience. He realized that, in the history of philosophy, attempts to construct knowledge about transcendent objects resulted in *Antinomies*, i.e., dilemmas where the arguments for contradictory theses are equally good. So it was not just that God, freedom, and immorality transcended experience, it was the problem that a coherence test had no way to really prefer one view over its opposite. That the limitation apparently applies to what are conceived as transcendent *objects*, and not just the abstractions of Platonic Forms or Aristotelian essences, is significant if we think that the only objects are phenomenal objects, which may exist transcendently, in themselves, but which nevertheless are not separate from the world. Another feature of the problem, however, is that even phenomenal objects can be conceived with characteristics that could only apply to them in themselves — and that is to be in some way *unconditioned*. Moral freedom is to possess an unconditioned causal agency. Even the Skeptic Hume rejected such freedom, because he thought it violated causality. Kant thought that it does not violate causality as such, but depended on the *way* that causality is applied: whether as a continuous series, going into indefinite past and future, or as a series with an origin, i.e., as the causal act of free will. The problem with unconditioned realities applies to the world itself, since the universe *as a whole* is an independent, unconditioned object. Thus Kant described the finitude or infinitude of space and time as itself an Antinomy. This was a dilemma only seriously addressed by Einstein's theory that space is finite but unbounded (using a non-Euclidean geometry); but then at present, the observational evidence, for a finite observable (but actually infinite) universe, is against it. So to most astronomers now, Einstein's solution to the Antinomy is no longer available. Kant thought that Antinomies over God, freedom, and immorality might be resolved in morality, but Fries and Nelson continued to take seriously the limitations of *theoretical* knowledge.

This limitation produced in Fries another epistemological feature. Kant had viewed religion as purely a function of morality (not itself a function of a possible experience), and Fries was going to agree with Kant that we certainly have no *perceptions* of transcendent objects. Fries, however, believed that there was more to religion than morality, and that religious feelings were about something real, even if they were neither perceptions nor intuitions. He therefore distinguished *Wissen* (knowledge), which is empirical and phenomenal knowledge, and *Glaube* (belief), which is derived from nonintuitive immediate knowledge, from *Ahndung* (intimation, *Ahnung* is modern German), which is a feeling that can be associated with possible transcendent or unconditioned objects, even while providing no knowledge of them. But for Fries, the feelings are esthetic, i.e., of the beautiful and the sublime. A stronger theory was put forward by an early associate of Nelson, *Rudolf Otto* (1869–1937). Otto pointed out that religion contains a characteristic category of valuation, the *sacred* or *holy*. The character of the holy and our feelings about it are distinct from either the moral or the merely esthetic. It is, in Otto's neologism, *numinous*. In these terms, religion is not just about the moral or the beautiful, the majestic or awesome (i.e. sublime), it is about something *uncanny*, something that can be supernaturally frightening. In its rawest and simplest form, this is the feeling one gets walking past a cemetery after dark. That is a moment largely free of esthetic value, though people often try to invoke it in art. But it also ranges to the curious "fear of God": an expression that can refer neither to fear of harm from an evil being, nor fear of damage from some overpowering and uncontrolled force, since God is thought of as good, whose intentions and powers can contain nothing to fear. God would be "feared" only because he would be unlike any natural object, something, in Otto's words, "wholly Other." Even in a religion without a God, like Buddhism, there is no lack of the sacred and even the uncanny.

The Friesian School is an obscure chapter in the history of philosophy. Yet some of those subsequently influenced by Fries and Nelson, like Karl Popper and Rudolf Otto, are usually familiar to those who study philosophy of science and philosophy of religion, respectively. What is unfamiliar is the larger context for their thought, where Friesian epistemology and metaphysics address issues that go from the earliest days of Western philosophy to the changes that heralded the beginning of Modern philosophy. The mainstream of academic philosophy in the 20th century ended up as little better than a triumph of Nihilism, using science to demolish religion, and then using sterile irrationalities like "deconstruction" even to demolish science. Having given up hope in knowledge, "Theorists" (as they like to be called) then head directly for uncritical and illiberal versions of political activism, whose Marxist roots are usually all too obvious. This process is ugly and dispiriting. But alternatives exist. Kant is still the watershed of Modern philosophy, but his wisest successors, like Fries and *Arthur Schopenhauer* (1788–1860), are overlooked or belittled, in favor of monsters like G.W.F. Hegel or Friedrich Nietzsche. It remains to be seen whether the tradition, a ship close to capsizing, will be able to right itself and go forward.

REFERENCES

"Epistemology," The Proceedings of the Friesian School, http://www.friesian.com/epistem.htm.
Aristotle, The Categories, On Interpretation, Prior Analytics. Harvard: Loeb Classical Library, 1962.
Aristotle, Posterior Analytics. Topica, Harvard: Loeb Classical Library, 1966.
Kant, Immanuel, Critique of Pure Reason. New York: St. Martin's Press, 1965.
Nelson, Leonard, Progress and Regress in Philosophy, Vol. 1 and 2. Oxford: Basil Blackwell, 1970.
Nelson, Leonard, Socratic Method and Critical Philosophy. New York: Dover, 1965.
Otto, Rudolf, The Idea of the Holy. Oxford, 1972.
Penrose, Roger. The Emperor's New Mind, Oxford University Press, 1990.
Popper, Karl. The Logic of Scientific Discovery, Hutchinson of London Ltd. 1959, 1977.
Schopenhauer, Arthur. The World as Will and Representation. Vol. 1 and 2. New York: Dover, 1966.

5 Reconsidering the Notion of the Organic

Robert E. Ulanowicz

CONTENTS

ABSTRACT

The advent of the Enlightenment entailed a radical shift in worldviews from one wherein life dominates all events to the perspective that all phenomena ultimately are elicited by encounters between lifeless, unchanging particles. A necessary casualty of this shift has been the notion of organic behavior. The possibility exists, however, that both the pre- and post-Enlightenment attitudes are extremes, and that a general, more complete description of nature might lie between these poles. One useful tool for such interpolation appears to be Karl Popper's definition of propensity as an agency that is intermediate to deterministic force at one end and pure chance at the other. Another is Robert Rosen's description of organic behavior as self-entailing. One way of interpreting self-entailment is to identify it with the influence of the aggregate configuration of processes upon the structure and composition of the organic system. To paraphrase Alfred North Whitehead, the creature is derivative of the creative process. One particular embodiment of the top-down influence of processes upon components might be akin to the action of autocatalysis among propensities, the effects of which can be measured using network and information theories. Once one expands the scope of consideration to include boundary conditions as well as dynamics, it then can be argued that the organic narrative is actually more compact than the conventional neo-Darwinian construct. Furthermore, the revised organic narrative provides a more appropriate metaphor for other self-organizing systems than did classical organicism, and it has the advantage of serving as a more congenial setting within which to portray the origin of life.

5.1 INTRODUCTION

Fascination with organic behavior held sway over much of Western thought for centuries preceding the dawn of the Enlightenment. One encounters the metaphor of human institutions in the guise of organisms as early as Aristotle and Paul of Tarsus. By way of contrast, Democritus and Lucretius, with their view on nature as comprising encounters between unchanging, lifeless atoms, were notable for being exceptions to the prevailing view that the universe was suffused throughout with life [1]. So dominant was the vision of the ubiquity of life that the overriding challenge to philosophical thought before the 17th century had been to explain somehow the existence of death in the world.

The situation changed radically with the advent of Newton and the publication of his *Principia*, which (by accident) [2] obviated the necessity to invoke living agencies to explain the movements of the spheres. Thereafter the fashion in natural philosophy quickly came to emphasize the material outlook of Hobbes in conjunction with the mechanical notions of Descartes. Lucretius would have found himself quite at home with the precipitating consensus that the world is dead in all its dimensions. Living beings became annoying exceptions in the world of the dead, and how to explain the origin of the "epiphenomenon" called life has become a vexing conundrum for contemporary thinkers.

One necessary casualty of the Enlightenment was the notion of organic behavior. True, remnants of the organic metaphor remained in the writings of Liebnitz, Comte, Spencer, and Von Betalanffy, but by and large organicism has been eschewed by modern thinkers (and not entirely without justification, it should be added. Any number of totalitarian leaders over the past two centuries have invoked the organic metaphor to justify their own oppressive agendas.)

The extremism of dictators, however, provides a clue to resolving the confusion left in the wake of the Enlightenment revolution. It is easy to recognize that the predilection of the ancients and medieval scholastics to see life everywhere was an extreme view, but one can also hold that it likewise remains a radical stance to view the world as dead, through and through. Organicism, because it focused primarily upon those rigid, almost mechanical aspects of organismal behaviors, also could be classified as extreme. Is there indeed no refuge from such radical caricatures of the world in which we live — some narrative of behavior that portrays a more balanced dialogue between the quick and the dead? [3] Here, the reader's attention is drawn to the domain of ecosystems. In a world where phenotypic characteristics appear to be the ineluctable consequences of genomic traits, the study of ecosystems affords a refreshing breath of freedom. As the renowned developmental biologist, Gunther Stent so neatly puts it,

> Consider the establishment of ecological communities upon colonization of islands or the growth of secondary forests. Both of these examples are regular phenomena in the sense that a more or less predictable ecological structure arises via a stereotypic pattern of intermediate steps, in which the relative abundances of various types of flora and fauna follow a well-defined sequence. The regularity of these phenomena is obviously not the consequence of an ecological program encoded in the genomes of the participating taxa [4].

When viewed in a Newtonian context, the relative independence of ecosystems from the mechanics of genomes leads some to regard the concept of "ecosystem" as irrelevant — they see the biotic community as nothing more than stochastic disarray [5]. Obviously, such conclusion was not what Stent had in mind. Rather, it appears he may have been making the rather bold suggestion that ecosystems come closer to the crux of organic behavior than do organisms themselves. That is, ecosystem behavior is ordered to a degree, but that degree falls pointedly short of mechanical determinism. Just perhaps, then, ecosystems offer the desired refuge from extremism.

Another inference we might draw from Stent's comments is that mechanisms are not the only causal entailments at work in ecosystems. Such a conclusion, however, conflicts with the meta-physics underlying all of scientific thought since *Principia*, which allows for only two types of causalities — material and mechanical. Robert Rosen, among others, has also questioned the assumed completeness of mechanical causal entailment in living systems [6]. He illustrates such incompleteness by referring to the rational numbers, which seem at first glance to be arbitrarily dense — between any two rational numbers it is possible to find an infinity of other rational numbers. Yet we know from number theory that the rational numbers are incomplete as regards our conception of what constitutes a number. Between any two rational numbers one can also find an infinity of irrational numbers. Kurt Gödel was able to show with his Incompleteness Theorem that any attempt to formalize completely a notion of mathematics will comprise a syntactic truth that is always narrower than the whole set of truths about the mathematical concept itself. (Some unformalized semantic residue always remains.) Whence, Rosen goes on to argue that the formalisms of physics describing the causal entailments in mechanical systems will perforce remain incomplete descriptions of organic behavior. It seems that in our zeal to apply Occam's Razor and avoid an "alpha type error" (the assumption of false agencies), we have gone to an extreme, thereby committing a "beta type error" by proscribing any forms of causal entailment other than those material and mechanical.

5.2 CHANCE AND PROPENSITIES

Yet another aspect of the radical minimalism inherent in the Newtonian worldview has been provided by Karl R. Popper [7]. Popper noted how mechanical causality exists only at one pole (extreme) of the causal spectrum: Either a cause has a precise effect or it has none. There is positively no room for chance in the Newtonian rendition of the world. Should any particle anywhere swerve from its lawful course, it would eventually throw the entire universe into shambles. Yet chance and contingency seem to be very much evident in natural phenomena in general and in biological systems *a fortiori*. Conventional wisdom has reconciled mechanism and chance via what is commonly referred to as the "Grand Synthesis." Upon closer inspection, this "synthesis" resembles nothing more than a desperate and schizoid attempt to adjoin two mutually exclusive extremes. That is, narrative is constantly switching back and forth between the realms of strict determinism and pure stochasticity, as if no middle ground existed. Once more, the picture is incomplete, and a more effective reconciliation, Popper suggested, lies with agencies that are intermediate to stochasticity and determinism. Toward this end, he proposed a generalization of the Newtonian notion of "force" that extends into the realm between strict mechanism and pure chance. Forces, he posited, are simple idealizations that exist as such only at the extreme of perfect isolation. The objective of experimentation is to approximate to the fullest extent possible the isolation of the workings of an agency from interfering factors. In the real world, however, where components are loosely, but definitely coupled, and all manner of extraneous phenomena interfere, he urged us to consider the more general notion of "propensities."

A propensity is the tendency for a certain event to occur in a particular context. It is Bayesian in that it is related to, but not identical to, conditional probabilities. Consider, for example, the hypothetical "table of events" depicted in Table 5.1, which arrays five possible outcomes, b_1, b_2, b_3, b_4, and b_5, according to four possible eliciting causes, a_1, a_2, a_3, and a_4. For example, the outcomes might be several types of cancer, such as those affecting the lung, stomach, pancreas or kidney, while the potential causes might represent various forms of behavior, such as running,

TABLE 5.1

Frequency Table of the Hypothetical Number of Joint Occurrences that Four "Causes" (a_1 ... a_4) Were Followed by Five "Effects" (b_1 ... b_5)

	b_1	b_2	b_3	b_4	b_5	Sum
a_1	40	193	16	11	9	269
a_2	18	7	0	27	175	227
a_3	104	0	38	118	3	263
a_4	4	6	161	20	50	241
Sum	166	206	215	176	237	1000

TABLE 5.2
Frequency Table As in Table 5.1,
Except that Care Was Taken to Isolate
Causes from Each Other

	b_1	b_2	b_3	b_4	b_5	Sum
a_1	0	269	0	0	0	269
a_2	0	0	0	0	227	227
a_3	263	0	0	0	0	263
a_4	0	0	241	0	0	241
Sum	263	269	241	0	227	1000

smoking, eating fats, etc. In an ecological context, the b's might represent predation by predator j, while the a's could represent donations of material or energy by host i.

We notice from the table that whenever condition a_1 prevails, there is a propensity for b_2 to occur. Whenever a_2 prevails, b_5 is the most likely outcome. The situation is a bit more ambiguous when a_3 prevails, but b_1 and b_4 are more likely to occur in that situation, etc. Events that occur with smaller frequencies, e.g., [a_1,b_3] or [a_1,b_4] result from what Popper calls "interferences."

We now ask how might the table of events appear, were it possible to isolate phenomena completely — to banish the miasma of interferences? Probably, it would look something like Table 5.2, where every time a_1 occurs, it is followed by b_2; every time a_2 appears, it is followed by b_5, etc. That is, under isolation, propensities degenerate into mechanical-like forces. It is interesting to note that b_4 never appears under any of the isolated circumstances. Presumably, it arose purely as a result of interferences among propensities. Thus, the propensity for b_4 to occur whenever a_3 happens is an illustration of Popper's assertion that propensities, unlike forces, never occur in isolation, nor are they inherent in any object. They always arise out of a context that invariably includes other propensities. That is, propensities are always imbedded in a configuration of processes.

The interconnectedness of propensities highlights an unsung aspect of the role of contingency in systems development — namely, that contingencies are not always simple by nature. Chance events can possess distinctive characteristics and can be rare, or possibly even unique in occurrence. Inculcated as we are in the atomic assumption that undergirds Newtonianism, we almost always consider chance events as generic in nature, point-like in extent, and instantaneous in duration. Thus, when Prigogine writes about macroscopic order being elicited by microscopic fluctuations, it is implicit that the latter are generic and structure-less [8]. Chance perturbations, however, happen to come in an infinite variety of forms, and any given system may be very vulnerable to some categories of disturbance and relatively immune to others.

Even if disturbances come in different flavors, we have been conditioned by Francis Bacon to expect all phenomena to repeat themselves in due time. Repeatability is, after all, a cornerstone of normal science. Once we open the door to complex contingencies, however, we must be prepared to face the possibility that some contingencies might be *unique* for once and all time. In fact, we should even gird ourselves to encounter a world that is absolutely rife with one-time events. The key to this possibility is that contingencies be more generally regarded as configurations or con-stellations of both things *and processes*. That many, if not most, such configurations are probably unique for all time follows from elementary combinatorics. For, if it is possible to identify n different things or events in a system, then the number of possible combinations of events varies roughly as n factorial.

It does not take a very large n for $n!$ to become *immense*. Elsasser called an immense number any magnitude that was comparable to or exceeded the number of events that could have occurred since the inception of the universe [9]. To estimate this magnitude, he multiplied the estimated number of protons in the known universe (ca. 10^{85}) by the number of nanoseconds in its duration (ca. 10^{25}) to yield a maximum of about 10^{110} conceivable events. It is often remarked how the second law of thermodynamics is true only in a statistical sense; how, if one waited long enough, a collection of molecules would spontaneously segregate themselves to one side of a partition. Well, if the number of particles exceeds 80 or so, the *physical* reality is that they would *never* do so (because $80! >> 10^{110}$).

Because propensities always exist in a context, and because that context usually is not simple, it becomes necessary to consider the reality and nature of complex contingencies. To capture the effects of chance and history, it will no longer suffice simply to modulate the parameters of a mechanical model with generic noise [10]. In a complex world, unique events are continually occurring. They are by no means rare; they are legion! Perhaps fortunately, the overwhelming majority of one-time events simply happen and pass from the scene without leaving any trace in the more enduring observable universe. On occasion, however, a singular contingency can interact with a durable system in such a way that the system readjusts in an *irreversible* way to the disturbance. The system then carries the memory of that contingency as part of its *history*. Again, no amount of waiting is likely to lead to an uncontrived repetition of what has transpired.

The efficacy of Popper's generalization is that the notion of propensity incorporates under a single rubric law-like behavior, generic chance and unique contingencies. We note for reference below that, irrespective of the natures of any eliciting interferences, the transition depicted from Table 5.1 to Table 5.2 involves proceeding from less-constrained to more constrained circumstances. It is the progressive appearance of constraints that we have in mind when we speak of the "development" of an organic system. We now ask the questions, "What natural agency might contribute to the transition from Table 5.1 to Table 5.2?"; or, in a larger sense, "What lies behind the phenomena we call organic growth and development?," and "How can one quantify the effects of this agency?"

5.3 THE ORIGINS OF ORGANIC AGENCY

If Rosen deconstructed the notion that an organic system can exist exclusively as a collection of mechanisms, he was at least responsible enough to hint at a direction in which one might seek to construct a new theory of organic behavior. Rosen claimed that, in contradistinction to mechanical systems, organic systems are self-entailing. That is, all of the processes necessary for the continued existence of the organic structure are present *within* the system. Various forms of anabolism, catabolism, reproduction, and maintenance all exist within an organic system and causally entail one another. Although material and energy are required from the external world, such resources resemble more objects than processes. In keeping with Popper's emphasis on context and arrangement of propensities, we choose to identify the configuration of constituent processes as the very kernel of the organic system, from which most behaviors derive. In Popper's own words: "Heraclitus was right: We are not things, but flames. Or a little more prosaically, we are, like all cells, *processes* of metabolism; *nets* of chemical pathways [11]" [Emphases mine].

The priority given to the configuration of processes over the structure of stable components in organic systems is likely to sound either exotic or transcendental to many readers, but it is neither. Its roots lie in the "process philosophy" of Alfred North Whitehead and Charles Saunders Peirce Whitehead, for example, was inclined to identify any "creature [as being] derivative from the creative process [12]." As for a nontranscendental exemplar, I offer the dynamics of autocatalysis, or indirect mutualism.

To make clear exactly what I mean by "autocatalysis," I take it to be a particular manifestation of a positive feedback loop wherein each process has a positive effect upon its downstream neighbor [13]. Without loss of generality, we will confine our discussion to a serial, circular conjunction of three processes A, B, and C. Whenever these processes involve very simple, virtually immutable chemical forms or reactions, then the entire system functions in wholly mechanical fashion. Any increase in A will invoke a corresponding increase in B, which in turn elicits an increase in C, and whence back to A.

Matters become quite different, however, as soon as the configuration of processes grows combinatorically complex. Such entities, by virtue of their rich continuum of variations, are capable of small, contingent changes in structure, each variation of which will continue to function as an autocatalytic loop as before, albeit with somewhat more or less effectiveness. In keeping with Popper's ideas, it would also be more appropriate to say that the action of process A under such circumstances has a *propensity* to augment a second process B. That is, the response of B to A is not prescribed deterministically. Rather, when process A increases in magnitude, most (but not all) of the time, B also will increase. B tends to accelerate C in similar fashion, and C has the same effect upon A.

Autocatalysis among propensities gives rise to several system attributes, which, as a whole, distinguish the behavior of the system from one that can be decomposed into simple mechanisms [14]. Most germane is that such autocatalysis now becomes capable of exerting *selection* pressure upon its ever-changing, malleable constituents. To see this, we consider a small spontaneous change in process B. If that change either makes B more sensitive to A or a more effective catalyst of C, then the transition will receive enhanced stimulus from A. Conversely, if the change in B either makes it less sensitive to the effects of A or a weaker catalyst of C, then that perturbation will likely receive diminished support from A. We note three things about such selection: (1) that it acts on the constituent processes or mechanisms as well as on the elements (objects) that comprise the system, (2) that it arises *within* the system, not external to the system, and (3) that it can act in a positive way to select *for* a particular system end (greater autocatalysis), rather than always *against* the persistence of an individual material form. The first attribute defeats all attempts at reductionism, while the latter two distinguish autocatalytic selection from the "natural selection" of evolutionary theory.

It should be noted in particular that any change in B is likely to involve a change in the amounts of material and energy that are required to sustain process B. Whence, corollary to the selection pressure is the tendency to reward and support those changes that serve to bring ever more resources into B. As this circumstance pertains to any and all members of the feedback loop, any autocatalytic cycle of propensities becomes the epicenter of a *centripetal* configuration, toward which as many resources as available will converge. Even in the absence of any spatial integument (as required by the related scenario called autopoeisis) [15], the autocatalytic loop itself defines the focus of flows.

Centripetality implies that whenever two or more autocatalytic loops exist in the same system and draw from the same pool of finite resources, *competition* among the foci will ensue. In particular, whenever two loops share pathway segments in common, the result of this competition is likely to be the exclusion or radical diminution of one of the nonoverlapping sections. For example, should a new element D happen to appear and to connect with A and C in parallel to their connections with B, then if D is more sensitive to A and/or a better catalyst of C, the ensuing dynamics should favor D over B to the extent that B will either fade into the background or disappear altogether. That is, the selection pressure and centripetality generated by the autocatalytic configuration can guide the replacement of elements.

Of course, if B can be replaced by D, there remains no reason why C cannot be replaced by E and A by F, so that the cycle A, B, C could eventually transform into D, E, F. This possibility implies that the characteristic lifetime of the autocatalytic form generally exceeds those of most of its constituents. The incipience of the autocatalytic form before, and especially its persistence beyond, the lifetimes of most of its constituent objects imparts causal priority to the agency of the configuration of processes. True, the inception of the feedback loop can be interpreted as the consequence of conventional mechanistic causes. Once in existence and generating its own selection pressures, however, those instigating mechanisms become accidental to the selection agency that arises. Any argument seeking to explain the behavior of the whole system entirely as the result of shorter-lived constituent objects erroneously ignores the ascendant agency of the configuration of processes, which winnows from among those ephemeral and transitory mechanisms.

We recall that with Democritus, the aim of rational explanation was to portray all processes as the consequence of universal laws that act on eternal and unchanging fundamental atoms. At atomic scales, this reductionist agenda has worked reasonably well, but, as soon as one encounters complex configurations of processes, temporality becomes reversed. At the mesoscales, it is the configurations of processes that are most enduring in comparison with which their constituents appear merely as transients. Therefore, the most natural direction for causality to act at intermediate scales is from the persistent configurations of processes toward their transient constituents, whose creation the former mediates.

5.4　THE INTEGRITY OF ORGANIC SYSTEMS

Although autocatalytic systems appear to be potent and durable, their continued existence is not without its own contingencies. As Popper noted, any configuration of processes can be considered as a network, and the overall topology of networks leads one to some striking conclusions about some inherent limitations possessed by organic systems. Earlier, Rashevsky had also described the constituent processes of an organism in terms of a network wherein the various forms visible in the system (nose, brain, kidney, etc.) [16] are connected to each other by processes. Whence, the nose transmits olfactory signals to the brain via the electrical process of neural transmission. In ecosystems one might link a lower species, such as an anchovy, to a higher trophic taxon, such as a striped bass, via the process of predation. (Striped bass consume anchovies.)

Rashevsky's student, none other than Robert Rosen, was able to prove a theorem about the vulnerabilities of organisms (particular instances of our more general "organic system") using nothing more than the general topology of their networks [17]. He identified two classes of processes in organisms — those that can generally be called metabolism and those that are restituent or effect repairs. Rosen proved that it is impossible for every restituent process in an organism to be reestablishable. That is, there is at least one restituent process in each organism, which, if rendered inoperative, cannot be reconstituted. He went on to prove a second theorem that if the nonreestablishable process happens to connect with external sources, then the destruction of that process will cause the failure of the whole system. Organisms are mortal, it seems.

To prove his theorems, Rosen had to consider what he called a "closed circuit of restituents." He was able to show how, if one process in a closed circuit of restituents fails, all the others will fail as well. Rosen's reasoning applies as well to autocatalytic loops, so that one might view as a corollary of his theorem the fact that autocatalytic systems are in general nondecomposable. For, if one process of an autocatalytic cycle is obviated, there is no guarantee that its antecedent and subsequent processes will link with one another in catalytic fashion, whence autocatalysis breaks down, and all its constructive and integrative powers vanish with it. We see how Rosen's conclusion for one subclass of organic systems (organisms) bears implications for the more general category: Organic systems are nondecomposable. They differ radically from purely mechanical systems, which can be broken apart into disjoint objects and reassembled again without any discernible loss of function.

5.5　FORMALIZING ORGANIC DYNAMICS

As Rosen and Gödel demonstrated, formalizing any set of entailments always carries with it the risk of incompleteness. Such danger did not deter Rosen himself from engaging in his own formalizations — to good end. Furthermore, the scientific enterprise demands at least some degree of quantification before a proposition can be legitimately entertained. Accordingly, we reconsider the effects that autocatalysis works upon a system in the hopes of discerning some pattern among

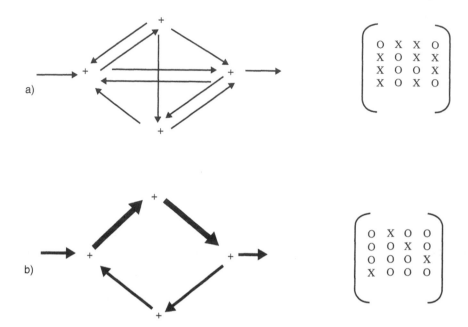

FIGURE 5.1 Schematic representation of the major effects that autocatalysis exerts upon a system. (a) original system configuration with numerous equiponderant interactions. (b) same system after autocatalysis has pruned some interactions, strengthened others, and increased the overall level of system activity (indicated by the thickening of the arrows). Corresponding matrices of topological connections indicated to the right.

them. There appears to be two major facets to its actions: Autocatalysis serves to increase the activities of all participating constituents; and it prunes the network of interactions so that those links that most effectively contribute to autocatalysis come to dominate the system. These facets of behavior are extensive (size-dependent) and intensive (size-independent), respectively. Schematically this transition is depicted in Figure 5.1. The upper figure represents a hypothetical, inchoate four-component network before autocatalysis has developed, and the lower one, the same system after autocatalysis has matured. The magnitudes of the flows are represented by the thickness of the arrows. To the right appear the matrices that correspond to the pattern of flows. One recognizes that the differences between the matrices in Figure 5.1 resemble those between Tables 5.1 and 5.2, and we recall how that transition was associated with the appearance of progressive constraints.

There is not sufficient space to detail how these two facets of autocatalysis came to be quantified [18]. Suffice it here merely to convey the results. We begin by designating T_{ij} as the amount of some conservative medium transferred from compartment i to some other compartment j within the organic system. The sum of all such magnitudes for all the processes in the system becomes what is known in economic theory as the "total system throughput" — a measure of a system's total extent, or activity. In symbols,

$$T = \sum_{i,j} T_{ij}$$

where T is the total system throughput. Growth thereby becomes an increase in the total system throughput, much as economic growth is reckoned by any increase in Gross Domestic Product.

As for the "pruning," or development effected by autocatalysis, it will be related to changes in the probabilities of flow to different compartments. We note, therefore, that the joint probability that a quantum of medium both leaves i and enters j can be estimated by the quotient T_{ij}/T, and that the conditional probability that, having left i, it then enters j can be approximated by the quotient

$$T_{ij} \Big/ \sum_k T_{ik}.$$

One can then use these probability estimates to calculate how much information is provided by the increased constraints. The appropriate measure in information theory is called the "average mutual information" or AMI,

$$AMI = \sum_{i.j} \left(\frac{T_{ij}}{T} \right) \log \left(\frac{T_{ij}T}{\sum_p T_{pj} \sum_q T_{iq}} \right).$$

The AMI has recently been shown to be the logarithm of the number of distinct roles in the system that is sculpted by its various constraints [19]. To demonstrate how an increase in AMI actually tracks the "pruning" process, I refer the reader to the three hypothetical configurations in Figure 5.2. In configuration (a) where medium from any one compartment will next flow is maximally indeterminate. AMI is identically zero. The possibilities in network (b) are somewhat more constrained. Flow exiting any compartment can proceed to only two other compartments, and the AMI rises accordingly. Finally, flow in schema (c) is maximally constrained, and the AMI assumes its maximal value for a network of dimension 4.

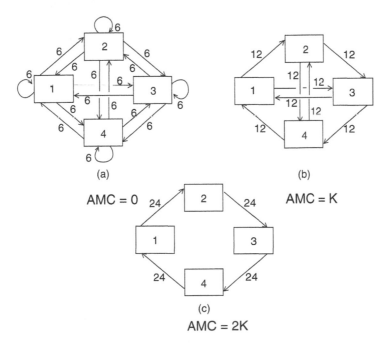

(a)

AMC = 0

(b)

AMC = K

(c)

AMC = 2K

FIGURE 5.2 (a) The most equivocal distribution of 96 units of transfer among four system components. (b) A more constrained distribution of the same total flow. (c) The maximally constrained pattern of 96 units of transfer involving all four components.

Because autocatalysis is a unitary process, we can incorporate both factors of growth and development into a single index by multiplying them together to define a measure called the system ascendency, $A = T \times \text{AMI}$. In his seminal paper, "The Strategy of Ecosystem Development," Eugene Odum identified 24 attributes that characterize more mature ecosystems [20]. These can be grouped into categories labeled species richness, dietary specificity, recycling, and containment. All other things being equal, a rise in any of these four attributes also serves to augment the ascendency. It follows as a phenomenological principle of organic behavior that "in the absence of major perturbations, organic systems have a propensity to increase in ascendency." Increasing ascendency is a quantitative way of expressing the tendency for those system elements that are in catalytic communication to reinforce and mutually entail each other to the exclusion of nonparticipating members.

I should hasten to emphasize in the strongest terms possible that increasing ascendency is only half of the dynamical story. Ascendency accounts for how efficiently and coherently the system processes medium. Using the same mathematics, one can compute as well an index called the system overhead, Φ, that is complementary to the ascendency [21].

$$\Phi = -\sum_{i,j} T_{ij} \log\left(\frac{T_{ij}^2}{\sum_k T_{kj} \sum_q T_{iq}} \right).$$

Overhead quantifies the inefficiencies and incoherencies present in the system. Although these latter properties may encumber overall system performance at processing medium, they become absolutely essential to system survival whenever the system incurs a novel perturbation. At such time, the overhead comes to represent the degrees of freedom available to the system and the repertoire of potential tactics from which the system can draw to adapt to the new circumstances. Without sufficient overhead, a system is unable to create an effective response to the exigencies of its environment. The configurations we observe in nature, therefore, appear to be the results of a dynamical tension between two antagonistic tendencies (ascendency vs. overhead).

5.6 UNDER OCCAM'S RAZOR

As mentioned above, some readers will regard the description of the development of organic systems to be unnecessarily complicated in comparison to the very simplistic scheme that emerged from the Grand Synthesis and usually is referred to as the neo-Darwinian scenario of evolution. Neo-Darwinians will immediately grasp for Occam's Razor to excise the offending notion of organic behavior. But I wish to rejoinder with the question, "Which description is really simpler? [22]" Howard Pattee (personal communication) emphasizes how any description of change in nature must consist of two elements — the dynamic by which those elements interact and the boundary conditions or context within which the dynamic transpires.

Charles Darwin consciously followed the prevailing Newtonian approach toward describing nature by explicitly externalizing all the agencies that elicit change under his rubric, "natural selection [23]." The remaining dynamics are rather easy to codify and describe. Natural selection, however, remains an enormously complicated "boundary condition" that at times is described in almost transcendental tones. We implicitly are urged to keep our eyes focused intently on the simple dynamical description and to pay no attention whatsoever to the overwhelming complexity within which that dynamic occurs. Subsequent emendations to Darwin's scheme have not altered his basic Newtonian separation and focus.

According to Pattee, however, any natural description should be judged in its entirety. By this standard, neo-Darwinism, due to the encumbrance of the complexity hidden in natural selection, does not fare well under Occam's razor. By contrast, self-organization theory (of which this organic narrative is an example) *includes* far more agency into its dynamics. A slightly more complicated description of the operative dynamics ensues, but the cost of such complication is more than repaid by the degree to which it simplifies the boundary value problem. That is, a significant amount of biological order, which under the Darwinian scheme had to be explained by arbitrary and innumerable manifestations of "natural selection," is now consolidated as the consequences of dynamics that are *internal* to system description. Kauffman called this class of phenomena "order for free," in the sense that the given pattern did not have to be encoded into the genome of the organism [24].

As regards overall problem description, the jury is still out as to whether neo-Darwinism or self-organization theory better satisfies Occam's criterion, but proponents of the latter description have every reason to be optimistic that the scales are beginning to tip in their favor.

Apropos simplicity of description, one may well ask, which metaphor more simply and appropriately pertains to *biological* phenomena — the mechanical or the organic? The answer should be tautologically obvious, but the minimalist/mechanists would have us think otherwise. Daniel Dennett, for example, bids us imagine the progressive complexity of biological entities as analogous to "cranes built upon cranes," whereby new features are hoisted on to the top of a tower of cranes to become the top crane that lifts the next stage into place [25]. He specifically warns against invoking what he calls "skyhooks," by which he means agencies that create order but have no connection to the firmament.

Once one is convinced that the organic metaphor does have a legitimate place in scientific narrative intermediate to the mechanical and the stochastic, then organic analogs for biological phenomena become the simplest and most natural possible. This point was brought home to me via one of the few "Eureka!" experiences I have ever encountered. I was working distractedly in my garden, pondering why I thought Dennett's analogy was inappropriate, when my eye was drawn to a muscadine grapevine that has grown on the corner of my garden fence for the last 25 or so years. In the initial years after I had planted it, the lead vine had become a central trunk that fed an arboreal complex of grape-bearing vines. Eventually, the lateral vines had let *down* adventitious roots that met the ground some distance from the trunk. Then in the last few years, the main trunk had died and rotted away completely, so that the arboreal pattern of vines was now being sustained by the new roots, which themselves had grown to considerable thickness.

No need for skyhooks here! The entity always remains in contact with the firmament, and bottom-up causalities continue to be a necessary part of the narrative. Yet it is the processes within the later structures that *create* connections, which eventually replace and displace their earlier counterparts. Top-down causality, the crux of organic behavior, but totally alien to mechanistic-reductionistic discourse, fits the developmental situation perfectly. Evolution is like a muscadine grapevine. As strange as that analogy might seem at first, it fits the description far better than Dennett's mechanical construct. In striving for causal simplicity by concocting mechanical metaphors for what is more inherently and legitimately organic phenomena, mechanists wind up complicating the picture unnecessarily and lead us astray from a more natural perspective on the living world.

5.7 THE ORGANIC PERSPECTIVE

To summarize, the realm of ecology appears to require a metaphysics that differs in essence from the conventions of the Enlightenment [26]. This ecological metaphysics can more readily accommodate a looser statement of the organic metaphor. To wit:

1. The causal entailments in organic systems are more replete than those of simple mechanical systems. In the organic context, Newtonian causal closure does not impart the virtue of simplicity, but rather is seen as a procrustean minimalism. Newtonian closure cannot be as easily reconciled with stochastic events without resorting to a schizoid view of nature. Causal agency is more naturally apprehended as a continuum of propensities, rather than an admixture of the extremes of mechanical determinism and pure chance.

2. An organic system is the culmination of a *history* of chance events that have been embedded into their functioning. While some of the entrained events may have been of a generic nature, others most assuredly were complex, unique, and nonrepeatable.

3. Organic systems, although they decay after they have ceased to function organically, are strongly *nondecomposable* while they are functioning (living).

4. The key feature of organic systems appears to be the top-down influence that combinations of processes exert upon their constituent elements (objects) and mechanisms (constituent processes.) The strength of such top-down influence can vary among the types of organic systems, approximating the deterministic in ontogenesis, but allowing for considerable plasticity in ecosystems.

As regards organicism, heretofore the major problem has been that organic behavior was miscast as almost deterministic or mechanical. Even early ecologists, like Frederic Clements could conceive of organic agency only in this fashion. They can hardly be faulted for such assumption, however, when the most obvious manifestations of organic behavior, organisms themselves, lie against one extreme of the range of organic behavior and *do* develop in close approximation to rigid determinism. But the crux of organic behavior is neither mechanism nor determinism. These are only peripheral limits toward which organic behavior, if left unimpeded, could drift. The kernel of organic systems lies rather in the coherence and direction that system-level configurations of processes impart to their components. This "ecological" form of organicism seems far less threatening a notion than its earlier counterpart. To convince oneself of this, one need to only consider the absurdity of a dictator trying to describe his/her totalitarian state in terms of an ecosystem with its loose connections and manifold feedbacks.

As regards any similarity of ecological organicism with vitalism, we note as how the former concept does not involve any independent "elan-vital" that enters the system from without. The contributing processes are all internal to the system itself. Furthermore, like the muscadine grape, autocatalytic configurations of processes are all inexorably connected with their material resources.

Finally, to return to the radical shift in attitudes toward life and death that opened this essay, it should now be apparent that the notion of creativity in organic systems can afford an appealing resolution. We need simply to place more emphasis upon the creative role of configurations of processes, and the conundrum evaporates. Physicists, for example, relate how the conjunction of processes that gave rise to the elementary forms of matter bear marked formal similarity to those that have just been described. No longer must dead matter and living forms be juxtaposed in our minds across an unbridgeable chasm, but rather now both should be viewed as outgrowths of a common form of creative agency.

ACKNOWLEDGMENTS

The author was supported in part by the National Science Foundation's Program on Biocomplexity (Contract No. DEB-9981328) and by the Across Trophic Systems Simulation Program of the United States Coast and Geodetic Survey (Contract 1445CA09950093). The author wishes to express his gratitude to Dr. Andrzej Konopka for soliciting this manuscript and for his helpful suggestions for revisions.

REFERENCES

1. Haught JF. Science, Religion, and the Origin of Life. AAAS Seminar, Washington, D.C., September 13, 2001.

2. Ulanowicz RE. Beyond the material and the mechanical: Occam's razor is a double-edged blade. Zygon 1995; 30:249–266.

3. Ulanowicz RE. Ecology, a dialogue between the quick and the dead. Emergence 2002; 4:34–52.
4. Lewin R. Why is development so illogical? Science 1984; 224:1327–1329.
5. Simberloff D. A succession of paradigms in ecology. Synthese 1980; 43:257–270.
6. Rosen R. Life Itself. New York: Columbia University Press, 1991.
7. Popper KR. The Open Universe: An Argument for Indeterminism. Totowa, NJ: Rowman and Littlefield, 1982.
8. Prigogine I, Stengers I. Order out of Chaos: Man's New Dialogue with Nature. New York: Bantam, 1984.
9. Elsasser WM. Acausal phenomena in physics and biology: a case for reconstruction. Am Sci 1969; 57:502–516.
10. Patten BC. Out of the clockworks. Estuaries 1999; 22:339–342.
11. Popper KR. A World of Propensities. Bristol: Thoemmes, 1990.
12. Whitehead AN. Process and Reality. New York: The Free Press, 1978.
13. Ulanowicz RE. On the ordinality of causes in complex autocatalyic systems. In: J.E. Earley and R. Harre, eds. Chemical Explanation: Characteristics, Development, Autonomy. New York: Annals of the New York Academy of Science, 2003; 988:154–157.
14. Ulanowicz RE. Ecology, the Ascendent Perspective. New York: Columbia University Press, 1997.
15. Maturana HR, Varela FJ. Autopoiesis and Cognition: The Realization of the Living. Dordrecht: D. Reidel, 1980.
16. Rashevsky N. Mathematical Principles in Biology and their Applications. Springfield, Illinois: Charles C. Thomas, 1961.
17. Rosen R. A relational theory of biological systems. Bull Math Biophys 1958; 20:245–260.
18. Ulanowicz RE. Growth and Development: Ecosystems Phenomenology. New York: Springer-Verlag, 1986.
19. Zorach AC, Ulanowicz RE. Quantifying the complexity of flow networks: how many roles are there? Complexity 2003; 8:68–76.
20. Odum EP. The strategy of ecosystem development. Science 1969:262–270.
21. Ulanowicz RE, Norden JS. Symmetrical overhead in flow networks. Int J Syst Sci 1990; 1:429–437.
22. Ulanowicz RE. The organic in ecology. Ludus Vitalis 2001; 9:183–204.
23. Depew DJ, Weber BH. Darwinism Evolving: Systems Dynamics and the Geneology of Natural Selection. Cambridge, MA: MIT Press, 1995.
24. Kauffman S. At Home in the Universe: The Search for the Laws of Self-Organization and Complexity. New York: Oxford University Press, 1995.
25. Dennett DC. Darwin's Dangerous Idea: Evolution and the Meanings of Life. New York: Simon and Schuster, 1995.
26. Ulanowicz RE. Life after Newton: An ecological metaphysic. BioSystems 1999; 50:127–142.

6 The Metaphor of "Chaos"

Wlodzimierz Klonowski

"What is Chaos? It is this order that was destroyed during Creation."

S.J. Lec 'Unkempt thoughts'

CONTENTS

6.1 THEORIES OF CHAOTIC BEHAVIOR

The metaphor of "chaos" is more convoluted than other metaphors (e.g., hierarchy), because the emotional meaning of the word "chaos" greatly differs from its formal meaning in contemporary science.

It is often mistakenly thought that the word "chaos" comes from *The Bible*, but in its original Hebrew version, the word "chaos" does not appear. The word *chaos* has its roots in Greek reflexive verb *chaino*, meaning "to open," and when used as a noun, meaning "something open," especially an abyss. In Greek mythology, descriptions of the "birth of the world" usually start with the statement like "In the beginning there was Chaos." In ancient Egypt, the most powerful god was Nun (Nu) — the God of Chaos, "father" of all other gods.

Superficially, *chaotic systems* behave in completely random fashion. Yet, close analysis shows that they show a subtle structure, an underlying order. Chaos is seemingly unpredictable, indecomposable, and yet is fully determined, given a specific set of untold conditions. (Note: The seeming unpredictability is not an artifact of the structure of chaos. It is an indication of our ignorance of initial conditions.) Chaos may be observed only in *nonlinear systems*, i.e. in which the influence of a change of any state variable depends on the actual state of the system and/or the reaction to any external stimulus is not directly proportional to the strength of this stimulus. We will come back later to the differences between chaoticity and stochasticity.

Linear systems have *states of equilibria* — a system that reached equilibrium is "dead"; "Old chemists never die, they just reach equilibrium" says a dictum. Nonlinear systems have so-called *attractors* — attractor is a subset of system's states to which the system "is attracted" when time runs; a system that reached a state that belongs to its attractor remains "alive" — different processes in the system are not stopped, but state of the system remains confined to the states belonging to this system's attractor. The very essence of chaotic system is the presence of a *chaotic attractor*. It used to be called *strange attractor*, but it was shown that while all chaotic attractors are strange, not all strange attractors are chaotic [1]. It is the presence of chaotic attractors that makes nonlinear dynamic of chaotic systems extremely sensitive to initial conditions.

The wide range of fields has been touched by the metaphor of "chaos," between them is the System Biology. Biology is the discipline distinguished among all sciences by the interplay of seems to equate *randomness* with chaos. Chaos is not random, it is fully determined. This plays an essential role in the creation of forms, and *complexity*, which is the presence of irreversible, dissipative fractal structures, playing a determining role in generating and shaping life forms.

6.1.1 CHAOS AND GENERAL SYSTEMS THEORY

The aim of General Systems Theory is to study all complex systems, independent of type of components, focusing instead on concepts such as boundaries, input–output, state space, feedback, control, information, hierarchies, and networks [2,3]. The organization of a system is the set of relations that define it as a unity; the structure of a system is the set of relations among its components. Step by step, people have universally recognized that a system is more than just the sum of its components, and has its own hierarchy and functional structure. Although superficially different, many kinds of biological systems display patterns of connectivity that are essentially similar. In most models of biological systems (for example, dynamical systems, cellular automata, and state spaces), the patterns of connectivity between elements are isomorphic to directed graphs. These isomorphisms suggest that the above sources of emergent behavior may be universal. In no area of science Aristotelian saying "The whole is greater than the sum of its parts" is more evident than in biology.

Ashby [4] formulated the law of requisite variety for adaptive systems, the principle that every dynamic system will self-organize, and the requirement that every regulator of a system must also be a model of that system. Aulin [5] proposed a similar law of requisite hierarchy governing both control systems and societies. Simon [6] has studied the problem-solving techniques that adaptive systems (people, organizations, computers, etc.) use to cope with complexity and proposed an evolutionary explanation for hierarchical organization [7].

The principles of variation and selection were first formulated by Charles Darwin [8] to explain the origin of biological species. Modern overviews of evolutionary theory in biology can be found in Dawkins [9]. Gould and Eldredge [10] have proposed the theory of *punctuated equilibrium*, according to which evolution is a largely chaotic, unpredictable process, characterized by long periods of stasis, interspersed by sudden bursts of change. Kauffman [11,12] has tried to understand how networks of mutually activating or inhibiting genes can give rise to the differentiation of organs and tissues during embryological development. This led him to investigate the properties of Boolean networks of different sizes and degrees of connectedness — he proposed that the self-organization exhibited by such networks of genes or chemical reactions is an essential factor in evolution, complementary to Darwinian selection by the environment. Kauffman's mathematical model is based on the concept of "fitness landscapes" (originally introduced by Sewall Wright). A fitness landscape is a distribution of fitness values over the space of genotypes. Evolution is the traversing of a fitness landscape; peaks represent optimal fitness.

Langton [13], inspired by works of Dawkin and Kauffman, initiated a new approach called *artificial life* that tries to develop technological systems (computer programs and autonomous robots) exhibiting life-like properties, such as reproduction, sexuality, swarming, and coevolution. Langton [14] has also proposed the general thesis that complex systems emerge and maintain on the *edge of chaos*, the narrow domain between frozen constancy and chaotic turbulence. The state between order and chaos (the "edge of chaos") is sometimes a very "informative" state, because the parts are not as rigidly assembled as in the case of order and, at the same time, they are not as loose as in the case of chaos. The system is stable enough to keep information and unstable enough to dissipate it. The system at the edge of chaos is both a storage and a broadcaster of information. At the edge of chaos, information can propagate over long distances without decaying appreciably, thereby allowing for long-range correlation in behavior: ordered configurations do not allow for information to propagate at all, and disordered configurations cause information to quickly decay into random noise. Living organisms dwell "on the edge of chaos," as they exhibit order and chaos at the same time, and they must exhibit both in order to survive.

Prigogine's "Brussels School" has demonstrated that physical and chemical systems far from thermodynamical equilibrium tend to *self-organize* and thus to form *dissipative structures* [15–17]. Irreversible processes and nonequilibrium states turn out to be fundamental features of the real world. Prigogine distinguishes between "conservative" systems (which are governed by the three conservation laws — for energy, translational momentum, and angular momentum — and which give rise to reversible processes) and "dissipative" systems (subject to fluxes of energy and/or matter). The latter give rise to irreversible processes — most of nature is made of such "dissipative" systems. Dissipative systems conserve their identity, thanks to the interaction with the external world, nonequilibrium becomes a source of order. Self-organization is the spontaneous emergence of ordered structure and behavior in open systems that are in a state far from equilibrium, described mathematically by nonlinear equations.

Haken [18–20] has suggested the label of *Synergetics* for the field that studies collective patterns emerging from many interacting components. Synergetics applies to systems driven far from equilibrium, where the classic concepts of thermodynamics are no longer adequate. It expresses the fact that order can arise from chaos and can be maintained by flows of energy/matter. Systems at instability points (at the "threshold") are driven by a "slaving principle": Long-lasting quantities (the macroscopic pattern) can enslave short-lasting quantities (the chaotic particles), and they can force order on them (thereby becoming "order parameters"). The system exhibits a stable "mode," which is the chaotic motion of its particles, and an unstable "mode," which is its macroscopic structure and behavior of the whole system. Close to instability, stable modes are "enslaved" by unstable modes and can be ignored. Instead of having to deal with millions of chaotic particles, one can focus on the macroscopic quantities. *De facto*, the degrees of freedom of the system are reduced. Haken shows how one can write the dynamic equations for the system, and how such mathematical equations reflect the interplay between stochastic forces ("chance") and deterministic forces ("necessity").

Eigen has introduced the concepts of natural self-organization through *hypercycle*, an autocatalytic cycle (a cycle of cycles of cycles) of chemical reactions [21]. Then he proved that life can be viewed as the product of a hierarchy of such hypercycles. Hypercycles are capable of evolution through more and more complex stages. Hypercycles compete for natural resources and are therefore subject to natural selection.

Bak has shown that many complex systems will spontaneously evolve to the critical edge between order and chaos; this phenomenon, called *self-organized criticality*, may provide an explanation for the punctuated equilibrium dynamics seen in biological evolution [22].

The modeling of nonlinear systems in physics has led to the concept of chaos as a deterministic process characterized by extreme sensitivity to its initial conditions [23]. *Cellular automata* (created by von Neumann [24] in his investigation of complexity of self-reproducing automata), mathematical models of distributed dynamical processes characterized by a discrete space and time, have been widely used to study phenomena such as chaos, attractors, and the analogy between dynamics and computation through computer simulation [25], while *catastrophe theory* proposed a mathematical classification of the critical behavior of continuous mappings [26], extending the much older works by the biologist D'Arcy Thompson [27].

Holland [28] founded *genetic algorithms* — parallel, computational representations of the processes of variation, recombination, and selection on the basis of fitness that underlies most processes of evolution and adaptation.

Foerster formulated the order from noise principle [29] and formulated the "second-order" cybernetics or the cybernetics of observing systems, according to which models of systems change the very systems they intend to model [30]. His emphasis on circular, self-referential processes has been elaborated in works on autopoietic systems — autopoiesis (self-production) denotes the fact that organisms produce their own components, while the environment is a source of perturbations that need to be compensated in order to maintain the system's organization [31]. *Cybernetic systems* are merely a special case of self-organizing systems.

Between common characteristics of complex systems, there are causal networks that contain cycles, in particular, self-reinforcing cycles. Forrester's [32] approach to systems dynamics is based on the analysis of positive and negative feedback cycles in networks of many interacting variables; this approach was at the base of the famous world model presented in a report to the Club of Rome. It is worth noting that the Club of Rome's model did not predict what actually happened in the environment. Boulding [33] proposed a related, flow-based model of the evolution of society, integrating ecology and economics. This transition from biological evolution to social exchanges naturally leads to the modeling of economic and social processes, to new interdisciplinary fields of *Econophysics* and *Sociodynamics* [34–36].

Mandelbrot, who was born in Poland in 1924 into a family with an extremely academic tradition, has founded the field of *fractal geometry* [37], which models the recurrence of *self-similar structures* (*patterns in space*) at different scales, exhibiting power laws. Fractals are geometric objects characterized by some form of self-similarity; that is, parts of a fractal, when magnified to an appropriate scale, appear similar to the whole. Coastlines are approximate fractals. Fractals are closely related to chaos — *the geometry of strange attractors is fractal geometry*. So, the nonlinear dynamics of chaotic systems can show itself as *fractality in time domain — oscillations, with similar smaller timescale oscillation imposed on those, with still smaller similar timescale oscillation imposed on these, etc.* Paar et al. [38] demonstrated that a simple mechanism of coupled oscillators with weak dissipation can lead to a complex coexistence of various modes involving *truncated fractals*, i.e., their self-similarity extends at most over a few orders of magnitude — biological fractals, unlike mathematical ones, are necessarily truncated. Such fractality may play a role in generating some basic features of biological systems.

Complex adaptive systems are ubiquitous in nature. They include brains, ecosystems, and even economies. They share a number of features: each of these systems is a network of agents acting in parallel and interacting; behavior of the system arises from cooperation and competition among its agents; each of such systems has many levels of organization, with agents at each level serving as building blocks for agents at a higher level.

6.1.2 Deterministic Chaos

In our everyday language "chaotic" means "complete absence of order or shape; confusion" [39]. It has been said "Chaos is a name for any order that produces confusion in our minds" (attributed to George Santayana, the author of *Realms of Being*). We are really confused while learning about *deterministic chaos* (it does sound like an oxymoron, doesn't it?) — the complex systems and processes that are governed by the deterministic laws in a form of the first-order ordinary differential equations (ODEs); however, when one observes these systems, they look completely disordered, like they were *stochastic*, i.e., governed by the laws of probability.

A system that evolves in time governed by the laws that may be expressed by ODE has been called *deterministic*, because knowing the equations and the initial conditions one is supposed to be able to calculate all system's characteristic in any other, future moment, and even at any moment in the past. General analytical methods (algorithms) of solving exist only for so-called linear ODEs, i.e., equations that contain neither a second nor any higher power, nor a product of any variables or differentials. Solutions of linear ODEs are typically given in a form of combinations of exponential and trigonometric functions.

Only a limited set of ODEs may be solved analytically. Most ODEs may be solved only approximately, using numerical methods and contemporary computers. However, since in a computer only integers may be exactly represented while real numbers are always represented with only finite accuracy, the result of solving an ODE may be wrongly treated as being a good representation of the real process. In reality, the obtained numerical solution may have nothing in common with the real system one wants to model. But even more important, since in practice the initial conditions may be reproduced only with limited accuracy, the behavior of the real system, despite starting with "the same initial conditions" may, in reality, be completely different each time a process is repeated. That is, the behavior of the system itself, and not just of its numerical, computer model, may be intrinsically chaotic.

ODEs containing only the first powers of the derivatives, but still some nonlinear algebraic terms such as squares, cross-products, or higher powers of the dependent variables and/or the independent variable are called quasilinear ODEs (QLODEs); since computers came into use, it is possible to solve QLODEs using numerical methods. QLODEs are but a special case of *differential systems*, and if time is taken as the independent variable one speaks about *dynamical systems*. QLODEs are extremely important in science and engineering. For example, the whole chemical kinetics is based on QLODEs. It has not, however, been really understood until the 1970s that even quite simple QLODEs may have extremely complicated solutions, including chaotic solutions. Anybody using computers for modeling dynamical systems should be fully aware of possible pitfalls.

To illustrate "strange" behavior of chaotic systems, let us consider the QLODEs proposed to model some meteorological phenomena [40]. However, the same equations may be used for modeling completely different systems, for example, lasers (of course, with very different physical interpretation of variables and parameters involved) [41]. Here we are not interested in physical interpretation, but only in the behavior of solutions obtained by numerical integration of Lorenz equations:

$$dx/dt = -ax + ay \tag{6.1a}$$

$$dy/dt = bx - y - xz \tag{6.1b}$$

$$dz/dt = -cz + xy \tag{6.1c}$$

We have chosen two sets of parameters:

$$a = 10;\ b = 28;\ c = (8/3) \tag{6.2a}$$

$$a = 10;\ b = 28;\ c = (8008/3000) \tag{6.2b}$$

For each set of parameters (Equation 6.2), three different sets of initial conditions have been used:

$$x(0) = 0.999;\ y(0) = 0.999;\ z(0) = 0.999 \tag{6.3a}$$

$$x(0) = 1.000;\ y(0) = 1.000;\ z(0) = 1.000 \tag{6.3b}$$

$$x(0) = 1.001;\ y(0) = 1.001;\ z(0) = 1.001 \tag{6.3c}$$

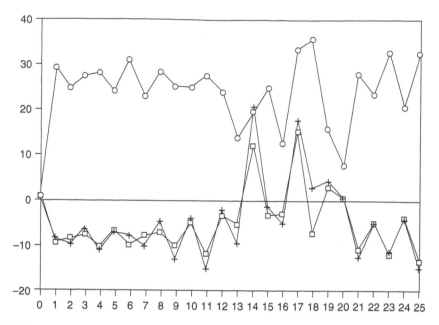

FIGURE 6.1 Example of a solution of system (Equation 6.1) [for parameters (Equation 2b) and initial conditions (Equation 3c)]; x (squares), y (crosses), and z (diamonds) vs. t.

The first result, surprising for an "uninitiated" person, is that the system shows very irregular behavior. In Figure 6.1, solution of Equation 6.1 (a–c) with parameters (Equation 6.2b) and initial conditions (Equation 6.3c) is shown. While inspecting Equation 6.1 (a–c), no one could directly predict such a strange behavior of the solution — because of the simplicity of the equations, one would rather expect smooth, regular behavior.

Figure 6.2 shows one "face" of behavior of chaotic system — its *extreme sensitivity to small changes in system parameters*. When parameters are changed from Equation 2a to Equation 2b, that is one of the parameters, c, is increased one tenth of a percent, while the initial conditions remain unchanged, at the beginning the solutions remain practically unchanged. Then, suddenly, a big difference appears. All three dependent variables show similar chaotic behavior.

Another "face" of the behavior of chaotic system — its *extreme sensitivity to initial conditions* — is shown in Figure 6.3. When the system parameters (Equation 6.2) remain unchanged, but initial conditions are changed from Equation 3a to Equation 3b, i.e., one tenth of a percent, again at the beginning nothing seems to change, but then suddenly big differences appear.

So, small changes of initial conditions and/or of system parameters bring dramatic changes in a long-time behavior of the solution of Equation 6.1. The obtained results are not some artifacts introduced by numerical integration, but are really the inherent properties of the considered system.

Deterministic chaos sounds like an oxymoron, something like "Ugly Beauty" — a nice piece of music, composed by Theolonius Monk or "Progressive Conservative" — one of the main political parties in Canada. Deterministic chaos has been observed in different physical and chemical systems. It is impossible to predict chaotic behavior by a simple inspection of QLODEs that model a given system, since chaotic behavior may take place only for some values of the system parameters; for other values the behavior may be quite regular.

Anybody using such models (in chemistry, biology, engineering, economics, business, and humanities) should be aware of possible pitfalls hidden in even simple QLODEs. Using computers to store and retrieve information does not require any knowledge of mathematics, nor understanding

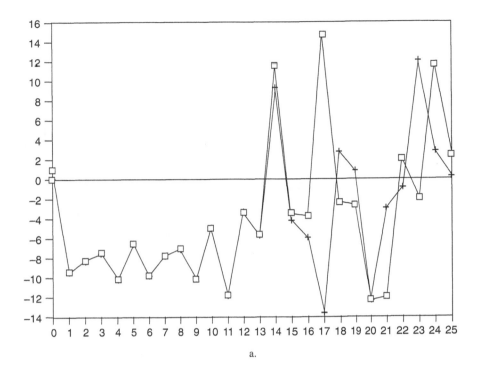

a.

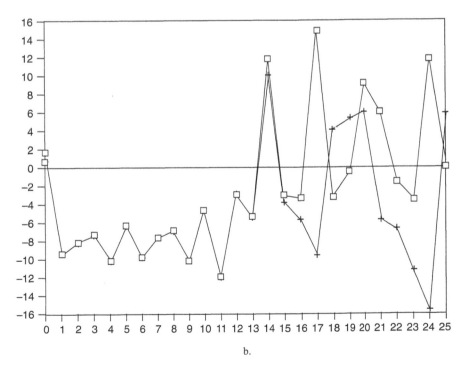

b.

FIGURE 6.2 Example of deterministic chaos — sensitivity of solution of system (Equation 6.1) to system parameters — *x* vs. *t* for parameters (Equation 2a) and (Equation 2b), respectively; a. for initial conditions (Equation 3a); b. for initial conditions (Equation 3b); *y* and *z* show similar chaotic behavior.

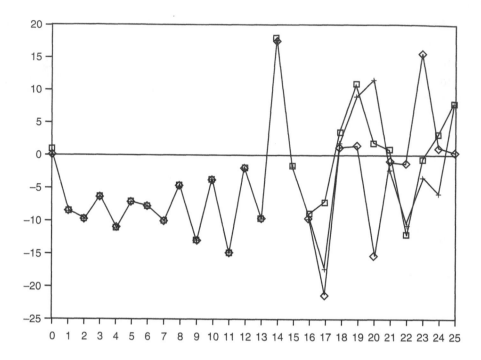

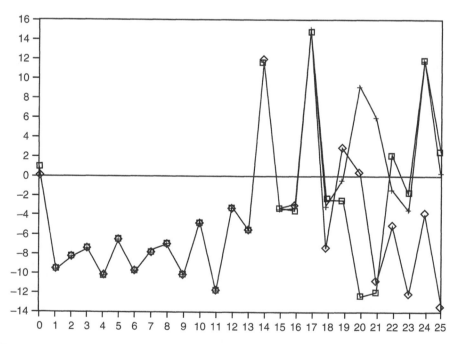

FIGURE 6.3 Example of deterministic chaos — sensitivity of solution of system (Equation 6.1) to initial conditions — x vs. t for initial conditions (Equation 6.3a), (Equation 6.3b), and (Equation 6.3c), respectively. a. For parameters (Equation 6.2a); b. for parameters (Equation 6.2b); y and z show similar chaotic behavior.

of how computer works, like watching TV does not require knowledge of Maxwell equations of electromagnetism. However, a person who applies a computer for problem solving in which numerical information is processed and transformed does need to have a really good mathematical background and enough knowledge about the considered system or process, so to be able to choose a proper mathematical method and not to misinterpret the results. Unacquaintance with ODEs is an example of *innumeracy* — an inability to understand numbers — the mathematical equivalent of illiteracy.

6.2 NONLINEAR DYNAMICS — CHAOS AT WORK

The "strange" properties of Lorenz equations (Equation 6.1) arise due to the nonlinear cross-terms — xz and xy — without these terms, it is a simple linear dynamical system that does not show extreme sensitivity to parameters or to initial conditions. In linear systems, the principle of superposition applies. One consequence of that principle is seen in the Newtonian dynamics example. In Newtonian dynamics, if the x component of the force applied to a body is a times greater, y component b times greater, z component c times greater, the components of acceleration of the body will also be correspondingly a, b, c times greater, because the mass of the body is assumed to be constant, and the components of acceleration are independent one from another.

Traditional reductionistic view trying to explain properties of a system by properties of its components may be fundamentally flawed because of *emergent properties* of complex chaotic systems. On a so-called *mesoscopic scale* (smaller than accessible to direct observation but big enough that the number of particles involved in the organized pieces of matter is sufficiently large to be beyond explicit calculation from the constituent parts), biological "miracles" — phenomena that can be categorized and described but not completely explained in terms of detailed microscopic mechanisms — happen.

Examples are biomolecules such as enzymes and intracellular structures, which are responsible for dramatically complex biological processes that are reproducible and can be modified but not derived from the properties of their constituents. There exist several families of so-called *conformons* — molecular structures that also carry information — driving distinct biological functions at the molecular level [42–44], for example timing processes within proteins (*Klonowski–Klonowska conformon*) [42, 44].

Another example are spatiotemporally organized networks of physicochemical processes that may be identified with *intracellular dissipative structures* [42], defined generally as the dynamics organization of matter in space and time kept far from equilibrium by a continuous dissipation of free energy. Nonequilibrium structures such as subcellular sol–gel dissipative network structures [45, 46] may be built up even of subunits that taken separately are in thermodynamic equilibrium.

Also brain activity may be explained as chaos on mesoscopic scale [47]. What distinguishes brain chaos from other kinds is the filamentous texture of neural tissue called neuropil, which is unlike any other substance in the known universe [48]. Neural populations stem ontogenetically in embryos from aggregates of neurons that grow axons and dendrites and form synaptic connections of steadily increasing density. At some threshold, the density allows neurons to transmit more pulses than they receive, so that an aggregate undergoes a state transition from a zero-point attractor to a nonzero point attractor, thereby becoming a population. Such a property has been described mathematically in random graphs, where the connectivity density is an order parameter that can instantiate state transitions [49]. Accordingly, state transitions in neuronal populations can be interpreted as a kind of sol–gel transition progressing in the neuropil medium.

6.2.1 ENERGETIC AND INFORMATIONAL INTERACTIONS

One may use a mechanical analogy to better understand the issue. A pack of cards lying on a tabletop is in equilibrium. Its potential energy shows a deep minimum, and a substantial force like a gust of wind is necessary to scatter the cards. The same pack, if arranged into a house of cards with the potential energy much higher, becomes extremely sensitive to very small forces, whether acting on the whole system or even on a single card. Building such a nonequilibrium structure requires application of forces as well as an energy input much higher than those needed to just scatter a pack of cards. But the response time of the system becomes extremely short, and system sensitivity becomes extremely high.

It is obvious that changing the state of a system in equilibrium requires a significant force, an *energetic interaction*, whereas the state of a system far from equilibrium may be changed by a very small force, an *informational interaction*.

Extreme sensitivity to physiological parameters enables very quick regulatory response.

The system of macromolecular network structures that occupy and organize the interiors of all living cells, the so-called *cytoskeleton* structures, which are of nanometer size scale and undergo conformational oscillations in the nanosecond timescale, are the basis of what Hameroff calls *ultimate computing*; in the motto to his book he says "Billionth scale activities in biomolecular assemblies could define life itself, and provide a frontier for the evolution of technology" [50].

The cytoskeleton, within neurons and all living cells, is a parallel-connected network, which can utilize coherent cooperativity of its own collective phenomena to organize and process subcellular information. Dynamic structural activities of the cytoskeleton are responsible for all cytoplasmic rearrangements including formation and regulation of dendritic spines and synapses in the nervous system, which, in turn, are probably the very basis of learning and memory.

The key word here, which leads to both neural and computing mechanisms, is the word "collective." Collective mechanisms can exert long-range cooperativity and organization within parallel arrays. Collective phase transitions in brain parallel arrays could be "a fabric of consciousness," where an "idea" emerges like the property of superconductivity from a large number of simple, "aligned" subunits. In computer simulations of parallel networks, Choi and Huberman [51] observed collective effects manifested as diffuse reverberation, sustained oscillation, phase transition, and deterministic chaos. All collective phenomena but equilibrium phase transitions (for example, freezing of water in 0°C, 32°F) take place in systems far from equilibrium.

Thus, as suggested by Conrad, *each protein is a rudimentary computer* and converts a complex analog input to an output state of its own conformation. The evolution of form and information "from chaos" has been termed *morphogenesis*. It is now apparent that biological cells are complex entities whose action depends on collective functions of intracellular structures, first of all the cytoskeleton.

With equilibrium, there is no change, no self-organization, no self renewing, and no learning. That is why living systems should foster some complexity and some nonequilibrium. We may say that complex systems, from atomic level to social level, show a kind of physical fractal structure, despite the fact that similarity of different levels have rather quantitative character instead of a qualitative one, as it is in the case of mathematical fractals. Any complex system has to exchange energy and information with the environment to keep its structure and function. Living systems show *autopoiesis* — they continuously renew and regulate themselves to maintain the integrity of their structure.

6.2.2 OPEN AND CLOSED SYSTEMS — FROM A SINGLE MOLECULE TO METAMAN

It is interesting that the word *chaos* has its roots in Greek "*chaino*," meaning "to open" or "something open," because nonlinear phenomena, such as deterministic chaos may be observed only in open systems (i.e., systems that exchange with their environments both energy and mass), while classic thermodynamics considers closed systems consisting of many elements.

The three basic principles that are the basis of structure and function, as well as of modeling of the living system, are [52]:

Simplicity Principle — if any phenomenon is possible in a simple system, it is likely that a more complicated system will exhibit phenomena of the same type

Economy Principle — more complicated systems demand greater input of free energy, and so they are less economical

Actuality Principle — the present properties of the system depend only on its present structure

This is why discussions of simple model systems, for example, considerations concerning structure–property relationships in homogenous solutions of polymeric macromolecules [45], are important for understanding complex biological systems. Especially useful are theories in which there is no assumption of equilibria nor of the complete randomness of the system, because living systems are not in thermodynamic equilibrium. For example, the probabilistic theory of crosslinked systems and the resulting Topological Gelation Criterion [47] may be applied to cytoskeleton. Critically branched systems play probably a very important role in living cells. Since sol–gel transitions occur practically in an all-or-none fashion, gelation is potentially a powerful mechanism for regulation of cytoplasmic structure, and so, as the Actuality Principle states, cytoplasmic properties.

Nonequilibrium states are the "living" states of any system. In those states, some properties (such as concentration, electrical charges, etc.) are unequally distributed, and those differences drive all movements and changes. Equilibrium, the unique state when all properties are equally distributed, is the state of "death." It is true not just for a single cell or an organism. The socio-economical system in which there are no inequalities between the people may not exist for a longer period of time — it decomposes like a dead organism. If those who "invented" communism knew thermodynamics, they would understand that communism must necessarily fail, as we all observed recently. *"New is born"*— emerging properties occur as a result of nonequalities and nonlinear interaction between different parts of the system.

Stock comes to a conclusion that, in our social system, a new type of superstructure is being formed, that as a group we are becoming a new type of being, which he dubs *Metaman* [53]. This superstructure will have the same relationship to all individuals as a single individual has to all his or her cells. With the growth of the global economy with its world-spanning technologies, our machines are turning us into a gigantic creature whose needs and goals exceed the needs and goals of an individual's as much as an individual's exceed those of a single cell. Ultimately, the new superstructure, the new creature, will be composed of a blend of people and machines.

The notion of human society as a living organism goes back at least to ancient Greece. But it seems no longer to be a poetic metaphor; it is an evolving reality. Metaman has brains — our brains, augmented and linked by computer networks. Metaman has senses, like, for example, weather satellites providing a global, expanded vision. Metaman has memory — all databanks, with stored information growing literally each minute. And one may say that Metaman has a will — the sum of all our wills.

Metaman's responses to external stimuli, to challenges, will be much slower than those of a single organism, because the greater is a structure, the longer is the characteristic timescale. But when seen from a perspective of many decades, on the one side, Metaman's responses may seem to be as directed and purposeful as the behaviors of a single organism. On the other hand, Metaman may show patterns characteristic for autonomous big systems, like, for example, chaotic behavior.

6.3 CHAOS AND FRACTAL GEOMETRY OF NATURE

Methods of nonlinear dynamics and deterministic chaos theory provide tools for analyzing and modeling chaotic phenomena and may supply us with effective quantitative descriptors of underlying dynamics and of system's fractal structure. Fractal geometry, a new branch of geometry, has evoked a fundamentally new view of how both nonliving and living systems result from the coalescence of spontaneous self-similar fluctuations over many orders of time and how systems are organized into complex recursively nested patterns over multiple levels of space. Fractal geometry has proven to be a useful tool in quantifying the structure of a wide range of idealized and naturally occurring objects, from pure mathematics, through physics and chemistry, to biology and medicine.

But *fractals* were known long ago, despite the fact that they were not named fractals. The first fractal was proposed by a Polish mathematician Wladyslaw Sierpinski — it is now well-known as "Sierpinski's gasket." In my opinion, fractals appeared in medicine and biology much earlier than in mathematics, e.g., in the Chinese notion of a *homunculus* saying that one may find a kind of mapping of the whole organism practically on each organ — earlobe, foot, retina of the eye, etc.; this is a base of different branches of acupuncture, such as auriculotherapy.

Mandelbrot [37] who invented the word says: "I coined *fractal* from the Latin adjective *fractus*. The corresponding Latin verb *frangere* means 'to break': to create irregular fragments. It is therefore sensible — and how appropriate for our needs! — that, in addition to 'fragmented' (as in fraction or refraction), *fractus* should also mean 'irregular,' both meanings being preserved in fragment." So, the term *fractal* applies to objects in space or fluctuations in time that possess a form of *self-similarity* and cannot be described within a single absolute scale of measurement. Fractals are recurrently irregular in space or time, with themes repeated like the layers of an onion at different levels or scales. Fragments of a fractal object or sequence are exact or statistical copies of the whole and can be made to match the whole by shifting and stretching — spatial structures of many living systems are fractal, while sequential fractal scaling relationships are observed in many physiological processes.

Nonlinear dynamical system often have chaotic attractors. Here the metaphor of chaos theory intimately relates to fractal geometry since strange attractors turn out to have fractal structure.

6.3.1 FRACTAL DIMENSION

Fractal dimension is a measure of how "complicated" a self-similar figure is. In a rough sense, it measures "how many points" lie in a given set. A plane is "larger" than a line, while Sierpinski triangle sits somewhere in between these two sets. On the other hand, all three of these sets have the same number of points in the sense that each set is uncountable. Somehow, though, fractal dimension captures the notion of "how large a set is."

Fractal object has a property that more fine structure is revealed, as the object is magnified, similarly like morphological complexity means that more fine structure (increased resolution and detail) is revealed with increasing magnification. Fractal dimension measures the rate of addition of structural detail with increasing magnification, scale, or resolution. The fractal dimension, therefore, serves as a *quantifier of complexity*. The fractal dimension of an object provides insight into how elaborate the process that generated the object might have been, since the larger the dimension, the larger the number of degrees of freedom likely have been involved in that process.

Objects considered in Euclidean geometry are sets embedded in Euclidean space, and object's dimension is the dimension of the embedding space. One is also accustomed to associate what is called *topological dimension* with Euclidean objects — everybody knows that a point has dimension of 0, a line has dimension of 1, a square is two dimensional, and a cube is three-dimensional (3D). However, collection of real points has dimension greater than 0, real lines greater than 1, real surfaces greater than 2, etc. At each level, as the dimensions of an object move from one integer to the next, the complexity of the object increases; it becomes more area filling from 1 to 2, more volume-filling from 2 to 3, etc. One familiar example of naturally occurring fractal curves is a coastline. One cannot use topological dimension for fractals, but instead has to use what is called Hausdorff–Besikovitch dimension, commonly known as *fractal dimension*.

In fact, a *formal definition of a fractal* says that it is an object for which the fractal dimension is greater than the topological dimension. One consequence of that definition uses the concept of

TABLE 6.1
Dimensional Properties of Abstract Constructs

Object	Dimension	No. of Copies
Line	1	$2 = 2^1$
Square	2	$4 = 2^2$
Cube	3	$8 = 2^3$
Any self-similar figure	D	$P = 2^D$
Sierpinski's triangle	1.58	$3 = 2^D$

self-similarity — a fractal is an object made of parts similar to the whole. The notion of self-similarity is the basic property of fractal objects.

Taking advantage of self-similarity is a way to calculate fractal dimension. For example, one can subdivide a line segment into m self-similar intervals, each with the same length, and each of which can be magnified by a factor of n to yield the original segment. A square or a triangle may be subdivided into n^2 self-similar copies of itself, each of which must be magnified by a factor of n to yield the original object. Similarly, a cube can be decomposed into n^2 self-similar copies of itself, each of which must be magnified by a factor of n to yield the original cube (Table 6.1). If onc takes the magnification, n, and raise it to the power of dimension, D, one will get the number of self-similar pieces in the original object, P:

$$P = n^D \tag{6.4}$$

Solving this equation for D one easily finds that

$$D = \log(P)/\log(n) \tag{6.5}$$

Using this formula, one can calculate a fractal dimension of some fractals. A mathematical fractal has some infinitely repeating pattern and can be made by the iteration of a certain rule. For example, Sierpinski's triangle S may be decomposed into three congruent figures, each of which is exactly half the size of S — if we magnify any of the three pieces of S by a factor of 2, we obtain an exact replica of S; that is, S consists of three self-similar copies of itself, each with magnification factor 2 (Figure 6.4).

a.   b

FIGURE 6.4 Sierpinski's triangles — concept of self-similarity and calculation of D. a. The rule for creating the Sierpinski triangle is as follows: connect the midpoints of the sides of a triangle (this way four smaller triangles are created) and then delete the middle triangle; this procedure leaves three smaller triangles similar to the initial one whose midpoints of the sides could then be connected, and so on — after ten iterations one has as many as 29,524 triangles. In theory, the procedure may be repeated infinitely many times; b. different types of Sierpinski's triangles.

So, fractal dimension of Sierpinski's triangle is (Table 6.1)

$$D = \log3/\log2 = 1.58 \qquad\qquad (6.6)$$

In principle, a theoretical or mathematically generated fractal is self-similar over a limitless range of scales, while natural fractal images have only a limited range of self-similarity.

6.3.2 NATURAL FRACTALS

A mathematical fractal has an infinite amount of detail. This means that magnifying it adds additional details, so increasing the overall size. In nonfractals, however, the size always stays the same, no matter the applied magnification. If one makes a graph log(*fractal's size*) vs. log(*magnification factor*), one gets a straight line. For nonfractals, this line is horizontal since the size (e.g., length of a segment, area of a triangle, and volume of a cylinder) does not change. For fractal object, the line is no longer horizontal since the size increases with magnification. The geometric method of calculating fractal dimension finds that fractal dimension can be calculated from the slope of this line.

Both the similarity method and the geometric method of calculating fractal dimension require measuring of fractal size. For many fractals, it is practically impossible. For these fractals, one applies the so-called box-counting method [54].

Living bodies have many systems that have fractural structure and the pulmonary system is the best example. It is composed of tubes, through which the air passes into microscopic sacks called alveoli. Trachea, the main tube of the system, splits into two smaller tubes, called the bronchi, which lead to different lungs. Bronchi, in turn, split into smaller tubes, which are even further split. This splitting continues further and further until the smallest tubes, called the bronchioles, which lead into the alveoli. Another supporting evidence that lungs are fractal comes from measurements of the alveolar area, which was found to be 80 m^2 with light microscopy and 140 m^2 at higher magnification with electron microscopy.

The increase in size with magnification is one of the properties of fractals. Fractal structure of the bronchial tree is imposed by optimization of resource utilization requirements in the lung, such as efficient distribution of blood and air. Thus, the morphology of the lung is directly related to its function, and changes in its structure can be linked to dysfunction. Since the bronchial tree is a fractal structure, its fractal dimension can be used as a tool for the detection of structure changes and quantification of lung diseases.

Similar splitting can also be found in blood vessels. Arteries, for example, start with the aorta, which splits into smaller blood vessels. The smaller ones split as well, and the splitting continues until the capillaries, which, just like alveoli, are extremely close to each other. Because of this, blood vessels can also be described by fractal canopies.

Biological fractal structures such as lungs and the vascular bed represent probably optimal design for their particular functions, like air flow or blood flow, respectively. It may well be that the fractal, dendritic trees of neurons are also optimally designed, but, in this case, for the flow of the most important commodity: information. The surface of the human *brain* contains a large number of folds. Its fractal dimension is 2.73 to 2.79, the highest in the animal kingdom.

Folding of the nasal membrane allows better sensing of smells by increasing the sensing surface. However, in humans, this membrane is less folded than in other animals, so making humans less sensitive to smells. The membranes of cell organelles such as mitochondria and the endoplasmic reticulum are also folded. Fractal dimensions of some biological structures are given in Table 6.2.

Fractals can also be found in various biomolecules such as DNA and proteins.

The similarity method for calculating fractal dimension works for a mathematical fractal, which, like Sierpinski's triangle, is composed of a certain number of identical versions of itself. Natural objects like coastlines or roots do not show exactly the same shape, but look quite similar when

TABLE 6.2
Fractional Dimension of Anatomical Structures

Anatomical Structure	D
Bronchial tubes	Very close to 3
Arteries	2.7
Human brain	2.73–2.79
Alveolar membrane	2.17
Mitochondrial membrane (outer)	2.09
Mitochondrial membrane (inner)	2.53
Endoplasmic reticulum	2.72

they are scaled down. Due to their statistical scaling invariance they are called *statistically self-similar*. The miniature copy of a structure may be distorted, e.g., skewed; for this case, there is the notion of *self-affinity*; within a self-affine structure, the scaling factor is not constant [54, 55].

Certain structures like quasicrystals or network glasses display at any magnification scale similar, although never strictly identical images. Even in the apparently totally disordered systems such as glasses or polymer networks, we may observe statistical self-similarity (repetition of characteristic local structures and certain typical correlations between them), if we use probabilistic description of the network. This is the most characteristic feature of the self-similarity: the fundamental information about the structure of a complicated system is contained already in quite small samples, and we can reproduce all the essential features by adding up and repeating similar subsets *ad infinitum* even if they are not strictly identical like in crystalline lattice, but just very much alike like in quasicrystals or network glasses.

It should be emphasized that *D is a descriptive, quantitative measure*; it is a statistic, in the sense that it represents an attempt to estimate a single-valued number for a property (complexity) of an object with a sample of data from the object.

6.4 CHAOS AND FRACTALS IN MODELING OF NATURAL PHENOMENA

So, fractal structures are intimately associated with nonlinear dynamic processes and chaos, because chaotic attractors have fractal structure, which, in some sense, is the very essence of a chaotic system, since it makes a system sensitive to initial conditions and not conservative. It has been observed that fractal models are essentially unresponsive to errors, while tolerant to the variability in the environment. This is due to the broad-band nature of the distribution in scale sizes of a fractal object.

Microscopic observation of subcellular structural elements — microtubules — shows individual microtubules that interconvert between phases of growing and shrinking. It was suggested that a dynamic instability, a road to chaos, can function as a control mechanism for the spatial organization of microtubule arrays [56], which, in turn, controls important cellular functions such as cell division (mitosis). On the other hand, a whole organism may be treated as bioreaction, with the law of metabolism interpreted in terms of heterogeneous catalysis and fractal structure [57].

On the cellular level, it was observed [58] that when large arrays of strange attractors are coupled diffusively through one of the variables, the structure of the dynamics of the system from response changes chaotic to periodic and form large Archimedean spirals or concentric bands. This observation seems particularly relevant to the question of the formal temporal structure of the biological clock in metazoan organisms. In particular, although individual cellular oscillators, as manifested in the cell cycle, appear to be more or less periodic, cells oscillate with chaotic dynamics. Only when large aggregates of these cells are tightly coupled can a precise circadian clock emerge. Immune system also shows both oscillatory and chaotic behavior [59].

One of the most important practical aspects of fractal analysis may be its use in *quantitative cellular morphometry* [54]. It may give deeper insight into the development of complex biological structures and the processes that contribute to structure forming. One of the advantages of fractal analysis is the ability to describe irregular and complex objects.

In cellular morphometry, cells and nuclei can be *quantitatively* described by measuring their fractal dimension. For example, evaluation of fractal dimension of nuclear outline of lymphoid cell could be a useful tool to distinguish between benign and malignant cases. One may use fractal dimension of imaged contours to directly infer that of 3D surface — the surface's dimension, D_S, is simply one plus the contours' dimension, D_C:

$$D_S = 1 + D_C \tag{6.7}$$

It was shown that increase in measured D correlates with perceived increase in morphological complexity — fractal dimension is a good quantitative measure of the degree of morphological differentiation; it is also a useful measure for comparative studies across and among species, as they relate to cellular evolution.

The use of fractal geometry in microscopic anatomy is now well established and it will be increasingly useful in establishing *links between structure and function.*

Another very important practical aspect of fractal analysis is its use in biosignal analysis. Biosignals are generated by complex self-regulating systems. That is why physiological time series may have fractal or multifractal temporal structure, while being extremely inhomogenous and nonstationary. A characteristic feature of nonlinear (as opposed to linear) process is the interaction (coupling) of different modes, which may lead to nonrandom signal phase structure. Such collective phase properties of the signal cannot be detected by linear spectral methods. For example, until quite recently, the basic approach to analysis of brain signals involved the assumption that electro-encephalogram (EEG)-signal is stochastic. A fundamentally different approach is to view EEG as the output of a deterministic system of relatively simple complexity, but containing nonlinearities. When some central nervous system activities stop to be chaotic and become oscillating, a state that may be called *pathological order* may emerge in the brain. Since nervous system, starting from a single neuron, shows fractal structure, it has become more and more obvious that neural activity may demonstrate deterministic nonlinear dynamics (i.e., chaos).

An important question emerges: Is it healthy to be chaotic [60]. It seems that "active desynchronization" could be favorable to a physiological system. Variations of fractal dimension according to particular mental tasks were reported; for example, during an arithmetic task, this dimension increased. The neural dynamics underlying EEG seems to depend on only a few degrees of freedom. Also during anesthesia, fractal dimension of EEG signal is decreased — when patient becomes more anesthetized, EEG becomes less chaotic. *Consciousness may thus be described as a manifestation of deterministic chaos somewhere in the brain/mind.*

Fractal analysis of EEG signals and the development of neurophysiologically realistic models of EEG generation may produce new successful automated EEG analysis techniques and can have important diagnostic implications. For example, fractal dimension of EEG changes during sleep, diminishing from waking state, through stages 1, 2, and 3 to stage 4, and increases again in rapid eye motion (REM) sleep [61].

Biosignal record is a pattern called *waveform* — a planar curve that proceeds resolutely forward — it does not go backward and it does not crossover itself [i.e., it is a collection of (tx) point pairs, where t values increase monotonically]. To characterize a biosignal like EEG, fractal analysis in time domain (that is analysis of its waveform on a computer screen or a sheet of paper) may be more applicable that fractal analysis of the signal represented by embedding data in a multidimensional phase space; different algorithms have been proposed for such analysis, of which the method proposed by Higuchi [62] seems to be the best [61].

In Higuchi's method, the time series representing the analyzed waveform (in a computer memory time by necessity is a discrete variable) is subdivided into k subseries; each subseries represents again a waveform. If the length of such a curve may be expressed as proportional to k^D, then D is the curve's fractal dimension. It measures *complexity* of the curve — for a simple curve D equals 1, for a curve, which nearly fills out the plane D is close to 2.

Supposed that in a computerized data-acquisition system, the signal recorded on a selected channel is represented by the time series

$$x(1), x(2), x(3), \ldots, x(N) \tag{6.8}$$

where $x(i)$ is the signal's amplitude at the i-th moment of time ($i = 1, \ldots, N$) and N is the total number of points; from this one constructs k subseries $x(m,k)$:

$$x(m,k): x(m), x(m + k), x(m + 2k), \ldots, x(m + int[(N{-}m)/k]* k) \quad (m = 1, 2, \ldots, k) \tag{6.9}$$

where $int[\ldots]$ denotes the greatest integer not exceeding the number in the brackets; m and k are integers indicating the initial time and the time interval, respectively. For example, if $N = 100$ and $k = 4$ one obtains four subseries:

$$x(1,4): x(1), x(5), \ldots, x(97)$$

$$x(2,4): x(2), x(6), \ldots, x(98)$$

$$x(3,4): x(3), x(7), \ldots, x(99)$$

$$x(4,4): x(4), x(8), \ldots, x(100)$$

In Higuchi's method, the length of the curve represented by $x(m,k)$ is defined as:

$$L_m\left(k\right) = \frac{1}{k} \cdot \left[\left(\sum_{i=1}^{int\left(\frac{N-m}{k}\right)} \left| x(m + i \cdot k) - x(m + (i-1) \cdot k) \right| \right) \cdot \frac{N-1}{int\left(\frac{N-m}{k}\right) \cdot k} \right] \tag{6.10}$$

where the term $(N - 1)/\{int[(N - m)/k]_* k\}$ is a normalization factor. $L_m(k)$ are then averaged for all values of m,

$$L\left(k\right) = \frac{\sum_{m=1}^{k} L_m\left(k\right)}{k} \tag{6.11}$$

giving the mean value of the curve length, $L(k)$, for given value of k. The procedure is repeated for several k ($k = k_1, k_2, \ldots, k_{max}$) and then from the log–log plot of log(L) vs. log(k) using the least-square method one obtains Higuchi's fractal dimension of the signal, D:

$$D = -\log[L(k)]/\log(k) \tag{6.12}$$

Higuchi's fractal dimension is a very useful measure for characterization of EEG signals [61]. Fractal dimension, which is a measure of the signal complexity, helps to differentiate between EEG traces corresponding to different physiopathological conditions. It needs much shorter signal epochs than those needed for obtaining reliable values of fractal dimension of the signal in phase space by means of the attractor dimension or other correlated parameters. Fractal analysis in time domain allows investigating relevant EEG events shorter than those detectable by means of other linear and nonlinear techniques. For example, fractal dimension of EEG-signal clearly demonstrates an influence of magnetic field on the brain, while no influence of magnetostimulation could be noticed while inspecting the same EEG-recordings with the naked eye or analyzing with linear methods like the fast fourrier transform (FFT). Higuchi's fractal dimension was also used to assess the influence of phototherapy on patients suffering with seasonal depression [Seasonal Affective Disorder (SAD)]; again, EEG-signal assessing with naked specialist eye or using linear methods did not reveal any evident changes, but calculation of D demonstrates influence of applied phototherapy.

We have to stress that fractal dimension calculated in time domain as well as "classical" fractal dimension used in quantitative morphometry (like that of a coastline or of Sierpinski's gasket) should not be confused with fractal dimension of a chaotic attractor calculated in the phase space of a dynamical system. Attractor dimension, e.g., correlation dimension, is usually fractal, but it may be significantly greater than 2, it may provide some measure about how many relevant degrees of freedom are involved in the dynamics of the system under consideration. Calculation of attractor's fractal dimension requires previous embedding of the data in phase space, using, e.g., Taken's time delay method. Higuchi's fractal dimension is always between 1 and 2 since it characterizes complexity of the curve representing the signal under consideration on a two-dimensional plane. Higuchi's algorithm works on raw data — one does not need to embed data in a phase space.

6.5 EXAMPLES OF ORDER, CHAOS, AND RANDOMNESS IN NATURAL PROCESSES

In nature, one may observe ordered, chaotic, and random systems. It seems that in biology, chaotic systems are the most important. Ordered equilibrium systems form structures like crystals that are inert and unreceptive — they are not able to change when conditions in their environment change. On the other hand, processes in completely random systems like gases have so many degrees of freedom that they are not able to built any structure that may support any purposeful function for a certain sufficiently long period. Chaotic system are between these two extremenesses — chaotic processes may form structures that are lasting and at the same time are able to react on changes in the environment, i.e., to evolve. Ability of easy adaptation to changing external conditions is more important for system surviving than a stiff adjustment to some strictly defined conditions — that is why chaotic systems, which adapt extremely easily, are so important, and a decrease of chaocity of a system may lead to serious pathology, and so one may formulate.

Chaos has many different faces. One may observe chaos in chemical and mechanical systems, in living organisms and in inanimate nature, in man-made structures and in planetary systems. Chaos affects our bodies and our senses. And there are different "*roads to chaos*," roads leading through oscillating patterns and/or spatially organized nonequilibrium structures. In biology, chaos is present on all levels of organization — from single cell to human brain to population.

Cytoplasm of a living cell is in perpetual movement that may be observed under microscope. Mechanical properties of cytoplasm are changed by undergoing sol–gel transitions, similar to those observed in colloidal systems or in producing a jelly. Such sol–gel transition processes in microvolumes are reversible and may be chaotic or quasiperiodic [45,52] — microgels are formed and then dissolved in different places again and again, as long as the cell remains in a far-from-equilibrium state. Irreversible gelation of the whole cytoplasm is probably connected with cell death [63].

Is human brain activity chaotic? It cannot be fully random because then we would not be able to perform behaviors such as moving our limbs purposefully. It neither cannot be ordered, i.e., unambiguousity determines (in the classical mechanics' sense) because then it could not be creative. Fundamental overall properties of the brain, up to now almost taken for granted, are perhaps maintained/exercised by a nonlinear deterministic chaotic mode of brain function. If this is the case, borderline psychopathological phenomena, when they flare up, can be explained as resulting from sudden reduction of such a deterministic chaotic mode. In brain activities, chaos may have different functions:

1. Owing to the inseparability of the initial conditions from the equations of motion for nonuniform strange attractors, mutual information can be transmitted without attenuation. This is in opposition to the Shannon–Weaver rule for transmission in nonchaotic or uniform systems that there is always attenuation, and it reflects the fact that chaotic dynamics *creates* as well as *destroys* information during transmission and other operations.
2. Chaos "prevents falling into obsession," meaning degeneration to fixed-point attractors.
3. Chaos can retrieve patterns by rapid search through phase space.
4. Chaos can control a system dynamically and "buffer" it against unexpected stimuli, replacing "homeostasis" with "homeodynamism."
5. Chaos can generate the patterns (which Freeman calls "carrier" waves) that encode information (and as Freeman noted decode as well).
6. Chaos can facilitate the reorganization of functional units in brains that are needed for flexibility, insight and creation.

Also on the level of population dynamics, there is neither a full order nor a complete randomness. Populations of many species, especially those quickly reproducing, may show deterministic chaos. Despite the model's simplicity, over time the total number of individuals fluctuated wildly. It could remain practically steady for thousands of generations, then without any warning suddenly (like we saw in the case of Lorenz's systems. (Figure 6.1 to Figure 6.3) boom or crash. They could cycle up and down for several generation or show a completely chaotic behavior, and then suddenly become steady again. It is evident that studies of a population over several years may be absolutely useless to predict population numbers for next years. Also failing to find an environmental cause of a sudden, mysterious decline in a population may be result of the simple fact that there is no such a cause there. It is "a hopeful fact" that nature is so unpredictable. It probably helps to maintain the biological diversity of the planet. A kind of instability in population dynamics may be what keeps one species that might otherwise completely dominate from doing so, thus allowing the others to persist. Conrad [64] notes five possible functional roles for chaos in population dynamics:

1. The generation of diversity, as in the predator-prey species where the exploratory behavior of the animal is enhanced.
2. The preservation of diversity; for example, the diversity of behavior is used by the prey to act unpredictably and thereby elude being caught by the predator.
3. Maintenance of adaptability, that is to disentrain processes — in populations this would correspond to keeping a broad age spectrum.
4. Cross-level effect, for example, chaos on genetic level would contribute to the diversity and adaptability on the population level.
5. The dissipation of disturbances, phenomenon observed in synchronization of chaotic systems, is achieved by the sensitivity of orbits on the chaotic attractor to initial conditions; the attractor ensures dynamical stability of the chaotic system like a heat bath ensures that temperature of a system is kept constant.

One used to think about chaos as something commensurate with a pathology. It turns out that the exact opposite may be true. For example, increase of order (decrease of chaos) in the brain that manifests in decrease of fractal dimension of EEG signal may lead to epileptic seizures [61]. Similarly, in heart, healthy states seem to be more chaotic than sick ones — when the heart patient is about to die, the heart rhythm becomes actually more periodic. Different diseases were considered to be just disturbed biological processes of the organism. Some diseases, especially those which are characterized by the disorders in the differentiation (malformations and malignant tumors), may be understood as processes of the disturbed self-organization. Mackey and Glass [65] introduced a revolutionary concept of a *dynamic disease*, defined as one that occurs in an intact physiological control system operating in a range of control parameters that leads to abnormal dynamics. Congestive heart failure [66], fetal distress syndrome [67], neurological disorders, including epilepsy and movement disorders [68], and aging processes [69] can be considered and modeled as dynamic diseases.

It is noticeable that the same phenomena, which may be observed in medicine and physiology on the level of a single organism, may also be observed on other levels of organization, from subcellular "suborganisms" to "superorganisms" of whole populations. Cohen and Stewart in *The Collapse of Chaos* present the thesis that *explaining complexity is easy, it's explaining simplicity that's hard* [70].

6.6 EXAMPLES OF SYSTEMS AND PROCESSES THAT ARE NOT EASILY MODELED WITH CHAOS

Our world is mostly nonlinear. The science of *nonlinear dynamics* was originally christened "*chaos theory*" because from nonlinear equations, unpredictable solutions emerge. In classical Physics, linear models were successfully used — both classical Newtonian dynamics and Maxwellian electrodynamics are *linear*. But all models are only approximation of real systems and processes, and linear models may serve as good approximations of reality only if long-term influence of interactions with other (sub)systems may be neglected. Long-term even movements of astronomical bodies may become chaotic.

On the other hand, the difference between classical *random (stochastic)* systems and *deterministic-chaotic* systems and processes is in my opinion rather quantitative than qualitative. A stochastic system has, in theory, an infinite number of degrees of freedom, so if a system has really a big number of degrees of freedom, it behaves randomly, i.e., it is governed rather by laws of probability and statistics than by dynamic equations; the problem arises of what that "big number" really is. If a system under consideration has many "irreducible" (equally important) degrees of freedom, it is still considered stochastic. In practice, even using modern computers, we cannot deal with more than approximately 20 to 30 nonlinear dynamical equations. If a nonlinear system has only a few (three or more) degrees of freedom or if the number of degrees of freedom may be reduced to only a few "really important" ones, the system is considered to be deterministic-chaotic. In time domain, instead of speaking about degrees of freedom, one should rather speak about subprocesses with different characteristic timescales — really important are those subprocesses that show characteristic timescale comparable with the time of observation. Chemists have for a long time used the so-called quasistationary approximation — the chemical species with characteristic timescales much shorter than the time of observation reach their stationary concentrations and so they may be eliminated from kinetic dynamical equations by using Tikhonov's Theorem [71], while those species with characteristic timescales much longer than the time of observation have constant (practically initial) values. Haken's [18] concept of order parameters is based on similar principles — due to the differences in characteristic timescales, the degrees of freedom of the system are reduced. We need to stress that even trajectory of a Brownian particle (despite the fact that Brownian motion has always been used as a classical example of a stochastic process) may be characterized by its fractal dimension [54].

Not all nonlinear dynamical systems are chaotic. First, on a "road to chaos" complex system shows quite regular behavior, like a simple stable point or a few critical points — stable, unstable, and saddle; then when system parameters change, one may observe limit-cycle types of oscillations, then period doubling, etc. Nonlinear dynamics of such systems may be modeled using cellular automata, graph theory, neural networks. But, on the other hand, in linear systems, branching processes and fluctuations (noise) may produce signals that look very, very similar to those produced by nonlinear dynamics in chaotic systems, and it is often not easy to demonstrate that the system under consideration shows nonlinear dynamics.

Nonlinear phenomena are well known in physics. If one connects together two balloons — one inflated to be like an apple, another like a small watermelon — the bigger takes out the air from the smaller one, which collapses; it is so because the volume–pressure characteristics for air closed in a balloon is nonlinear. In Medicine, awareness of nonlinear effects is rather little, despite the fact that such effects are very well known from everyday life. For example, e.g., small amount of alcohol acts as a stimulator, while in a large dose, alcohol is poisonous; in general, due to nonlinear effects, a small dose of a drug or radiation may have a completely opposite effect than a greater dose — this phenomenon of nonlinearity in dose–response dependence is called *hormesis* [72].

However, deterministic chaos governs only a small subset of chaotic systems. Another class of nonlinear dynamics models that are not modeled with "classical" chaos theory are *reaction-diffusion systems* (RDS). These systems are modeled not by quasilinear ordinary differential equations (QODE) but by quasilinear partial parabolic differential equations, QPPDE, i.e., diffusion equations with nonlinear source–sink term, which is due to a chemical reactions. In RDS, one observes interesting processes of pattern formation [15].

Scientists hoped that development of the chaos theory would suggest explanations of brain activity — a resolution of the discrepancy between mesoscopic global order and aperiodic seemingly random activity at microscopic levels. Brains work with large masses of neurons. Each neuron typically receives synaptic input from thousands of other neurons within the radius of its dendritic arbor, and it gives synaptic output to thousands of others within the radius of its axon, and not the same thousands because each neuron connects with less than 1% of the neurons within its arbors, owing to the exceedingly high packing density of cortical neurons. These properties of dense but sparse interconnection of immense numbers of otherwise autonomously active nonlinear neurons provide the conditions needed for the emergence of mesoscopic masses, ensembles, and populations, which have properties related to but transcending the capacities of the neurons that create them. The most significant property of ensembles is the capacity for undergoing rapid and repeated global state changes [48]. However, models of low-dimensional deterministic chaos have failed to explain brain activity — one still knows too little about mesoscopic brain states. Freeman [48] suggests that what happens in brain and what manifests itself through EEG signal will be called "*stochastic chaos*," because it arises from and feeds on the randomized activity of myriads of neurons, and it provides the basis for self-organization.

6.7 CONCLUSIONS AND OPEN PROBLEMS

Chaotic systems are defined as those showing irregular (nonperiodic) behavior with extreme sensitivity to initial conditions and to small changes in system parameters due to their nonlinear nature; although chaotic behavior appears to be erratic or random, it is governed by deterministic equations of motion, in contrast to stochastic motion which, like noise, is described by probabilistic laws [73].

In everyday language, the word "chaotic" is used to denominate either "*completely unordered*" systems or "*absolutely unpredictable*" processes. Since we often think in terms of words rather than of pictures, words "chaos, chaotic" ring in our mind associations with something unordered and unpredictable. Chaos is apparently unpredictable behavior arising in a deterministic system because of great sensitivity to initial conditions. Chaos arises in a dynamical system if two arbitrarily close starting points diverge exponentially, strictly speaking this is not true. If the divergence is exponential, it is still bounded. The only circumstance for which it is unbounded is for time duration.

Recent years have seen an extraordinary growth of interest in complex systems and deterministic chaos. Together they point to the emergence of new paradigms, cutting across traditional disciplines. The overall approach is revolutionary since it requires us to adopt a new frame of reference. Even apparently simple deterministic systems can behave in a seemingly unpredictable manner. There already exists a variety of procedures for delineating whether chaotic behavior results from a nonlinear dynamical system with a few degrees of freedom, or whether it is caused by a stochastic noise.

We all know that if we need to keep something in order, it usually induces anxiety and tension. To disconcert the order seems to be much more effortless and makes us feel better. To "put something in order" means to decrease the entropy, and according to the Second Law of Thermodynamics, requires work to be done on the system. Disorder increases "by itself." But what we learned from theory of chaos is that disorder does not have to mean a lack of any governing dynamic laws — that chaos may be deterministic. But it still may be much easier to keep a system in such a chaotic state rather than in the ordered state. Chaos is more natural then order.

Many scientists do believe that functional complexity and behavior of living systems are due to an interaction of an immense number of a very few simple basic mechanisms on molecular or even submolecular level, which combined together in a nonlinear way, are responsible for even the most complicated phenomena. One does observe some very similar phenomena in simple chemical systems as in a living organism. But what is the difference between animate and inanimate, between biological systems and physicochemical systems? Critically, the metaphor of "chaos" does not give a clue how to answer this fundamental question.

But more and more scientists come to the conclusions that behind matter lies energy but also that behind energy lies wisdom that is expressed as information that life has a direction and purpose, which cannot be explained in purely physical terms. "The age-old theological view of the universe is that all existence is the manifestation of a transcendent wisdom, with a universal consciousness being its manifestation. If I substitute the word information for wisdom, theology begins to sound like quantum physics" — writes Schroeder [74]. And he adds: "To call the phenomenon of life complex trivializes the reality ... the ordered, information-containing complexity found in life is of a type qualitatively different from that found in the sub-structures from which it arose...." Systems can give rise to secondary systems that are more complex, but that complexity is a fractal extension, an increase in amount but not in type, of the original systemís complexity. With life, the increase in complexity seems to be one of a new type as well as amount." As the Pope John Paul II said: "Providence does not know a notion of pure chance."

As in practically any scientific problem, various authors agree (and even overlap) on certain points concerning chaotic behavior, appear diametrically opposed in some instances, and have mutually inclusive stances in other cases; certain tenets from *each* model may ultimately prove relevant to the whole. We often are like a Monsieur Jourdain from Moliere's "Le Bourgeois Gentilhomme" who says (II.iv): "Good heaven! For more than forty years I have been speaking prose without knowing it." The most likely reasons for failing to observe some phenomena and processes where they exist appear to be due to experimental design. We often do not see what we do not look for.

REFERENCES

1. Grebogi C et al. (1984). Physica D 13:261–268.
2. Bertalanffy L. von (1973). General System Theory (Rev ed). New York: George Braziller.
3. Wiener N (1961). Cybernetics: or Control and Communication in the Animal and the Machine. New York: M.I.T. Press.

4. Ashby WR (1964). An Introduction to Cybernetics. London: Methuen; Aulin A (1982) The Cybernetic Laws of Social Progress. Oxford: Pergamon.
5. Aulin A (1982). The Cybernetic Laws of Social Progress. Oxford: Pergamon.
6. Simon HA (1981). The Sciences of the Artificial (2nd ed). Cambridge, MA: MIT Press.
7. Heylighen F ed. (1997). The Evolution of Complexity. Dordrecht: Kluwer Academic.
8. Darwin C. The origin of species by means of natural selection or the preservation of favoured races in the struggle for life (Edited with and introduction by Burrow JW). Penguin classics, 1985. (First published by John Murray, 1859).
9. Dawkins R (1989). The selfish gene (2nd ed). Oxford: Oxford University Press.
10. Gould SJ and Eldredge N (1977). Punctuated equilibria: the tempo and mode of evolution reconsidered. Paleobiology 3:115–151.
11. Kauffman SA (1993). The Origins of Order: Self-Organization and Selection in Evolution. New York: Oxford University Press.
12. Kauffman SA (1995). At Home in the Universe: The Search for Laws of Self-Organization and Complexity, Oxford University Press.
13. Langton CG (1992). Life on the edge of chaos. *Artificial Life II*. New York: Addison-Wesley, 41–91.
14. Langton CG ed. (1997) Artificial Life: An Overview (Complex Adaptive Systems), Bradford Books.
15. Nicolis G and Prigogine I (1977). Self-Organization in Non-Equilibrium Systems. New York: Wiley.
16. Nicolis G and Prigogine I (1989). Exploring Complexity. New York: Freeman.
17. Prigogine I and Stengers I (1984). Order out of Chaos. New York: Bantam Books.
18. Haken H (1983). Synergetics — An Introduction. (3rd ed). Berlin: Springer.
19. Haken H (1987). Advanced Synergetics. (2nd ed). Berlin: Springer.
20. Haken H (2002). Synergetics — past, present, and future. Attractors, Signals, and Synergetics. In: Klonowski W, ed. Froniers on Nonlinear Dynamics Vol. 1. Lengerich, Berlin: Pabst Science Publishers, 19–24.
21. Eigen M and Schuster P (1979). The Hypercycle: A Principle of Natural Self-Organization. Berlin: Springer.
22. Bak P, Tang C, and Weisenfeld K (1988). Self-organized criticality. Phys Rev A 38:364–374.
23. Gleick J (1987). Chaos: Making a New Science. New York: Penguin Books.
24. Neumann J von (1966). Theory of Self-Reproducing Automata. In: Burks AW, ed. Champaign: University of Illinois Press.
25. Wolfram S (1994). Cellular Automata and Complexity: Collected Papers. Reading, MA: Addison-Wesley.
26. Thom R (1975). Structural Stability and Morphogenesis. Reading, MA: Benjamin.
27. Thompson D (1917). On Growth and Form. Cambridge: Cambridge University Press.
28. Holland JH (1992). Adaptation in Natural and Artificial Systems: An Introductory Analysis with Applications to Biology, Control and Artificial Intelligence. Cambridge, MA: MIT Press.
29. Foerster H von and Zopf G, eds. (1962) Principles of Self-Organization. New York: Pergamon.
30. Foerster H von (1996). Cybernetics of Cybernetics (2nd ed). Minneapolis: Future Systems.
31. Maturana HR and Varela FJ (1992). The Tree of Knowledge: The Biological Roots of Understanding (rev ed). Boston: Shambhala.
32. Forrester JW (1973). World Dynamics (2nd ed). Cambridge, MA: Wright-Allen Press.
33. Boulding KE (1978). Ecodynamics: a new theory of societal evolution. London: Sage.
34. Anderson PW, Arrow KJ, and Pines D, eds. (1988). The Economy as an Evolving Complex System. Redwood City, CA: Addison-Wesley.
35. Arthur WB (1994). Increasing Returns and Path Dependence in the Economy. Ann Arbor: University of Michigan Press.
36. Weidlich W (2003). Sociodynamics — a systematic approach to mathematical modeling in the social sciences In: Klonowski W, ed. From Quanta to Societies, Frontiers on Nonlinear Dynamics Vol 2. Lengerich, Berlin: Pabst Science Publishers, 15–46.
37. Mandelbrot B (1983). The Fractal Geometry of Nature. New York: WH Freeman and Co.
38. Paar V, Pavin N, and Rosandic M (2001). J Theor Biol 212(2):47.
39. Hornby AS, Gatenby EV, and Wakefield H (1963). The Advanced Learners Dictionary of Current English, (2nd ed). London: Oxford University Press.
40. Lorenz EN (1963). Deterministic nonperiodic flow. J Atmosph Sci 20:130.

41. Harrison RG and Uppal JS (1988). Instabilities and chaos in lasers. Europhysics News 19(6):84.
42. Ji S (1991). Biocybernetics: A machine theory of biology. In: Ji S, ed. Molecular Theories of Cell Life and Death. New Brunswick, NJ: Rutgers University Press, 41.
43. Ji S (1999). The linguistics of DNA: words, sentences grammar, phonetics, and semantics. Ann N Y Acad Sci 870:411–417.
44. Ji S (2000). Free energy and information contents of conformons in proteins and DNA. Biosystems 54:107–130.
45. Klonowski W (1979). A new criterion for critical branching in polymer system. Presented by Prigogine I and Glansdorff P to Bull Cl Sci Acad R Belg LXIV, 568–577.
46. Klonowski W (1984). Probabilistic theory of crosslinked macromolecular systems with applications in chemical physics and biophysics. Books on Demand AU00326. Ann Arbor, MI: University Microfilms International.
47. Klonowski W (1988). Random hypergraphs and topological gelation criterion for crosslinked polymer systems. In: Kennedy JW and Quintas LV, eds. Applications of Graph Theory in Chemistry and Physics. North Holland; Discrete Applied Math 19, 271–288
48. Freeman WJ (1995). Societies of Brains. Mahwah NJ: Lawrence Erlbaum Associates.
49. Erdos P and Renyi A (1960). On the evolution of random graphs. Publ Math Inst Hung Acad Sci 5:17–61.
50. Hameroff SR (1987). Ultimate Computing, Biomolecular Consciousness and Nano Technology, North-Holland.
51. Choi MY and Huberman BA (1983). Phys Rev A 28(2):1204; Cohen B. (1997). The Edge of Chaos: Financial Booms, Bubbles, Crashes and Chaos, Wiley.
52. Klonowski W (1988). Representing and defining patterns by graphs: applications to sol-gel patterns and to cytoskeleton. Biosystems 22:1–9.
53. Stock G. Metaman: The Merging of Humans and Machines into a Global Organism. Princeton, NJ: Princeton University Press.
54. Klonowski W (2000). Signal and Image Analysis Using Chaos Theory and Fractal Geometry Machine Graphics and Vision 9(1/2), 403–431.
55. Kraft R. Fractals and Dimensions. HTTP Protocol at: http://www.edv.agrar.tumuenchen.de/dvs/idolon/dimensions/dimensions.html
56. Bayley PM and Martin SR (1991). Microtubule dynamic instability: some possible physical mechanisms and their implications. Biochem Soc Trans 19:1023–1028.
57. Sernetz M et al. (1985). J Theor Biol 117:209–230.
58. Klevecz et al. (1991). Cellular Oscillators.
59. De-Boer, Perelson, and Kevrekidis (1993). Bull Math Biol 55(4):781–816.
60. Pool, R. (1989). Is it healthy to be chaotic? Science 243(4891): 604–607.
61. Olejarczyk E. (2003). Analysis of EEG signals using fractal dimension method, PhD Thesis, Warsaw: Institute of Biocybernetics and Biomedical Engineering Polish Academy of Sciences.
62. Higuchi T (1988). Approach to an irregular time series on the basis of the fractal theory. Physica D 31:277–283.
63. Winchester AM (1985). Irreversible Gelation.
64. Conrad M (1986). Five possible functional roles for chaos in population dynamics.
65. Mackey and Glass (1977) Dynamic Diseases.
66. Goldberger et al. (1986). Congestive heart failure.
67. Rapp P (1979). Neurological disorders, including epilepsy.
68. Modanlon and Freeman (1982) Fetal distress syndrome.
69. Mandel (1988) Movement disorders and aging processes.
70. Cohen and Stewart (1994). The collapse of chaos: Discovering simplicity in a complex world. New York: Viking Press.
71. Klonowski W (1983). Simplifying principles for chemical and enzyme reaction kinetics. Biophys Chem 18:73–87.
72. Klonowski W (1999). Nonlinearity and statistics — implications of hormesis on dose-response analysis. Biocyb Biomed Eng 19(4):41–55.
73. Arecchi FT and Farini A (1966). Lexicon of Complexity (1st ed). Firenze: Studio Editoriale Florentino.
74. Schroeder G (2001). The hidden face of God. How Science Reveals the Ultimate Truth. Boston: Free Press.

7 Biological Complexity: An Engineering Perspective

Stephen W. Kercel

CONTENTS

7.1 WHEN ENGINEERING DECISIONS INVOLVE LIVING PROCESSES

We live in "the biological century." To an extent never previously imagined, engineering and economic decisions are being influenced both by the biological paradigm, in general, and by specific biological issues in particular [1]. As Friedman notes, "All fields of engineering will be players in the biological century, enriching and complementing one another, and fulfilling a promise we can hardly imagine" [2]. Biological questions affect practical decisions about issues both exotic and mundane.

At the exotic end of the spectrum, serious decisions are influenced by the possibility of extraterrestrial life. For example, on September 21, 2003, NASA deliberately flew the Galileo spacecraft into the atmosphere of Jupiter [3]. The reason for this decision was that the spacecraft was running out of maneuvering fuel, and there was a risk that the uncontrolled spacecraft might eventually crash into Europa. Since there is a possibility that Europa has liquid water, and that it may harbor life, a crash of a terrestrial spacecraft into Europa posed the risk of contaminating indigenous organisms with terrestrial microbes.

The characterization of extraterrestrial life is an awkward one. At the most fundamental level, would we recognize it if we see it? The difficulty is shown by the search for life on Mars. "The three biology experiments discovered unexpected and enigmatic chemical activity in the Martian soil, but provided no clear evidence for the presence of living microorganisms in soil near the landing sites" [4]. Because the Viking experiments to detect life were based on terrestrially biased preconceptions, they produced inconclusive results. "In the case of Mars, the issue has been complicated by the emotional belief in an Earth-like Mars, which has largely been shown to have been a myth" [5]. The debate as to whether or not available data prove or disprove the existence of life on Mars continues to the present time [6,7].

It is not surprising that the problem of identifying extraterrestrial life is difficult, when one considers how difficult it is to distinguish living from nonliving processes even on Earth. The disease carrier for "Mad Cow Disease" is a protein called a prion. Because it behaves in a way that suggests a capacity to replicate and preserve itself in a hostile environment, the prion was formerly (before its identification) assumed to be a living organism. However, despite the fact that a brain into which one pathogenic (bad) prion is introduced rapidly becomes infested with bad prions, they do not reproduce. Through a process similar to enzyme catalysis, bad prions cause the refolding of the chemically identical good prions into a conformation in which they are pathogenic. It has been shown that no new macromolecules are synthesized during this apparent change of conformation [8–10].

The point illustrated by these extreme examples is that preconceptions can lead to misguided practical decisions. For instance, several fabulously expensive interplanetary missions have led to inconclusive experimental data regarding the existence of extraterrestrial life, because the experiments were not designed to recognize the relevant distinctive features of living processes. If engineers are to be serious "players in the biological century," they must begin by appreciating how living processes *differ fundamentally* from traditional engineering systems.

7.2 THE ROLE OF CAUSATION

The distinction between living and nonliving processes is prototypical of all the problems the engineer faces when dealing with organisms. Determining whether or not the process is an organism appears to be a fundamental classification problem that arises in biological and chemical research. It is prototypical of all classification problems, the identification of whether a process is a member of a given class. The solution of a classification problem typically informs the solution to the generic engineering problem, deciding what to do next in the face of a given situation.

Feature identification is believed to be the most difficult aspect of the classification problem [11]. As Robert Rosen has noted [12], the most reliable features of any system (process) are the entailment structures that constrain its behavior. He writes: "Any system, be it formal or natural, is characterized by the entailments within it. Thus, a *formal* system is characterized by its inferential structure, which entails new propositions from given ones. And a *natural* system is likewise characterized by its causal structure" [12].

7.2.1 Characterization by Causation

To appreciate how a natural system is characterized by its causal entailment structure, we must first understand the Modeling Relation (MR). This particular relation has been known for well over a century. "We form for ourselves images or symbols of external objects; and the form which we give them is such that the logically necessary (denknotwendigen) consequents of the image in thought are always the images of the necessary natural (naturnotwendigen) consequents of the thing pictured" [13].

In describing the properties of the MR, Rosen tells us "modeling is the art of bringing entailment structures into congruence [14]. However, this statement seems to leave us no wiser than when we started. How does art enter into the discussion; are we not instead supposed to be scientific? What is an entailment, much less an entailment structure? What does it mean that two different entailment structures are congruent? In what sense are they not identical? If they are not identical, what similarity between them causes us to declare them congruent? Why do these questions matter?

The first point to appreciate is that the MR is a relation in the mathematical sense [15, 16]. Suppose that A and B are sets, and that there exists a set, R, of ordered pairs, where the first element of each pair in R is an element of A, and the second element of each pair in R is an element of B. In mathematical notation: $a \in A$, $b \in B$, $(a,b) \in R \Leftrightarrow aRb$. In the MR, the members a and b of each ordered pair in R are entailments from two different systems.

Entailments are the relationships that constrain the organization of a process. There are two sorts of systems that might appear in the MR, natural systems and formal systems. Natural systems are systems in physical reality that have causal linkages; if certain causative events impinge upon a natural system, then the system will behave in a certain way, or produce certain events in effect. This consequential linkage of cause and effect in a natural system is a causal entailment. If event P causes event Q, the relationship $P \Rightarrow Q$ is a causal entailment [17].

Formal systems are conceptual systems that have inferential linkages; if certain hypothetical propositions impinge upon a formal system, then they will produce certain consequential propositions in conclusion. This consequential linkage of hypothesis and conclusion in a formal system is an inferential entailment. If proposition R implies proposition S, then the proposition $R \Rightarrow S$ is an inferential entailment [17]. In formal systems that are descriptive of processes of life and mind, impredicative inferential entailment structures and the semantics implied by them abound [18]. Despite the seeming similarity of terminology, a formal system in a MR is not to be confused with Hilbert's strictly syntactic formal system, more commonly called a formalism [19].

Entailment structures are inherent within a system; they are the distinguishing features that characterize the system [12]. They do not cross over from one system to another. In Figure 7.1, we see a natural system, N, distinguished by its structure of causal entailments, where causal entailment "a" is the relation "event h causes event j." Formal system, F, distinguished by its structure of inferential entailments, where causal entailment "b" is the relation "proposition p implies proposition q." The entailment structures of two distinct systems are distinct from one another; causes or hypotheses in one do not produce effects or conclusions in the other. In fact, this provides the answer to one of the questions posed above. Its self-contained entailment structure is what provides identity to a system and distinguishes it from other systems.

The fact that distinct systems are nonidentical does not preclude them from being regarded as being in some sense similar. Similar systems should have distinguishing features that closely

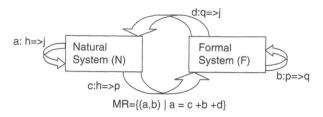

FIGURE 7.1 The Rosen MR. A natural system (N) is a system in ontological reality. A Formal System (F) is an abstract representation. The entailment $a{:}h \Rightarrow j$ is the representation of a causal entailment in N; the entailment can be read as "event h causes event j." The entailment $b{:}p \Rightarrow q$ is the representation of an inferential entailment in F; the entailment can be read as "proposition p implies proposition j." The entailment $c{:}h \Rightarrow p$ is the representation of a measurement; the entailment can be read as "event h corresponds to proposition p." The entailment $d{:}q \Rightarrow j$ is the representation of a prediction; the entailment can be read as "proposition q corresponds to event b." The MR is a relation between the set of events in N and the set of propositions in F. The "+" signifies concatenation. The condition for the MR to exist is that there must be an entailment in N, an entailment in F, a measurement and a prediction such that when the measurement is concatenated with the entailment in F is concatenated with the prediction, the same result is obtained as the result of the entailment in N. In other words, the MR exists between two sets of entailments if sets contain entailments that commute.

correspond to each other; similar systems belong to the same equivalence class. Dissimilar systems should have distinguishing features that do not closely correspond to each other; dissimilar systems belong to different equivalence classes. The partitioning into equivalence classes is based on the entailment structure. As already noted, the distinguishing feature of a system is its entailment structure. Thus, we would expect similar systems to have entailment structures in which there is some degree of correspondence between the entailments.

To establish this correspondence, consider a system of encodings and decodings [20]. For example, we might have a system of encodings that encodes a set of events in the natural system, N, in Figure 7.1, into a set of propositions in the formal system, F. We might also have a system of decodings that decodes a set of propositions in the formal system, F, into a set of phenomena in the natural system, N. Although the two systems remain independent in the sense that causes or hypotheses in one do not produce effects or conclusions in the other, encodings and decodings can link the two systems.

This linkage between entailment structures provides the means of determining the similarity between two systems. Suppose that an event, h, in N can be encoded to a proposition, p, in F; we can think of the encoding arrow, c, in Figure 7.1 as a measurement on a natural system. Suppose further that the proposition, p, when applied as a hypothesis in the inferential structure in F entails another proposition, q, in F as a conclusion. In other words, the two propositions are entailed as an implication, b = $(p \rightarrow q)$, in F. Suppose that this entailed proposition, q, in F can be decoded into an event, j, in N; we can think of the decoding arrow, d, in Figure 7.1 as a prediction by a formal system.

Rosen identifies the similarity as congruency between the entailment structures, and defines it in the following way [21]. Suppose that in the underlying reality, the event h in N causes event j in N. In other words, the two events are entailed as a causal linkage, a = $(h \Rightarrow j)$, in N. Suppose further that the linkages commute. Event h is encoded by c to proposition p, c = $(h \Rightarrow p)$. We already have b = $(p \Rightarrow q)$. Then suppose d = $(q \Rightarrow j)$, decoding proposition q to event j. Then we concatenate these entailments, c + b + d = $h \Rightarrow p \Rightarrow q \Rightarrow j$. We note that this concatenation of entailments exactly corresponds to the causal entailment a = $(h \Rightarrow j)$. In other words, the commutation is also described as a = c + b + d. If there exists no such entailment b in F, having a commutative relationship with some entailment a in N, then the two systems do not have congruent entailment structures. Entailment structures are congruent to the extent that such correspondences between entailments exist.

If such correspondences between the entailments in the two systems do exist, then we can learn something about one entailment structure by observing the other. This is the essence of the MR. When it is applied to a formal system to obtain predictions about a natural system, the inferential entailments in the formal system correspond to the causal entailments in the natural system. Where the relationship holds up and where it breaks down are both understood. This is where the MR differs from "black box" simulations. Construction of a "black box" is a simple curve fit. It makes no claim about the causal links in underlying reality, offers no understanding of the natural system it purports to describe, and offers no warning as to when the description will break down.

In contrast, for any valid MR, the identification of the encodings and decodings between two systems is an act of discovery based on insight or understanding. The benefit of this understanding is the awareness of the specific entailments so described, and a clear indication of the scope of applicability (or nonapplicability) of the formal system as a model. The cost of this understanding is that it is an art and not a science in the reductionist sense; there is no automatic or algorithmic method for determining either the encodings or decodings. In fact, there is not even any necessity or assurance that the system of decodings can be obtained from some straightforward inversion of the encodings.

This digression into modeling has laid the background for identifying the crucial distinguishing features of a process. The distinguishing feature of a natural process is the structure of causal entailment that constrains the behavior of the system. We can gain novel insight into that causal entailment structure by asking questions about a model of the process, where a model is a process with an entailment structure that commutes with the entailment structure of the process being modeled. However, the crucial point remains; *a natural process is characterized by its structure of causation.*

7.2.2 CLASSES OF CAUSATION

An effect or event occurs, and we ask, "Why did this happen?" Aristotle said that any event is the result of a transformation, and it has multiple kinds of causes [22]. Merely asking why an event happened is not specific enough. We can ask, "What is the input of the transformation?" This is important, because, in the absence of an input, a transformation produces no output. What we would call the input to the process is, in Aristotle's parlance, the *material cause.*

However, in the absence of other influences, a material cause does not spontaneously transform itself into something else. In an ordered Universe, the transformative process must operate according to constraints that we characterize by laws. The intersection of the constraints inherent in reality (that we characterize as Laws of Nature) and the morphology of a natural system result in a specific structure of constraints on that system, often characterized as a Law of Behavior specific to that system. These describe the constraints on the behavior of the input, the process, and the output. Aristotle called this constraint *efficient cause.* As a popular example, it is often said that the efficient cause of a house is the "builder's know-how." In a linear engineering system, the transfer function characterizes the efficient cause. It is important to note that "efficient cause" is not to be confused with the conventional concept of "efficiency"; the two terms refer to entirely different concepts.

To account for the specific form of a specific output, there must be a constraint on efficient cause. That constraint is *formal cause.* This is an abstraction of the output, rather than the output itself. "The form is the definition" [23]. For example, if the effect of a process is to produce a house, then the formal cause is the quality of "houseness," not the house itself. In the house analogy, it is often said that the formal cause of a house is the blueprint, the representation of the specific form of the house. The form or definition is seen in the parameters that would represent properties of specific components (such as inductance and capacitance values) in the transfer function and would affect specific constraints on the output of the particular system (such as resonant frequency) [24]. For example, the value of the coefficients of a finite impulse response (FIR) filter would be a formal cause and would be visible in specific consequences such as the actual location of the passband.

By far the most controversial of the Aristotelian causes is final cause, the answer to the question, "What purpose is the effect for?" For man-made productions, this may not be a big problem. A servo achieves whatever purpose its designer chooses; beyond that, it has no effect on final cause and is unaffected by it. In the case of naturally occurring mechanisms, such as planetary orbits, questions of final cause are disallowed. The perfectly mechanistic Universe of classical physics is not for anything; the Universe is simply a historic accident, unfolding without goal or purpose. The crucial point is that no mechanism, natural or man-made, produces its own final cause.

7.2.3 WHY MACHINES DO WHAT THEY DO

To appreciate how causation works in traditional engineering systems, consider the causal entailments illustrated by the linear servomechanism in Figure 7.2. The *effect* or output of the servo is characterized by $Y(S)$. What causes $Y(S)$? What the servo does is to transform $X(S)$ into $Y(S)$. One of the causes is material cause, $X(S)$, a set of present and previous states also known as input or initial conditions. When engineers speak of "causal systems," they are referring only to material cause. There is feedback in the system, but it occurs only on the level of material cause. It is closed to material cause. In other words, material cause is an integral part of the process.

Obviously, the transformation of $X(S)$ into $Y(S)$ does not proceed willy-nilly. The process is constrained or entailed, by $Y(S) = H(S) X(S)$. For a linear time invariant case, $H(S)$ is technically termed the *transfer function*. The transfer function is a special instance of a Law of Behavior or dynamical law, a description of how the mechanism constrains the transformation process. The Law of Behavior (here characterized by $H(S)$) is another answer to the question of why $Y(S)$ occurred. In a linear system, the transfer function characterizes the *efficient cause*.

The transfer function is determined by the interaction of the physical morphology of the servo and the constraints inherent in reality that are often characterized as the Laws of Nature (Maxwell's equations, Newton's Laws of Motion, and so on). Efficient cause forces behavior by constraining the possible allowable behaviors. Crucially, the constraint imposed by the efficient cause of any natural process *depends on the physical morphology of the process*.

It is worth reiterating the distinction between formal and efficient cause. To change the properties of a process by changing its morphology, and hence its dynamical laws is to update the efficient cause. To tune or modulate the parameters (diffusion rate, capacitance, etc.) of a process while leaving the morphology unchanged is to update the formal cause. This distinction is important in interpreting biological behavior.

A very simple concrete example of causation in machines is shown in Figure 7.3. The effect is the output V_2. The material cause is V_1. Efficient cause is the transfer function $H(S) = S^2LC/(S^2LC + 1)$, where S is a property of the material cause, and the values of parameters L and C represent formal cause. There is no final cause.

A consequence of the absence of final cause is that there is no entailment of the properties of the parts due to the influence of the whole. In the absence of final cause, the influence of the parts

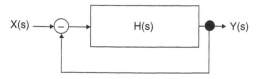

FIGURE 7.2 A linear servomechanism. $X(S)$ is the representation of the input signal. $Y(S)$ is the representation of the output signal. $H(S)$ is the representation of the "transfer function," a mathematical expression that completely characterizes the linear servomechanism. The "–" sign in the circle indicates negative feedback; a replica of the output signal is coupled in parallel but out of phase with the input signal.

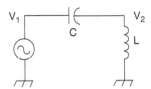

FIGURE 7.3 A series electrical circuit. The circle with a wave inside represents a voltage source whose value is V_1, the input. Element C is a lumped capacitance. Element L is a lumped inductance. The "three-pronged" symbol is ground. V_2, the voltage across the inductor, is the output.

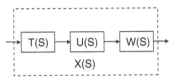

FIGURE 7.4 A linear system made up of linear subsystems. $T(S)$, $U(S)$, and $W(S)$ are the transfer functions subsystems in series. From the perspective of the world "outside the box," a composite transfer function $X(S) = T(S)\,U(S)\,W(S)$ is an exact equivalent.

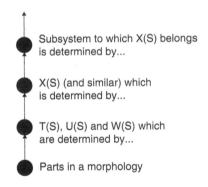

FIGURE 7.5 A linear hierarchy of linear systems. Transfer function $T(S)$ is determined by the properties and morphological layout of its internal components; likewise for $U(S)$ and $W(S)$. The transfer function of $X(S)$ is determined by the transfer functions and morphological layout formed by the subsystems. The system characterized by $X(S)$ and similar components constitute a larger system, which in its turn might be a subsystem in an even larger system. In principle, this regress of systems serving as subsystems of larger systems can increase without limit. Significantly, the properties of the larger system (represented by its transfer function) are completely determined by the properties its parts (represented by their transfer function), but the properties of the parts are unaffected by their membership in the whole. Abstractly, the transfer functions of the subsystem are unaffected by the context provided by the larger system of which they are members.

is presumed to entail all of the properties of the whole. An example is shown in Figure 7.4. We have three subsystems, whose efficient causes are fully characterized by $T(S)$, $U(S)$, and $W(S)$. Morphologically, they are connected in series, and constitute the parts of a larger system whose efficient causes is fully characterized by $X(S) = T(S)U(S)W(S)$.

This relationship is typical of the principle of "upward causation" illustrated in Figure 7.5. Efficient cause of machines is entailed in a linear hierarchy. The system characterized by $X(S)$ is fully entailed by subsystems characterized by $T(S)$, $U(S)$, and $W(S)$ in a relationship governed by their morphological relationship. The subsystems characterized by $T(S)$, $U(S)$, and $W(S)$ are fully entailed by subsystems of relationships governed by their morphologies. In the direction of increasing scale, the system whose efficient cause is characterized by $X(S)$ can serve as one of many subsystems in an even larger system, and the efficient cause serves as an entailment on the properties of that larger system.

The process is context independent. There is no provision for the possibility that $T(S)$, $U(S)$, or $W(S)$ might be influenced by their participation in $X(S)$. Neither is there any possibility to account for the influence of context by arbitrarily patching "context terms" into the transfer function. Such terms would represent the influence of more parts on the context, but not the influence of the context on the parts. One might argue that context is a part, but the influence of context on the efficient cause is to update the entire morphology of a process. As the next section indicates, it is not reasonable to treat it as patching in another part.

7.2.4 WHY EFFICIENT CAUSE IS WHAT IT IS

If we ask where the efficient cause of the system in Figure 7.2 came from, the answer is that the efficient causes of the subsystems are what they are, and by interacting in the morphology of the system they give rise to the overall efficient cause. The morphology is externally entailed, and typically no more is said about it. In the case of a man-made system, one might say that an engineer, knowing the laws of nature, synthesizes a morphology, which, constrained by the laws of nature, must lead to an efficient cause that is described by $H(S)$. A mechanism need not be designed by the Hand of Man in order to have an efficient cause. In the case of a naturally occurring mechanism, such as a planetary orbit, a particular morphology is given (why or how it is given is irrelevant), and that morphology, being governed by the Laws of Nature, produces an efficient cause. In either case, whether entailed by the Hand of Man or by some invisible hand in nature, a crucial property of mechanisms is that morphology, which gives rise to efficient cause, is externally entailed. In that sense, mechanisms are open to efficient cause.

Clearly, patching in more parts to represent context is unrepresentative of the influence of context. However, might we not be able to represent the influence of context by using a mechanism to entail another mechanism? In this case, the causal relations would be nested in a hierarchy. In Figure 7.6, at the lower level of the hierarchy, two transformations occur. Material cause $X(S)$ is transformed into effect $Y(S)$ through a process whose constraint is characterized by efficient cause $H(S)$. Material cause $X(S)$ is transformed into effect $Z(S)$ through a process whose constraint is characterized by efficient cause $J(S)$. However, there is also a metalevel transformation. Suppose some process in the environment could reach inside $H(S)$ and rearrange its morphology, converting it to the morphology that corresponds to $J(S)$. In that case, the process $H(S)$, which serves as an efficient cause on the lower level, also serves as a material cause on the metalevel. Likewise, the process $J(S)$, which serves as an efficient cause of a different process on the lower level, becomes an effect of the metalevel process. The transformation of metamaterial causes $H(S)$ into metaeffect $J(S)$ is constrained by a metaefficient cause.

As a crude example, suppose the transformation $Y(S) = H(S) X(S)$ (illustrated in Figure 7.3), where $X(S) = V1$, $Y(S) = V2$, and $H(S) = S2LC/(S2LC + 1)$. Suppose the metatransformation is performed by a circuit-rebuilding robot, and that the robot updates the morphology that underlies $H(S)$, by connecting an additional inductor L1 in parallel with C. The updated efficient cause is represented by transfer function $J(S) = (SL + S3LL1C)/(SL1 + SL + S3LL1C)$. In addition, the robot that entails $J(S)$ by updating $H(S)$ is constrained by its own externally entailed efficient cause.

At first glance, it appears that the nesting of Figure 7.6 solves the problem of context dependency. However, it has merely moved the problem up a level. If we look at Figure 7.6 and ask what efficiently caused the efficient cause, $J(S)$, the answer is that there exists an efficient cause at a metalevel. If we ask what caused the metaefficient cause, it is reasonable to suppose that there is

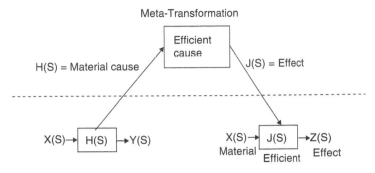

FIGURE 7.6 Entailment of efficient cause in mechanisms. In Aristotelian causation, material cause is transformed into effect. In a linear system, causation results in the transformation of a material cause represented by $X(S)$ into an effect represented by $Y(S)$. That transformation proceeds in a constrained manner; $H(S)$ completely describes that constrain in a linear system. The constraint here characterized by $H(S)$ is the efficient cause that entails the transformation. The constraining influence is modulated by properties characterized by the parameters of the transfer function; that modulating influence is the formal cause. There is no final cause in linear systems. If we can reach inside a linear system and rearrange the morphology of its parts, we can change the characteristics of the efficient cause; this is represented as a change in the transfer function. That can be seen as a transformation on a metalevel. In the entailment of an efficient cause, we can think of the "efficient cause" (viewed at the low level) characterized by $H(S)$ as the material cause at the metalevel. Likewise, we can think of the updated "efficient cause" (viewed at the low level) characterized by $J(S)$ as the effect at the metalevel. As on the lower level, the transformation at the metalevel is constrained, and that constraint is the metaefficient cause.

yet another level of the hierarchy, and that the metaefficient cause is entailed at a meta–meta level. Having seen this much of the pattern, a new problem immediately arises: how do we account for context dependency but avoid an infinite regress of efficient causation in a finite physical process?

It was to break that infinite regress that Rosen first discovered the conditions that enable the existence of an entailment structure that is "closed to efficient cause" [25]. In Figure 7.7, B represents the process of metabolism, in which material cause is taken from the environment and integrated into the substrate of an organism. f is the efficient cause of B. Also note that Φ is the efficient cause of f. Since the process is open to toxins as well as nutrients, f can be degraded by the environmental insults to which it is exposed. To preserve the integrity of f, a repair process is included; it has f as its effect, and Φ as its efficient cause. A hierarchy develops in which Φ efficiently causes f, which efficiently causes B, a straightforward linear hierarchy. The problem is that Φ is no less vulnerable to environmental insult than B and f. However, nothing fundamentally precludes the possibility of B's efficiently causing Φ, another linear hierarchy. Thus, when we look at the whole relationship, Φ efficiently causes f efficiently causes B efficiently causes Φ. Rosen avoids the infinite regress by implementing closure to efficient cause, leading to a hierarchical closed loop.

This is a key concept for understanding all that follows. In processes of life and mind, the distinguishing feature of Rosen complexity (equivalent to autopoiesis) [26] is a *hierarchical* closed loop of causal entailment [27]. It is both a loop and a hierarchy of metaefficient causes. The key limitation on the operation of a machine is that "there can be no closed path of efficient causation in a mechanism" [28]. Many theorists, such as Bateson, recognize the necessity of circular causation [29] and of hierarchical entailment [30] in complex processes. What Rosen showed is that both circularity and hierarchy are necessary in the same process *simultaneously*.

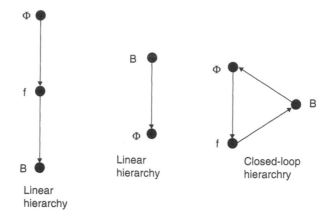

Linear
hierarchy

FIGURE 7.7 Linear and circular hierarchies. Consider hierarchy as a concept of "containment." If *B* is contained in f, then *B* is lower in the hierarchy than f. Likewise, if f is contained in Φ, then f is lower in the hierarchy than Φ. If Φ is contained in *B*, then Φ is lower in the hierarchy than *B*. In the closed-loop hierarchy, if we start from Φ, we see that *B* is contained in f is contained in Φ; in other words, *B* is contained two layers deep in Φ. However, if we start from *B*, we see that Φ is contained in *B*. Intuitively, it appears that the closed loop is logically incoherent. However, Aczel's proof of the existence and uniqueness of a solution to Ω = {Ω} demonstrates that, counterintuitive as it might be, a circular hierarchy is logically coherent.

7.2.5 ENTAILMENT OF DOWNWARD CAUSATION

The hierarchical loop leads to the radical departure from traditional engineering systems. It still has upward causation as a linear hierarchy does. In both graphs, in Figure 7.8, the efficient causes of the behaviors of the parts at Φ are combined to yield an efficient cause of bigger subsystems at f. However, when we close the loop in the right-hand graph, something else occurs. f is the efficient cause of *B*, which is the efficient cause of Φ. In other words, by traversing the loop the long way around, we get an effect similar to imposing an edge directed from f to Φ; f, the larger system, exerts a causative influence on Φ, the smaller components. In other words, the loop affords simultaneous upward (parts influencing the whole) and downward (whole influencing the parts) causation.

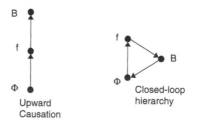

FIGURE 7.8 Simultaneous upward and downward causation. In a linear hierarchy of efficient causes, the path of causation is unidirectional; if Φ efficiently causes f, which efficiently causes *B*, we see that *B* has no influence on Φ. However, with a closed loop, despite the fact that we can only traverse the loop in one direction, we can see both influences. Starting from Φ and moving the long way around, we see that Φ influences *B*. However, starting from *B*, and moving the short way around, we see that *B* influences Φ. Thus, if traversing the loop between two nodes, the long way around represents upward causation (the influence of the parts on the whole), then traversing the loop between same two nodes the short way around represents downward causation (the influence of the whole on the parts).

Since they entail their own structure of upward and downward causation, physical processes possessing a closed-loop hierarchy of efficient cause are called *endogenous processes*. They have the quality of making themselves up as they go along. "Endogeny" is a conventionally used medical term that describes the property of a system that grows from within itself, or makes itself up as it goes along [31]. The same term is used for a similar concept in economics [32]. Some biologists consider this property to be a defining feature of living systems. For example, Margulis and Sagan [33] flatly state that a system is alive if and only if it is autopoietic. They identify autopoiesis as the process of life making itself; they capture the essential point, although they simplify a much richer concept [26].

The endogenous loop imparts a downward causation (or constraint of the whole on its components) that acts as an internally generated final cause. Final cause imparts an identity on a process; it determines its own goals acts in a manner to preserve itself. In other words, the closed-loop hierarchy produces an integrated whole that has an identity that can be perturbed by, but remains distinct from, its ambience. The endogenous causal entailment structure holds the process distinct and intact.

Based on approximately a half century of experimental observation, Freeman says that this is what brains do. Their endogenous nature requires a closed loop in their causal linkages. Freeman calls this circular causality [34]. Based on his observations of brain behavior, Freeman says that through neural activity, goals emerge in brains, and find expression in goal-seeking behavior [35]. He summarizes a lifetime of observation of brain activity, both in primitive animals and in humans, with the pithy observation that "brains are hypothesis driven."

Abduction of hypotheses is not merely the hallmark of what brains do; it is the thing that brains do that computers cannot do [36]. For a system to establish, update, and seek its own goals requires that it be able to abstract semantic meaning from streams of symbols. The process by which it does so is abduction, the abstraction of hypotheses from sensory data with far more effectiveness than random guessing. For a hypothesis to be useful, the system must test it. Again the process is semantic. Semantic behavior does not occur in mechanisms or algorithms. They have no internal semantics, and any data they use are strictly syntactic [37]. In fact, the ability to abstract meaning from a stream of signs and symbols is an externally observable effect that distinguishes endogenous processes from mechanisms such as servos and algorithms.

This concept of final cause differs from Aristotle's notion of Final Cause in that it is internally entailed and is updated due to perturbations of material and formal cause. In other words, final cause does not impart strict finality more than efficient cause imparts strict efficiency. Nevertheless, internally entailed final cause corresponds to Aristotelian Final Cause in some key particulars. First, it does impart identity or wholeness to the process. More crucially, it induces goal-seeking behavior for the purpose of preserving that identity; this corresponds to the teleological property of Aristotelian Final Cause. Also, it represents downward causation or the constraint of the whole on the properties of the components.

7.3 DO HIERARCHICAL LOOPS OF ENTAILMENT MAKE SENSE?

The concept of a hierarchical loop of entailment is not especially new. It is one of the oldest concepts in rational philosophy. It has been dismissed by philosophers both ancient and modern as being logically incoherent. In set-theoretic logic, it is seen as leading to Russell's Paradox, the incoherent notion of the set of all the sets, a construct that provably both is and is not a member of itself [38]. Objections to the idea that a system could be the cause of its own causes go back at least as far as Aristotle. "And so, in so far as a thing is an organic unity, it cannot be acted on by itself; for it is one and not two different things" [39]. However, the fallacy in both Russell's and Aristotle's objections to endogenous entailment is based on the presupposition that systems are syntactic things rather than semantic processes.

7.3.1 Impredicatives: Answering Russell's Paradox

Recall that one of the uses of the MR is that we can gain insights about a process by asking questions about a congruent model of the process. Is it possible to construct an abstract model that is congruent with an endogenous process? If so, and if the model is logically coherent, might we not reasonably conclude that the process being modeled is also coherent?

One solution to Russell's Paradox is the Foundation Axiom of set theory, the notion that no set may include itself as a member. This necessarily prohibits the notion of a set that contains all the sets as members. One of the consequences of the Foundation Axiom is a ban on ambiguity. "By formalizing the theory, the development of the theory can be reduced to form and rule. There is no longer ambiguity about what constitutes a statement of the theory, or what constitutes proof in the theory" [40].

However, ambiguity is not incoherence, and the Foundation Axiom is too restrictive a means of assuring noncontradiction. Coherence can be obtained with the less-restrictive antifoundation axiom (for the genuinely curious, the simplest form of the antifoundation axiom says "every tagged graph has a unique decoration"). The detailed discussion of this definition is beyond the scope of this chapter, but is readily accessible in Barwise and Etchemendy [41]. The more general hyperset theory afforded by the antifoundation axiom is provably no less coherent than conventional set theory; sets turn out to be a degenerate case of hypersets.

In set-theoretic jargon, the prototypical hyperset, Ω, is the set that contains itself as a singleton. More casually, Ω is that object, which solves $\Omega = \{\Omega\}$, where the brackets $\{\}$ denote a set. This definition is no more a conceptual stretch than the definition of i as that object which solves $i^2 = -1$. The existence and uniqueness of Ω is proven; there is one and only one object that satisfies $\Omega = \{\Omega\}$ [42].

$\Omega = \{\Omega\}$ is not a circular definition, or a claim that a thing merely is what it is. It is an impredicative definition. The identity of Ω is established by the specific constraint on its relationship with itself. The conventional meaning of the term "impredicative" is given by Kleene. "When a set M and a particular object m are so defined that on the one hand m is a member of M, and on the other hand, the definition of m depends on M, we say that the procedure (or the definition of m, or the definition of M) is impredicative. Similarly, when a property P is possessed by an object m whose definition depends on P (here M is the set of the objects, which possess the property P). An impredicative definition is circular, at least on its face, as what is defined participates in its own definition" [43].

Note that Kleene does not dismiss impredicativity because it is circular "on its face." It has depth below the face; the object on one level refers to its definition on the other, and there is a specific constraint on how that reference operates across the two levels. The distinguishing feature of an impredicative definition is the hierarchic closed-loop constraint. Kleene shows that the constraint in the impredicative definition can be used to define the least upper bound on a set of real numbers.

The easiest way to gain a sense of hypersets is to consider the graphical representation of sets as in Figure 7.9. A set is a sort of abstract container that can define a hierarchy of containmnet. 1 is a number; it is not a set. $\{1\}$ is the set whose only member is the number 1. Sets can be nested; $\{\{1\}\}$ is the set whose only member is the set whose only member is 1.

The hierarchy defined by set nesting is represented by a directed graph. A terminal node in the graph represents a member of a set, and not usually a set itself. A nonterminal node represents a set; directed edges (signifying the property, "is a member of") from the node that represents the set are connected to all the nodes that represent members of the set. The leftmost object in Figure 7.9 is the set containing three members; those members are the numbers 1 and 2 and the two-element set whose members are the letters a and b. The same members may appear at different levels in the hierarchy. The next leftmost object in Figure 7.9 is another set containing three members; those members are the numbers 1 and 2 and the two-element set whose members are the numbers 1 and 2.

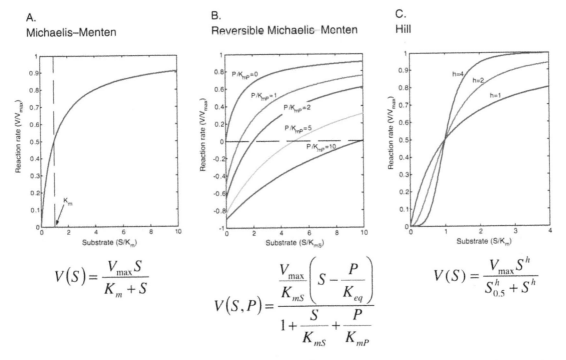

COLOR FIGURE 2.5 Examples of commonly used kinetic rate laws. Panels A, B, and C give the functional form of the Michaelis–Menten, reversible Michaelis–Menten, and Hill rate laws and graphs of reaction rate vs. substrate. The reversible Michaelis–Menten reaction schemes yield negative rates when $P/K_{eq} > S$.

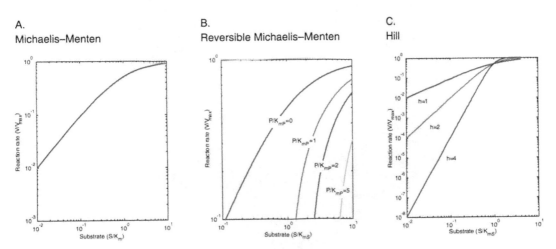

COLOR FIGURE 2.6 Examples of kinetic rate laws plotted in logarithmic coordinates. Nearly linear behavior is often found over wide ranges of substrate concentration. Linear regions in logarithmic coordinates indicate power law behavior in linear coordinates.

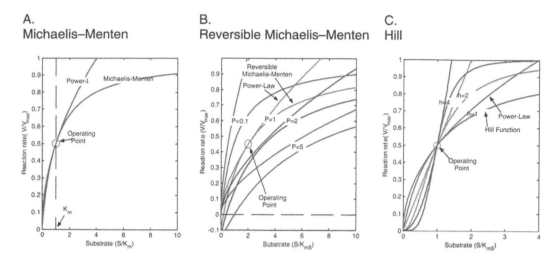

COLOR FIGURE 2.7 Power law approximations to the Michaelis–Menten, reversible Michaelis–Menten, and Hill rate laws. The power law approximation fits the given rate law exactly at the operating point and typically provides a good approximation to the rate law about that point.

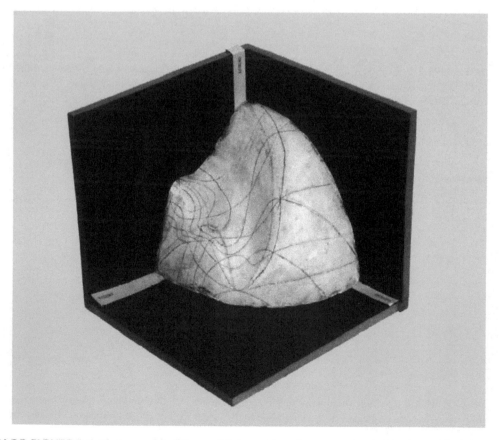

COLOR FIGURE 9.1 Plaster model of the equilibrium states of water constructed by James Clerk Maxwell and sent as a present to Josiah Willard Gibbs. (©The Cavendish Laboratory, University of Cambridge. Reproduced with permission.)

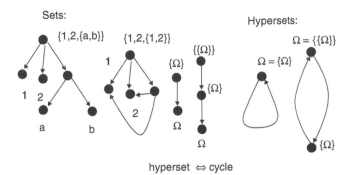

FIGURE 7.9 Sets and hypersets. In set notation, containment is represented by bracket notation or directed graphs. In the leftmost figure, the set {a, b} is contained in the set {1,2,{a, b}}. In traditional set theory, the same member might appear at multiple levels of the containment hierarchy, as shown in the first figure from the left. We can see hierarchies of sets as nested containers; for example set {{Ω}} represents a container that contains a container that contains Ω. In the rightmost figure, the structure of containment is folded into a loop; Ω itself represents a container that contains a container that contains Ω. In hyperset theory, it is easily proven that the structure Ω = {Ω} is identical to the rightmost structure; this is called the prototypical hyperset. A set is a hyperset if and only if its graph contains a cycle; it includes at least one node such that if one follows the directed edges, one eventually returns to the node.

The two objects in the middle of Figure 7.9 are regular sets. Ω is a Greek letter; it is not a set. {Ω} is the set whose only member is the letter Ω. {{Ω}} is the set whose only member is the set whose only member is the letter Ω. If the constraint Ω = {Ω} is imposed on the two-node one-edge graph forcing the two nodes in the regular set to become identical, the result is a hyperset. The constraint need not be imposed only on nodes at adjacent levels. The constraint Ω = {{Ω}} on the three-node two-edge graph forces the top and bottom nodes in the regular set to become identical, turning the regular set into a hyperset.

The graphical representation provides a convenient way to distinguish a set from a hyperset. If the graph includes a set of edges that form a closed cycle, anywhere in the graph, then the graph represents a hyperset. If the graph includes no set of edges that form a closed cycle, anywhere in the graph, then the graph represents a regular set.

Although the first impulse of mathematicians was to attempt to define impredicative structures out of existence, they could not do so. In fact, it did not even make sense for them to try; impredicative structures are neither incoherent nor vacuous as was once feared [44]. For example, the fundamental wavelet equation $\phi(t) = 2 \sum_n h_0(n) \phi(2t - n)$, a mathematical principle at the heart of modern signal processing engineering has an inherently and necessarily impredicative definition [45]. Remarkably, impredicative structures have been found to be indispensable to the description of the foundations mathematics itself [43].

There is a temptation to confuse impredicativity with recursion. However, they are not the same. Recursive definition of a class requires a specific instance of the class and a constraint upon the relationship between members of the class. Arguably, it is a more constrained version of the notion of impredicative definition. Impredicative definition of a class requires only a constraint upon the relationship between members and the class as a whole.

These definitions indicate a practical difference between recursion and impredicativity. Recursion is characterized by a starting point and finite bottom. The defined bottom corresponds to the initial conditions in a differential equation [46]. A recursive algorithm repeatedly calls lower-lever versions of itself until it "hits bottom." At the bottom of the recursion, it discovers the state vector characterizing the initial conditions, each step upward through the constraint of the recursive hierarchy entails a state transition. The fully evolved state vector is returned by the highest level

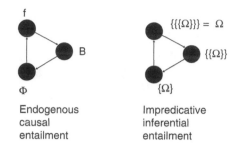

Endogenous
causal
entailment

Impredicative
inferential
entailment

FIGURE 7.10 Similarity of ontological structure of "closure to efficient cause" to the epistemological structure of impredicativity. The Rosen structure of efficient cause leads to the left-hand figure. The efficient cause of metabolism (f) is entailed by (or contained within) the efficient cause of the repair process (Φ), which is entailed by (or contained within) the efficient cause of the repair process, the replication process (B), which is entailed by (or contained within) the efficient cause, the metabolism process (f). In the hyperset $\Omega = \{\{\{\Omega\}\}\}$, exactly the same closed-loop hierarchy of containment is seen.

of the recursion to the function that called the recursion. In contrast, an impredicative process has neither a "bottom-out" condition nor any state vector at all. The stability of the process depends on the constraint on relationships between the dynamics at both the higher and the lower levels of the process.

If used carelessly, impredicatives can lead to incoherence, but they are not inherently incoherent. Thus, hyperset theory is a far less restrictive solution to Russell's paradox than that proposed by Russell. An axiomatic definition of "set" and a less restrictive axiomatic definition of a "proper class" that may include the "class of all the sets" are sufficient to avoid both Russell's paradox and his infinite regress of "types" [47].

Is endogeny incoherent? Consider the relationship between endogenous causes in the diagram in the leftmost graph of Figure 7.10. The causal entailments map to the inferential entailments of the three-node hyperset $\Omega = \{\{\{\Omega\}\}\}$, as shown in the rightmost graph of Figure 7.10. Similarly, the inferential entailments of the three-node hyperset map to the causal entailments. The entailment structures are congruent, and an MR exists between the two processes. The hyperset is coherent, and there is no reason to suppose that the corresponding structure of causal entailment should be incoherent.

7.3.2 FUNCTION: ANSWERING ARISTOTLE'S OBJECTION

Aristotle's objection to a process entailing its own causes was based on the notion that the process is a nonfractionable object. Although he saw that final cause imparts function, he did not take into account the possibility that the process might serve as a context in which subprocesses might have their own functions. Thus he did not consider the possibility that the functional components of a process might perturb each other's causal entailment structures, and consequently perturb the causal entailment structure of the whole process.

To appreciate the properties of a functional component, suppose we have a perceptibly heterogeneous system [48]. One part has different features than other parts. If we leave the system alone, it will exhibit some sort of behavior. If we remove or change a part, we get a change in behavior of the overall system. Crucially, the change in behavior that we get is unlikely to be the effect that we would predict by merely subtracting the behavior of the part from the overall behavior of the unmodified system. The effect of changing or removing one part is to replace the original system with a new system. The function of a part is the discrepancy in behaviors between the original system with the full complement of parts, and the new system with one part removed.

These parts must not be confused with the directly summable parts of a reducible system. Rosen avoids this confusion by offering a new term for a part that embodies function. He calls it *a functional component*. In this terminology, the difference between the two systems defines the component, and the difference between the two behaviors defines the function. In a complex system, a component with a function is the unit of organization.

A functional component is context dependent. It has inputs, both from the larger system of which it is a component and from the environment of the larger system. It also has outputs, both to the larger system and to the environment. If the environment, A, changes, then the function of the component, B, changes. A can typically be described by a family of mappings that carries a set (the domain X, where $x \in X$) to another set (the codomain Y, where $y \in Y$), such that, $y = a(x)$, or more formally, A: $X \rightarrow Y$. B can typically be described by another family of mappings that carries a set (the domain U, where $u \in U$) to another set (the codomain V, where $v \in V$), such that, $v = b(u)$, or more formally, B: U $\rightarrow$ V.

The functionality, F, of the functional component can be described as a mapping that maps a domain set of mappings (A, where mapping $a \in A$) to a range set of mappings (B, where mapping $b \in B$), such that $b = f(a)$, or F: A $\rightarrow$ B. The concept of a mapping that maps one set of maps to another set of maps is not unfamiliar to engineers. This is precisely what happens with a symbolic Laplace Transform.

A functional component differs from the idealized particle of Newtonian physics. The particle's identity (defined in terms of parameters such as mass) is unaffected by context. A particle does not acquire new properties by being associated with other particles. A functional component's context dependency requires that its identity be tied to its function in a larger system. Although it is a thing in itself, it both acquires and imparts new properties as a consequence of association with other functional components. Since the functional component provides a means of perturbing the entailment structure of its context, it is not the case that a process "is one and not two different things." It is a process with a function that can be perturbed by the operation of its functional components.

7.4 WHY ORGANISMS ARE NOT MACHINES

From the time of Descartes to the present day, it has been supposed that an organism is nothing but a particularly complicated machine. Csete and Doyle recount the modern version of the concept, observing that an organism is nothing but a mechanism with a hierarchy of modules, and feedback at various levels of the hierarchy [49]. They conclude with the following sage advice. "Biologists and engineers now have enough examples of complex systems that they can close the loop and eliminate specious theories. We should compare notes."

Perhaps engineers and mathematical biologists should compare notes. If they were to do so, biology would have much more to say to engineering than engineering would have to say to biology about new concepts. Engineering is rooted in the physical tradition of "upward causation," and denies the influence of the whole on the properties of the parts. For at least the past 60 years, it has been realized that this picture is incomplete. Speaking of the living process, Schrödinger says, "We must be prepared to find a new type of physical law prevailing in it, or are we to term it a non-physical, not to say a super-physical law?" [50].

The concept of downward causation resulting from a hierarchical closed loop of efficient cause is just such a superphysical concept. Traditional bottom–up causation is the degenerate case that arises when the hierarchical loop of causation is very large, the influence of the whole upon the parts is weak, and a section of the hierarchical closed loop can be reasonably approximated as a linear hierarchy. However, this is the *degenerate* case. In all but the most trivial biological processes, the influence of the closure of the loop of efficient cause cannot be ignored. Both the character of the loop and the character of modularity in biological processes are profoundly different from those in traditional engineering systems.

7.4.1 BOTH THE LOOP AND THE HIERARCHY ARE CRUCIAL

To appreciate the difference between endogeny and feedback, consider the two processes shown in Figure 7.11. The feedback servomechanism has no context dependency except the external entailment of its dynamical law of behavior (or transfer function if it is linear). That entailment is typically ignored; engineers typically assume that the transfer function is what it is without asking why. The input or material cause is not part of the context; it is part of the process. The input, transformation, and output are considered three parts of the process, isolated from the rest of the world. There is a self-referential loop, but it is nonhierarchical, operating strictly at the level of material cause.

The feedback servomechanism is isomorphic to recursion. The essentially nonhierarchical character of the recursion is seen in the fact that any program that calls the recursive algorithm will both enter and exit the recursive call at exactly the same nesting level. An essential point is that any such process is computable; it can be fully described by a recursive algorithm.

An organism, or a process entailed by an endogenous loop, open to material cause [53]. In other words, the input at the level of material cause is part of the context, not part of the process. In the process of metabolism, such inputs are integrated into the physical substrate of the organism. Since one effect of material cause is an update of the morphology of the subprocesses, updating the morphology incidentally updates the efficient cause. The updating of efficient cause is constrained by an efficient cause at a higher level, and every efficient cause in the process is the effect of another efficient cause within the process. In other words, the process is closed to efficient cause. The closed loop is at the level of efficient cause. Since each efficient cause serves as the context of the efficient cause that it entails, the self-referential loop is inherently hierarchical. This is the crucial difference between a feedback servomechanism and an endogenous process. The former is a flat loop of material cause, and the latter is a hierarchical loop of efficient cause, in which each efficient cause is contained within another efficient cause.

The endogenous loop is not isomorphic to recursion. Instead, it is homomorphic to impredicativity. The hierarchical character of the impredicativity has the effect that any program containing such a structure will deadlock, and is forbidden in computation [52]. An essential point is that any such process is incomputable. Except in the rare case that a bound on truncation error can be derived from an impredicative definition (as is the case for the Fundamental Wavelet Equation), there is no recursive algorithm that characterizes impredicativity.

If traversing the loop of efficient cause in one direction follows the entailment of efficient cause in the bottom–up sense, then traversing the same loop in the other direction follows the entailment of efficient cause in the top-down sense. This is the same as the influence of final cause, the influence of the whole upon the properties of the components. This imparts to the process self-determined identity, and self-determined goal seeking behavior to preserve that identity. There is no top–down causal entailment in a servomechanism.

In both an organism and a mechanism, the behavior of the process can be changed by *modulating* the formal causes. Recall that formal cause is a constraint upon efficient cause that results in the particular form of the outcome. In the case of a mechanism, this corresponds to properties of components characterized as parameter values. In the case of organisms, the genome is one of the formal causes [24]. In both the organism and the mechanism, both the environment and the internal working of the process can modulate the formal cause.

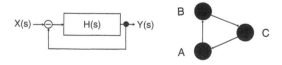

FIGURE 7.11 Two fundamentally different kinds of closure. The loop in a servomechanism is a nonhierarchic loop of material cause. The loop in Rosen complexity is a hierarchic loop of efficient cause.

When viewed from the perspective of causal entailment, the closed loop of endogeny and the feedback loop of a servomechanism are not merely different, they are practically orthogonal to each other. The servo is closed to material cause, open to efficient cause, and presumed to be unaffected by final cause. The feedback loop is nonhierarchical, *operating at the level of material cause*, and is isomorphic to recursion. It is computable. The endogenous loop is open to material cause, closed to efficient cause, and both entails and is entailed by its own final cause. The endogenous loop is hierarchical, *operating at the level of efficient cause*, and is homomorphic to impredicativity; thus it is incomputable.

7.4.2 AMBIGUITY IS INCOMPUTABLE

To appreciate that impredicativity is incomputable, it suffices to realize that ambiguity is absolutely forbidden to computation [53]. This is the reason that programs with hierarchical loops deadlock. They present ambiguous information to the next step in the algorithm, and the algorithm halts, waiting for disambiguating information that never arrives.

Both endogenous natural processes and impredicative abstractions have this key property not shared by the mechanisms commonly studied by traditional science or their formalistic analogs; they have ambiguous internal semantics. Speaking of brains Von Neumann says, "There exist here different logical structures from the ones we are ordinarily used to in logics and mathematics" [54]. In trying to find a starting point for his development, Von Neumann realized that he was being arbitrary. "By axiomatizing automata in this manner, one has thrown half the problem out the window, and it may be the more important half" [55].

When Von Neumann hoped he could throw semantics "out the window" but feared he could not, it is clear that he had in mind something other than the common notion that semantics is a simple this-for-that substitution of one sign for another. It is also clear that he was not referring to a property that applies only to natural languages. Instead of thinking of semantic meaning as substitution of signs, it is more instructive to think of the meaning of a sign as a relationship between a sign and proposition where the entailment of the proposition depends on the context-dependent relationship between a sign and its referent.

To appreciate how the relationship between a sign and its referent entails a proposition, consider a logical structure L, consisting of a set of sentences $\{this, that_1, \ldots, that_n\}$. The propositional demonstratives, $T = \{this, that_j| 1 \leq j \leq n\}$, refer to something in their situation, and are the domain of a mapping, $c: T \rightarrow Q$. Propositions, $Q = \{p, q_j| 1 \leq j \leq n\}$, are the codomain of the mapping. Given a situation, s, one can use a sentence φ to express a proposition p about s; within this structure, the propositional demonstrative **this** also refers to p. The *context*, c_s, of sentence φ is the pair $c_s = <s,c>$ consisting of situation, s, and mapping, c [56].

A statement Φ is a pairing of context, c_s, with sentence, φ. $p = Exp(\Phi)$ is a proposition expressed by Φ. $Val(\varphi)$ is a mapping from sentences to propositions defined by a particular logical structure L. The proposition $Exp(\Phi)$ is obtained from $Val(\varphi)$. There is a uniqueness theorem that says that for any sentence φ in context, c_s, there exists a unique proposition, $p = Exp(\Phi) = Exp(\varphi, c_s)$. $Exp(\Phi)$ is true if situation s, determined by the context, is of a particular type determined by the descriptive conventions of L, and false otherwise. (*Note*: The actual descriptive conventions, $Val(\varphi)$, can be set out in hyperset-theoretic notation. Barwise and Etchemendy show an example [56]).

Several fairly remarkable consequences flow from these ideas. They provide a coherent concept of falsehood. The Russellian concept of truth is that true propositions correspond to facts in reality. However, there are no false facts in reality that correspond to false propositions. In Austinian logic, used by Barwise and Etchemendy to ground this concept of semantics, ontological reality is what it is, and the values of *true* and *false* are strictly the properties of epistemological propositions commenting on that reality. More relevant to this discussion, the proposition, p, that corresponds to the sentence, φ, can, and usually does, include a hierarchic self-reference. Where $p = Exp(\varphi(this, that_1, \ldots, that_n), <s,c>)$, c includes the mapping of **this** to p.

The logical structure $p(\varphi) = Exp(\varphi, c_s,)$, which identifies the meaning of sentence φ, is inherently impredicative. There is no way of isolating φ on one side of the "=" sign. In other words, if the semantic meaning of a sign, φ, is a proposition, p, then the sign, φ, includes "internal semantics" [57]. The sign includes an implication that refers to its own internal structure; the structure means something to itself.

It is crucial to recognize that $\mathbf{L}$ is not a natural language. It is a logical structure that can serve as a model of utterances in a natural language. This means that there is no reason to suppose that there is a single $\varphi \in \mathbf{L}$ that corresponds to an utterance in a natural language. In fact, Korzybski notes that language semantics may require limitlessly many different instances equivalent to $\varphi \in \mathbf{L}$ to account for all the propositions implied by a natural language utterance. Terms such as 'yes,' 'no,' 'true,' 'false,' 'fact,' 'reality,' 'cause,' 'effect,' 'agreement,' 'disagreement,' 'proposition,' 'number,' 'relation,' 'order,' 'structure,' 'abstraction,' 'characteristic,' 'love,' 'hate,' 'doubt,' etc. are such that if they can be applied to a statement, they can also be applied to a statement about the first statement, and, so, ultimately, to all statements, no matter what their order of abstraction is. Terms of such a character I call *multiordinal terms*. The main characteristic of these terms consists of the fact that on different levels of orders of abstractions they may have different meanings, with the result that they have no general meaning; for their meanings are determined solely by the given context, which establishes different orders of abstractions" [58].

What is clear from Korzybski's observation is that if meaning is affected by context and if the context can change without limit, then meaning can change without limit. Ambiguity of meaning is inherent, and not an anomaly or error. The fact that a single expression might have many (even limitlessly many) different meanings, but nevertheless be properly understood in a specific context is what Barwise and Perry call "another aspect of the efficiency of language" [59].

Since ambiguity is inherent, there is no general procedure for disambiguating expressions or concepts. Perhaps the most compelling example of inherent ambiguity arises in logic itself; the concept of falsehood is inherently ambiguous [60]. A proposition is false if it is a *negation* or contains an inherent internal contradiction. It is also false if it is a *denial*, being internally coherent, but contradicting a true proposition. This does not mean that logic should have three truth values — true, negated, and denied; in all but a few pathological instances, false propositions are simultaneously negations and denials. However, it does mean that to avoid confusion, the logician must be aware of the possibility of the pathological instances, and determine from the situation which sort of falsehood is implied.

The fact that the situation *constrains* the meaning is crucial to this concept of semantics. In fact, it is these constraints that enable utterances to convey information about situations [61]. The situation determines how restrictive the constraints are. In the example given by Korzybski, there is a countable infinitude of meanings. In the more severely constrained case of falsehood in logic, there are only two meanings. In the even more severely constrained case of formalism, only one meaning is permitted. Significantly, it is not the case that ambiguity is a pathological instance of confusion arising from the careless use of language. Instead, it is the case that nonambiguity is an unnaturally restrictive instance of limiting a sign to a single meaning, so that it might conveniently be represented by an algorithm [53].

As Barwise and Perry note, these constraints are not limited to natural languages, but rather "the same basic pattern applies to all manner of meaningful items, in particular, to mental states" [61]. In other words, if a process can be put into an MR with an Austinian logical structure $\mathbf{L}$, then it has semantic properties, and those properties are defined by the constraints on the relationship between the sign and its referent.

As Bateson observes, semantic constraints have the property of nonlocality. "...the universe is *informed* by the message; and the 'form' of which we are speaking is not in the message nor is it in the referent. It is a correspondence between message and referent" [62]. There is no place where the correspondence or the constraints that govern it can be said to be located. They are everywhere, but not isolatable anywhere. "This matter of the localization of information has bedeviled communication theory and especially neurophysiology for many years" [62].

Neither nonlocality nor other properties of semantic constraint are limited to natural languages. **L** is not a natural language. It is a logical structure with its own internal semantic properties, including internal ambiguity. Assuming that we can discover entailments that enable us to map between the entailment structure in **L**, and the entailment structure of some other process, we can construct an MR between that process and **L**, and we can study the semantic properties of the process by asking questions about the semantic properties of model **L** [63].

Natural language is merely one of the processes that can be put in an MR with **L** [65]. If we investigate natural languages by modeling them with **L**, and then ask questions about **L**, quite often we do find an unsurprising result. It can happen that for one of the propositions that correspond to a comment about a sign in the natural language, there exists another sign that might be used in place of the first. In other words, "this-for-that" substitution (often mistaken for the entire scope of semantics) is a degenerate instance of the more general concept of semantics that arises from Austinian logic that *the meaning of a sign is the entailment of a proposition implied by the sign, where the constraints in the inferential entailment structure depend on the context of the sign and its referent.*

Because the context can change without limit, the meaning of a sign implied by its context can change without limit, and the task of disambiguation of a sign becomes a fools' errand. It is remarkable that logicians have tried to define impredicatives out of existence because they have inherent internal ambiguities. Those inherent ambiguities are neither a defect nor a weakness; they are a strength. As Barwise and his collaborators have shown, impredicatives such as hyperset theoretic constructs are an abstract means for representing the ambiguities that arise in endogenous natural processes.

7.4.3 CAN FUNCTION IN CONTEXT BE CHARACTERIZED BY DIFFERENTIAL EQUATIONS?

Despite the ambiguity and consequent incomputability of endogenous and impredicative entailment structures, the question often arises as to whether or not they can be represented by systems of differential equations. In particular, the question is whether or not they can be represented by a system of differential equations whose solution is a map of the form f: $R^n \rightarrow R^m$, a function that carries an n-dimensional vector of irrational numbers to an m-dimensional vector of irrational numbers. With rare exceptions, this strategy will fail.

To appreciate why, consider how context is taken into account. A differential equation is a map of the form $d{:}g \rightarrow h$, where g and h are representations of rates of change that impose a constraint on a process that is satisfied by another map of the form f: $R^n \rightarrow R^m$. One cannot simply add terms to d to account for the influence of context, since the terms represent the influence of parts rather than encompassing context. To account for the influence context on d, one would require an encompassing map e: $j \rightarrow k$, for which d is the solution. Once again, this is the first step down the road to infinite regress.

One might hope that the infinite regress could be sidestepped by the same sort of hierarchical loop as has already been discussed. In other words, suppose the process is represented by a hierarchy of maps, $d: g \rightarrow h$ solves e: $j \rightarrow k$ solves x: $y \rightarrow z$ ... solves $d: g \rightarrow h$, where each letter represents a category theoretic construct. Such a system of maps can probably be constructed. Indeed, this is precisely the sort of thing that situation semantics does.

The problem is that none of the maps are of the form f: $R^n \rightarrow R^m$. Such a map is not a constraint that is satisfied by another map. Numbers are constructible as Von Neumann ordinals from operations in set theory [65]. However, there is no means of constructing a hyperset by performing operations on regular sets. There are hyperset theoretic constructs that are orthogonal to number-based constructs, just as irrational numbers are orthogonal to imaginary numbers [66]. In other words, there are hyperset theoretic constructs that are independent of the concept of number. Thus, a category theoretic generalization of differential equations might be represented by a hierarchy of maps, $d: g \rightarrow h$ solves e: $j \rightarrow k$ solves $x: y \rightarrow z$ ... solves $d: g \rightarrow h$, and this might be represented as a loop of hyperset theoretic constructs, but nowhere in the loop would a solution fall out as lists of numbers.

7.5 WHY NOT REDUCTIONISM?

Despite the fact that we are dealing with processes whose entailment structures do not commute with the entailment structures that govern the relationships between lists of numbers, the question persists. Why do we not write a computer program to emulate the foregoing and do away with all the seeming hand waving? As already noted, there is no algorithm capable of implementing the hierarchical closed-loop entailment structure that is necessary to abduct the novel information necessary to deal with data that have context-dependent ambiguities. Ignoring the detail that ambiguity is forbidden to computing and that genuine ambiguity deadlocks an algorithm, thousands of brilliant minds have spent decades searching for an algorithm that exhibits "intelligent" behavior. In this section, we consider some of the reasons why that search has failed.

7.5.1 What Is Reductionism?

"When you measure what you are speaking about and express it in numbers, you know something about it, but when you cannot express it in numbers your knowledge about it is of a meager and unsatisfactory kind." — Lord Kelvin [67].

The above pearl of wisdom commonly appears on the opening page of scientific textbooks. It is fair to ask, why should we depart from it? After all, has not reductionism literally delivered the goods for 400 years? If that were not enough, has not the revolution in digital computing dramatically accelerated the rate at which those goods are being delivered? Why does the strategy discussed in this chapter take such a radical departure from a path that has proven to be so fruitful?

Before we consider the reasons why, it might be helpful to be clear just what it is that we are departing from. Terms like *algorithmic* and *reductionistic* are tossed around rather more casually by their users than is justified by the precision of thinking that they are supposed to imply. Just what *was* Lord Kelvin going on about, and why did he think it was so important?

For example, "algorithmic" has a precise meaning in computer science. It does not merely mean methodical. It is a process with five attributes [53]. It terminates after a finite number of steps. Most crucially, each step is unambiguously defined. It has zero or more input data. It has one or more output data. An algorithm must be effective.

That last attribute leads to the most confusion. What is it about a process that causes us to characterize it as *effective*? Whatever it is, it must be unambiguous. Otherwise it would contradict the second attribute, the ban on ambiguity. Church proposed "we define the notion ... of an effectively calculable function of positive integers by identifying it with the notion of a recursive function of positive integers (or of a lambda-definable function of positive integers)" [68,69]. Church had equated effectiveness with lambda-definability, but discovered that this was equivalent to the less awkward concept of the Turing machine. Hence, he defined *effective* as computable by a Turing machine, noting that "computability by a Turing machine ... has the advantage of making the identification with effectiveness in the ordinary (not explicitly defined) sense evident immediately" [70].

This definition of computational effectiveness as being calculable by a Turing machine was followed by the observation that nobody seems to be able to produce a calculable function that is not calculable by a Turing machine. There is hot debate in the computer science community as to whether or not the Busy Beaver problem [71] constitutes a counter-example to this observation. In any case, the Church–Turing thesis turns out to be a comment on effectiveness *as it relates to calculable functions*. "Every effectively calculable function that has been investigated in this respect has turned out to be computable by Turing machine" [69].

There is a variant on the Church–Turing thesis, which makes a rather grand claim about reductionism. The strong Church–Turing thesis (evidently never explicitly endorsed, nor rejected, by either Church or Turing) asserts that every physical process is effective and must be completely describable by a Turing machine. "I can now state the physical version of the Church–Turing principle: 'Every finitely realizable physical system can be perfectly simulated by a universal model computing machine operating by finite means'." This formulation is both better defined and more physical than Turing's own way of expressing it [72]. The claim is that every realizable physical system includes brains, and it must follow that intelligence can be captured in an algorithm.

This is a modern restatement of a concept that has dominated science for the past 400 years. Reductionism encompasses numbers and other systems that can be put in correspondence with numbers [73]. Suppose that x is a natural system or referent, and $P(x)$ is a proposition that asserts that some property of x is true. Rosen argues that an *essential attribute* of reductionism is that any such proposition can be algorithmically constructed by "ANDing" $P_i(x)$ where $i \in \{1, ..., N\}$, N is a natural number, and there are N true subproperties of x that are described by N independent propositions. Since infinity is not a number, this description limits a reducible system to a finite list of properties. This algorithm constitutes a list of conditions, each necessary, and all sufficient to establish the truth of $P(x)$. In Rosen's words, "... every property $P(x)$ of x is of this character."

Another *essential attribute* of reductionism is the context independence of the parts of a system [74]. Suppose that referent, x, can be fractioned into parts, x_j. The listable subproperties of x_j, described by $P_k(x_j)$, are independent of the fractioning process or any other context. $P(x) = \wedge_{j,k} P_k(x_j)\, j,k \in N$. This algorithm says that the largest property of the system can be found by ANDing all the subproperties of all the parts, and that doing so produces no information whatsoever about the context. In Rosen's words, "It is precisely this context independence that renders reductionism an entirely syntactic exercise, that gives it its algorithmic character, and embodies it in the kinds of lists and programs I described earlier."

These presumed attributes have some consequences [75]. Every proposition describing a property $P(x)$ has an algorithm for assessing its truth. Any natural system, x, can be constructed, given enough parts x_j, and an algorithm for constructing x from x_j. The process of analyzing a system x into its parts x_j is exactly reversible to a process of synthesizing x from its parts x_j. Correspondingly, the process of analyzing $P(x)$ into its subproperties and synthesizing it from its subproperties are reversible and algorithmic. If, as is asserted by its advocates, strict reductionism affords a complete description of all physically realizable processes, then as Rosen says, "everything is computable," including cognitive function.

In strict reductionism, "computable," "algorithmic," and "reducible" are three different terms describing the same kind of process. These processes are context independent. Their behavior is the direct sum of the behaviors of their parts. All elements of behaviors are localizable. Their causal entailments are strictly separable. Any closed loops of causality are strictly limited to material cause. At any instant in time, the entire state of the process can be characterized by a structure of irrational numbers (sometimes called a state vector). The entire biography of the process is subsumed in the evolution of values within that structure. As Lord Kelvin says, the knowledge is in the numbers; to raise distinction between the model and the process is to divert attention from the numbers or to obscure the thing that matters most to a true understanding.

7.5.2 WHERE DOES REDUCTIONISM FALL SHORT?

Recall Deutsch's claim: "I can now state the physical version of the Church–Turing principle: 'Every finitely realizable physical system can be perfectly simulated by a universal model computing machine operating by finite means'. This formulation is both better defined and more physical than Turing's own way of expressing it." Remarkably, Deutsch admits that this is not what Turing said.

Also, he characterizes his statement as a definition, and not as either a hypothesis or a proven conclusion.

However, is a "more physical" rendition of the Church–Turing thesis justified? The claim of the strong Church–Turing thesis seems to arise from the failure to notice a hidden ambiguity in the term *effective*. Church and Turing described a process as *effective* if and only if it was equivalent to a Turing machine. In the context of the five properties of an algorithm, this is the one and only admissible meaning of the term "effective." In physical ontology, *effective* has quite a different meaning; *effectiveness* is the property of producing events in reality.

If one ignores the ambiguity and treats these quite distinctly different meanings as if they were identical, then the syllogism follows quite naturally. All physical processes are *effective*. All *effective* processes are computable. Therefore all physical processes are computable. However, the conclusion remains unproved because *effective* has two different meanings in the two different premises.

The concept that reductionism explains everything predates by centuries the claims of Church, Turing, and those who retell Church's and Turing's story in their own way. The premises behind the conclusion that "reductionism explains everything" is that "reductionism explains everything in physics," the whole world is made up of physical particles governed by the laws of physics, and the behavior of the whole world is the direct sum of the behaviors of its parts. The problem is that none of these premises are provably true.

Contemporary reductionistic theories of physics do not explain all physical processes, even some seemingly mechanistic ones. For example, there is a law for describing the rate of decay of a large collection of radioactive nuclei. However, there is no law for identifying which nucleus will disintegrate next [76].

Furthermore, physics ignores endogenous causation. Despite the fact that discussion of endogenous causation and impredicative logic were typically disallowed in serious scientific discussion at the time, Schrödinger, one of the preeminent physicists of the last century, recognized 60 years ago that living processes must be entailed by something like endogeny. "To put it briefly, we witness the event that existing order displays the power of maintaining itself and of producing orderly events. That sounds plausible enough, though in finding it plausible we, no doubt, draw on experience concerning social organization and other events that involve the activity of organisms. And so it might seem that something like a vicious circle is implied" [77].

This led Schrödinger to an astounding conclusion. Reductionistic physics is inadequate to account for the properties of organisms. Speaking of living processes he says "we must be prepared to find a new type of physical law prevailing in it. Or are we to term it a non-physical, not to say a super-physical, law?" [50]. He was not referring to a mystical vitalistic Cartesian "ghost in the machine." He said that to understand living processes, there remains a larger physics to be discovered, and that larger physics is "something like a vicious circle."

The currently known laws of physics ignore endogenous causation, and, consequently, it is not valid to claim that a world that abounds in endogenous processes is fully governed by the currently known laws of physics. As Schrödinger shows, the premise that the whole world is the direct sum of the behaviors of its parts is an *assumption* that appears to ignore visible facts. Thus, the notion that a system is reducible to its parts is not proven; rather it is simply the converse of the previous assumption. The claim that the behavior of the whole world is the direct sum of the behaviors of its parts is an instance of the logical fallacy of "begging the question," considering that a proposition is proved because it is assumed.

7.5.3 THE MEASUREMENT PROBLEM

In his assertion that knowledge is expressed in numbers, Lord Kelvin mentions, but then quickly passes by, the notion that the numbers arise from a process of measurement. In fact, measurement is the foundation of physics. However, that foundation has a rather serious crack. About a century ago, it was realized that there was no practical way of validating what is measured, and this as yet unsolved "measurement problem" continues to perplex physicists even to the present day.

This is not a mere technicality of quantum mechanics. Irrespective of scale, the measurement problem is the question of why we should believe any measurement that we take. Rosen discusses the measurement problem in some detail. "Suppose we want to determine whether a predicate or proposition $P(x)$ is true about a given object, x. For instance, suppose we want to determine whether a particular molecule x has a certain active site on it. Or whether a particular pattern x manifests a certain feature. These seem to be nice, objective questions, and we would expect them to have nice, objective, cognitive answers. Basically these are measurement problems" [78].

He then notes that we might test the truth of $P(x)$ by measuring object x with a meter or feature-detector, y, an object different from the object x that we are trying to measure. However, can we believe the reading, $Q(y)$ produced by the meter or feature-detector, y? One way to test $Q(y)$ is to use another measuring device z (Rosen calls it a feature-detector detector) to measure y. The problem is why should we believe $R(z)$, the reading produced by the feature-detector detector? He then makes the point that this strategy is doomed because in order to work it would require an infinite regress of feature-detector-detector … detectors.

This is the classical measurement problem. Attempting to validate the validator starts down a path of an infinite regress of independent readings and validations. This is more than a philosophical puzzle. Since many measurements are made very near the technical limits of measurability, the question of whether or not the measurements are believable is crucially important.

We can truncate the regress at $Q(y)$, if we have some other reason for believing $Q(y)$ to be true. If we have a model of meter y, built from independent knowledge, we can use this model to infer the truth of $Q(y)$ without recourse to measurement. The problem with this strategy is where did the independent knowledge come from, to construct the model of the meter if no recourse was made to measurement?

Rosen offers an alternative strategy. He says we can use an impredicative structure to validate the measurement. "The only other possibility is to fold this infinite regress back on itself, i.e., to create an impredicativity" [79]. His strategy works as follows. We want to understand process x. We make a measurement with feature detector y that produces the proposition $P(x)$. We want to validate feature detector y. We make a measurement with feature-detector detector z that produces the proposition $Q(y)$. Suppose we break the infinite regress by letting the original process x serve the role of the feature-detector detector z (i.e., force z = x). In other words, he claims that we can validate $R(z)$ because within the impredicative structure (where $z = x$), we can validate $R(x)$, $Q(y)$, and $P(x)$ simultaneously.

Admittedly, Rosen does not describe the specific impredicative structure that works the wonder of validating $R(x)$, $Q(y)$, and $P(x)$ simultaneously. He implies that the structure is discoverable for specific instances of measurement. In other words, he points to a strategy for solving the measurement problem, but does not provide the solution.

An impredicative solution to the measurement problem is needed, because the measurement process is inherently impredicative, and consequently, cannot be described by reductionism. "The impredicativity that we have described, which is the only alternative to the unacceptable infinite regress, may be described in the following deceptively syntactic looking form: $x \leftrightarrow y$ (7.1). This is essentially the modeling diagram (Figure 7.1) in which the arrows represent what I call *encodings and decodings* of propositions about x into those of y, and which satisfy a property of commutativity (i.e., the path $x \rightarrow y \rightarrow x$ in the diagram is here the identity mapping). But Equation 7.1 represents an impredicativity and, hence, from the preceding discussion, it follows that both x and y themselves possess other models that are nonformalizable or incomputable. Stated otherwise, Equation 7.1 actually represents a multitude of inherently semantic interpretations; it has meanings that cannot be recaptured from it by purely syntactic means — by lists of other symbols and rules for manipulating them" [80].

Far from explaining everything in physics, we see that reductionism fails to explain the measurement process at the very foundation of physics. It is possible that a hyperset-theoretic model of valid measurement may be discovered. In fact, the necessity to do so is a significant open research question.

7.5.4 Brain Physiology

If reductionism fails to explain the measurement process at the very foundation of physics, would it be a great surprise to find that it fails even more spectacularly at explaining the observed processes of life and mind? The classic reductionistic theory of mental processing was Skinner's behaviorism. Chomsky punctured the idea by showing that behaviorism ignores the influence of the internal structure and organization of an organism on its own behavior [81].

As an alternative to the reflex arc structures of behaviorism, Miller, Galanter, and Pribram suggested that the mind might use an internal model of the world to execute plans was a better explanation of hierarchically structured behavior [82]. They recognized that most human behavior operates on different hierarchical levels simultaneously.

As Freeman and others have observed, brains do indeed construct internal models. However Miller, Galanter, and Pribram formulated their Test Operate Test Exit (TOTE) model as one level of a recursive algorithm that they supposed would describe the process. They failed to notice that a recursive algorithm executes one level at a time in sequence (and must, because each level waits for data from the next), and does not correspond to mental modeling, which they themselves note operates on different hierarchical levels *simultaneously.*

Had they posited an impredicative model instead of an algorithmic model, some fairly dramatic discoveries about mental processing might have been made decades earlier than they actually were. We form hypotheses by abduction, and test them by emotion. The "test" is the question of whether or not the knowledge feels right; we update our models if it feels wrong. When the knowledge feels right, we use it to proceed toward some internally defined goal or final cause, something an algorithm is incapable of forming. We perceive the result of the process as a comforting and satisfying *feeling of knowing* [83].

It is also instructive to consider what happens when the process fails. In autism, the sensory substratum is present and produces the right markers (chemical and electrical signals), though due to a neurological malfunction, the representation systems do not construct sensible percepts from them. As a result, the typical autistic child lives a life of unremitting terror [84].

With proper medication and training, autistic people can and do learn how to construct models most appropriate for their situation, exclusively "thinking in pictures" [85]. Verbal language is a pure abstraction into which high-functioning autistic people have learned to translate their visual concepts. This may be due to the fact that visual representation contains no emotions, smells, tastes, or sounds.

The highly developed visual representation systems of autistics overstimulate the neurological "fight or flight" response, while auditory and kinesthetic mental models are easily overwhelmed by sounds, touches, tastes, and smells. As a result, the daily life of the autistic adult is only slightly less frightening than that of the autistic child [86]. It is not surprising that if the abstraction of sensations into "knowing" feels right, then a meaningless jumble of sensations must feel terribly wrong.

Minds operate on different hierarchical levels simultaneously to abduct semantic emotions from sensory data. Where the meaning and emotion reside is nonlocal; it is in the relationship between the hierarchical levels of the mental model. The fact that one cannot point to a specific neuron or even a vector of neural weights, and say that "an emotion or meaning resides here" does not make them any less real. They have all the characteristics of impredicative processes.

The fact that minds operate nonlocally is no surprise given that brains, the substrate with which minds are inseparably entangled, are also largely nonlocal in their operation. Based on extensive observational and experimental research, Bach-y-Rita has found that "nonsynaptic diffusion neuro-transmission (NDN) is a major mechanism of information transmission in the brain and spinal cord. It includes the diffusion through the extracellular fluid of neurotransmitters released at points that may be remote from the target cells, with the resulting activation of extrasynaptic receptors as well as intrasynaptic receptors reached by diffusion into the synaptic cleft." He goes on to note that "NDN may be the primary information transmission mechanism in certain normal mass, sustained functions, such as sleep, vigilance, hunger, brain tone, and mood [87].

At its substratum, the brain/mind is organized by the relationship between synaptic firings, volume transmission, and glial activity. Each process organizes the pattern of events in the others. The processes are related in a bizarre hierarchy. Comparing these processes to electromechanical robots circuits, it is as if we have three devices that each continuously and simultaneously updates the morphology, and consequently, the dynamical laws of the others. Since either can be seen to be rebuilding the others, from its own perspective, each process appears higher than the other in a hierarchy. This same kind of hierarchy is found in a hyperset structure. Interpreted as a directed graph, the nodes in a hyperset form a hierarchy in which, from the perspective of any node in the hierarchy, that node is at the top. Even at the level of the physical substratum, the brain appears to be endogenously organized [88–90].

In summary, the brain/mind is observed to perform semantic processes nonlocally. Their operation is inseparable. In fact, what Damasio defines as Descartes' error is "the separation of the most refined operations of mind from the structure and operation of a biological organism" [91]. It is fundamentally incorrect to think of a brain as a computer and mind as the program running on it.

7.5.5 OTHER BIZARRE EFFECTS

What is remarkable about complex systems and impredicative models is that both produce bizarre effects. Note the distinction between absurd and bizarre. Absurd effects do not actually occur; they are incoherent with the causal entailment structure of reality. Bizarre effects do not contradict reality, but are unexpected. Three of the bizarre effects of complex systems are especially notable.

7.5.5.1 Determinism Vs. Freewill

Perhaps the most bizarre consequence of a system's having an endogenous entailment structure is that the system can produce an unlimited number of behaviors that are all consistent with the entailment structure. To express semantics in terms of syntax would require an infinite regress of syntax [92]. In other words, a complex model or an impredicative inferential entailment structure can produce an unlimited number of true propositions consistent with its inferential entailments. Similarly, a complex natural system or endogenous causal entailment structure can produce an unlimited number of events consistent with its causal entailments.

If an MR exists between the two systems, there must be partial congruency of entailment between them. There must be at least one causal entailment in the complex natural system that commutes with an inferential entailment in the model; if there is no congruency of entailment between them, then an MR does not exist between the two entities. Since the model and the process being modeled are not identical (i.e., the map is not the territory, and the distinction between the map and the territory is embedded in the territory), there cannot be total congruency of entailment between them. "There is accordingly no largest model of such a system" [93].

This partial congruency has a remarkable implication. To the extent that the entailment structures are congruent, the model predicts behaviors of the natural system, and conversely the natural system predicts behaviors of the model. Both systems have an unlimited number of different entailments and entailed behaviors (all possible, albeit not equally likely). Since the two systems are not fully congruent, each system must have entailed behaviors that are not entailed in the other. In other words, a complex natural system must have behaviors that are fully determined (consequent of the causal entailments that led to them), but are not predictable by a given complex model (including the mental model in the mind of any observer) of the system. In other words, the complex system is completely determined, but gives every appearance of freewill.

(*Note*: If the notion of overlapping infinities gives the impression of doubletalk, consider the following example. Imagine the set, $3\mathbf{Z}$, of natural numbers divisible by 3. Also, imagine the set, $2\mathbf{Z}$, of natural numbers divisible by 2. Most of the members of either set are not members of the other. The set, $6\mathbf{Z}$, of natural numbers divisible by 6 contains all the members common to both sets. All three sets are not only infinitely large but provably have identically the same cardinality or size, $\aleph_0$. In other words, $2\mathbf{Z}$ and $3\mathbf{Z}$ have infinitely many overlapping members, but just as many members that do not overlap. Infinitudes of congruent and noncongruent entailments work just the same way.)

The idea that a fully determined reality can produce unpredictable behaviors has a dramatic implication for the popular concept of random events in reality. There may not be any. Probability may not be a model of some hypothetical underlying randomness of nature. It may turn out as Jaynes suggests that probability is the measure of our limited knowledge of reality that necessarily arises from the fact that reality is inherently complex and we have no largest model of it. "Common language — or at least, the English language — has an almost universal tendency to disguise epistemological statements by putting them into a grammatical form, which suggests to the unwary an ontological statement. A major source of error in current probability theory arises from an unthinking failure to perceive this. To interpret the first kind of statement in the ontological sense is to assert that one's own private thoughts and sensations are realities existing externally in Nature. We call this the 'Mind Projection Fallacy,' and note the trouble it causes many times in what follows. But this trouble is hardly confined to probability theory; as soon as it is pointed out, it becomes evident that much of the discourse of philosophers and Gestalt psychologists, and the attempts of physicists to explain quantum theory, are reduced to nonsense by the author falling repeatedly into the Mind Projection Fallacy" [94].

7.5.5.2 Nonlocality

We have seen that nonlocal phenomena arise in both semantics and emotions. It also arises in impredicative representations. Consider the prototypical hyperset, that abstract entity, Ω, which solves $\Omega = \{\Omega\}$. Under the Zermelo–Fraenkel–Aczel axioms, the solution exists, is a unique entity, and the fact of its existence does not lead to contradictory propositions. Ω has a definition, but that definition is not formed by pointing to something else. The defining property of Ω is a specific constraint of the relationship of Ω with itself. However, you cannot point to Ω or $\{\Omega\}$ and say that the defining property lies in either place. The definition is nonlocal, or spread across the relationship. In other words it is the attribute of a nonfractionable process. It is this nonlocal nonfractional relationship that Rosen identifies as semantic. "Or to put it another way, real mathematical systems are full of impredicativities — semantic elements that endow these systems with enough inferential structure to close impredicative loops" [78].

Bateson speaks of the nonlocality of semantics in the same way. He does not limit meaning to something that must reside in the mind of a knower. He says rather that meaning lies in the correspondence between a symbol and its referent. "The universe, message-plus-referent, is given pattern or form — in the Shakespearean sense, the universe is *informed* by the message; and the 'form' of which we are speaking is not in the message nor in its referent. It is a correspondence between message and referent" [62].

Note the contrast with algorithmic processes which are reducible to open sequences of inferential linkages, and are only concerned with syntax or *structure* of events. Impredicative processes with closed loops of hierarchical inferential entailment are concerned with the semantics or *meaning* of events [95]. A crucial feature of impredicativity is that the distinction between the map and the territory is buried within the territory.

Bateson also notes the crucial distinction between *being right* and *not being wrong*. If we are trying to guess the missing characters in a garbled message, we do not algorithmically explore and eliminate all the wrong possibilities. We use redundancy afforded by context to focus directly on the right possibility. That is how people identify partial words on TV's *Wheel of Fortune*.

To appreciate how redundancy works, suppose that meaning is the essence of communication, and the isolated symbol is the special case. In other words, communication is the creation of redundancy. From the perspective of an observer of A talking to B, suppose that A has a specific message on his pad. No new information is added *about the message* by inspecting B's message pad after he receives the message from A. However, what the message means depends on a complex equivalence of what is being seen, felt, or heard. To B the message content itself is new information. To the outside observer, the meaning is in the rules of the game played by A and B. They explain to the observer an improbable coincidence, by what process did the same message appear on both of their pads?

To make a guess is to face a cut in a sequence of events and predict what will happen next. The discontinuity may be temporal or spatial. A pattern is an aggregation that will permit us to guess the exemplars of a set, given a small subset of exemplars. A message has meaning if it is about a referent. The "Universe of Relevance" consists of the message and its referent. The message "It is raining" conveys both the symbolic message and the ontological fact that raindrops are falling.

If you receive the message, then you have a better than random chance of guessing that it is raining without looking out the window. The universe is informed by the message. The "form" or meaning is not the string of symbols or the referent. It is in the *correspondence* between them. Note that the meaning is nonlocal. It is not located in either the string of symbols or the referent. It is entangled or a consequence of the relationship. The fact that you cannot say where the meaning resides does not make it any less real. Where the meaning resides is nonlocal. This nonlocality is a bizarre consequence of the endogeny of language. It is a perplexing problem to traditional thinkers in communications and neurophysiology.

7.5.5.3 Language

Semantic processes are nonlocal and ambiguous. They are found in many processes, not least of which is natural language. As Chomsky notes from his study of natural language, semantic processes have another bizarre property; they are generative [96]. Having mastered a language, one can understand indefinitely many expressions that are outside one's experience. In synthesizing understanding from streams of signs, the normal use of language is a creative activity.

He says that the motivation to study language is that it provides insights into the human mind. In particular, the notion of a Universal Grammar illustrates how the mind operates. In Chomsky's parlance, grammar is the answer to the question, what is the nature of the knowledge that makes it possible for a person to normally use it creatively? A generative grammar is the notion that the rules of grammar allow the construction of an infinitely large set of allowable structures according to a finite set of rules.

Although Chomsky does not make a causal interpretation of the Universal Grammar, such an interpretation is justified if one considers that Chomsky does see the process as a transformation, noting that rules relating deep structure to surface structure are called "grammatical transformations." He calls the relationship between surface structure (a set of syntactic rules) and deep structure (abducted semantic meanings) a "syntactic object." Deep structures in grammars appear to correspond to real mental constructs or activity. They are the effect produced by the transformation.

Deep structure is the meaning abducted from an incoming stream of symbols, and clearly this input to the transformation corresponds to Aristotelian material cause. One can consider the generative grammar at the heart of the Universal Grammar to be an efficient cause, characterizable by an as yet undiscovered "law of behavior," constraining the process of abductive transformation. The fact that the generative grammar can abduct limitlessly many meaningful outcomes from a finite entailment structure suggests that it is endogenous in character, and that the "law of behavior" describing it is impredicative.

Generative grammar appears the same way for every human language. What distinguishes one language from another is the surface structure, a formal cause, or syntactic set of parameters or finite set of rules that modulates generative grammar to operate in a particular way for a particular natural language. Chomsky sees generative grammar as a rich system of constraints upon possible surface structures (clearly this is typical of the relationship between efficient and formal cause; the possible properties of modulating parameters are necessarily constrained by the process that they modulate). That there are severe constraints on the properties of surface structure explains how it is that children are so quick to learn the surface structure of a given language from a sparse set of noisy exemplars? There is no predisposition to learn one language over another. If all are equally easy to learn, then at the foundation the process imposing the fundamental constraints must be the same. This hypothesis could be falsified if it failed to explain acquisition of some languages. However, Chomsky's own observational data bear out the hypothesis, and he has not found falsification of the hypothesis anywhere in the observational data.

Although surface structure is syntactic, by serving as the formal cause modulating the generative grammar that serves as efficient cause in the transformation of symbol streams into meaning, surface structure plays a crucial role in semantic interpretation. Surface structure subsumes the speaker's presuppositions about a sentence. Thus, minor changes in surface structure (e.g., word placement, passive vs. active) contribute to the semantics or meaning. Even a change in intonation or emphasis can convey different meanings, as the emphasis often goes to answer a particular question, while the other information merely adds detail or restates a presupposition.

Language has irreducible properties, not least of which are generative power, nonlocality of meaning, and context dependency. Chomsky says that the contributions of grammar and context to meaning *may* be inseparable. Given the impredicative character of language, they almost certainly are. The fact that the process leads to the abduction of meaning has one further implication, that meaning is subjective; it depends on the perspective of the user of language. Language serves and is entailed by the function that it serves for the user. In other words, subjective meaning serves as a final cause that entails the operation of language in the mind of the user (*Note*: The correspondence between Aristotelian causal classes and elements of Chomsky's theory is my observation. Chomsky may or may not endorse this particular interpretation of his work.)

Despite the irreducible properties of language, he claims that it is possible to learn something about it from reductionistic strategies. "Abstracted study" of language ignores context. It considers formal structures (syntax, deep structure, surface structure, and semantics) and relations between them. Chomsky considers that this is a legitimate hypothesis, although its presuppositions are not self-evident (in fact, by his own recognition of the context dependency of language, he himself cuts the ground out from under the presuppositions). The hypothesis is that linguistic competence is separable from how language is used. As in other forms of reductionism, its popularity is based on its ability to solve a certain class of problems. Since this strategy can solve *some* linguistic problems, might it not solve *all* linguistic problems?

Chomsky admits that the use of language is not explainable or understandable by such a reductionistic strategy. Consequently, he fears that it may be infeasible to study significant problems of language communication. His fear would be well founded if theoretical modeling were to continue to ignore final cause. Thus, even from a reductionist starting point, Chomsky allows the possibility that new intellectual tools are needed, and might be feasible.

The interaction of mind and language is so fundamental that Chomsky sees the need of a branch of psychology that has the goal of exhibiting and clarifying mental capacities to learn and use language. This capacity is apparently unique to humans. Language must be abductive, producing novel effects from a finite but endogenous causal entailment structure. This creativity or seeming "emergent complexity" makes it impossible to understand human behavior by stimulus–response or other reductionist methods. (*Note*: Again, this is my interpretation of Chomsky's words. He does not appear to believe that there is a serious alternative to reductionism. However, he does discuss the fact that traditional methods are not up to the task.)

7.6 DOES THE ENDOGENOUS PARADIGM IGNORE PAST INSIGHTS?

Remarkably, one of the objections to the notion that there exist incomputable processes that can only be represented by incomputable models is that it allegedly ignores decades of work by celebrated thinkers. That work typically falls into three classes — the computational theory of mind, the various accomplishments of the Artificial Intelligence community, and the seemingly more compelling instance of Von Neumann's self-replicating automaton. Despite the all-inclusive claims made by the champions of these ideas, it is clear that the ideas are far from all inclusive. In light of that disparity, rather than ignore their work, it could perhaps be more appropriate to ask, what principles have they ignored?

7.6.1 COMPUTATIONAL THEORY OF MIND

One of the more curious discussions of a computational theory of mind is that put forward by Kosslyn and Koenig. A superficial interpretation of Figure 9.1 in Chapter 9 of *Wet Mind* might be that the essential functions of the brain can be captured in a block diagram that is roughly as complicated as a digital signal-processing chip. However, in the context in which that diagram is given, they make the following qualification: "… we have been vague about the source of (free) 'will,' the ultimate source of decision making in the brain. What decides what to do next? We have inferred that there must be a "decision subsystem," but have not said much about it. We have eschewed talk of consciousness and emotion" [97].

Of these qualitative behaviors, they seem to find consciousness to be the least amenable to computational interpretation. "As we have demonstrated in previous chapters, the vocabulary of computation has proven useful for understanding mental activity. But there is no corresponding vocabulary for understanding the texture of experience" [98]. They then enumerate the requirements for an adequate theory of consciousness.

Their attributes of consciousness fly in the face of the notion of a brain as a computer. It is nonreducible. "Any theory of consciousness must describe a phenomenon that cannot be replaced by a description of brain events" [98]. It is inseparable from its physical substrate. "Thus, even though a description of consciousness cannot be replaced with a description of brain activity, brain activity is a necessary prerequisite to consciousness" [99]. It is nonlocalizable to brain events. "If the theory posits that consciousness has a function, then it must posit a function that cannot be accomplished by brain events *per se*" [99]. It does not serve as a largest model of mental activity. "A theory of consciousness must specify why some mental processes are accompanied by conscious experience whereas others are not" [99]. Finally, and, most remarkably, the hierarchy of consciousness (a behavior of the whole; organizational rather than physical) and brain activity (behaviors of the parts; purely physical) form closed loop in which each updates the other. "Thus, we assume that the relation between brain and consciousness cuts both ways: activity in the brain affects consciousness and vice versa. This requirement is daunting. Indeed, at first glance, it seems paradoxical: Something that is not a physical event must not only arise from a physical event, but also must be capable of feeding back and altering it" [99]. All these attributes are incomputable, but fully consistent with the interpretation of consciousness as an effect arising from an endogenous structure of causal entailment in the brain/mind.

Probably the most visible champion of the computational theory of mind is Pinker, a former student of Kosslyn's. Pinker does not exactly claim that the whole world (or even the whole human) is reducible. He admits the existence of consciousness (subjective), self (imponderable), free will (an enigma), meaning (a puzzle), knowledge (perplexing), and morality (a conundrum) [100]. The characterizations that he assigns to each process suggest that he agrees that none are amenable to computation.

How should we address such phenomena? Borrowing an idea from McGinn, he suggests that we should address them not at all [101]. Starting from the premise that the brain is a computer, it is incapable of comprehending incomputable processes. The suggestion has a fatal flaw. It ignores the fact that all those incomputable processes are themselves produced by the brain/mind. If the brain were a computer, it should produce none of them. It is clear from other parts of the book that the real reason that Pinker champions the computational theory of mind is a perfectly fair one. He has never seen what he regards as a compelling alternative.

Perhaps the most eloquent defense of the computational theory of mind is that by Churchland. Not surprisingly, she makes scant mention of ambiguity and no mention at all of the abductive resolution of ambiguous cues in a given context. She addresses semantics strictly in terms of Tarski's highly formalized semantics. In other words, she does not address the means by which minds use ambiguous sensory cues.

She does admit that "there is no algorithm for making wise choices" [102] and is skeptical of the notion that choices are uncaused, or are (equivalently) the result of some sort of quantum mechanical magic. That minds make choices, and do so nonalgorithmically remains to be explained (by somebody else), with the burden of proof resting (not unreasonably) upon the explainer. The idea that seemingly free (caused but not fully predictable) choices could be entailed by the inherent ambiguities in an endogenous causal entailment structure seems never to have come to her attention. In other words, she interprets brain function in terms of connectionism, even knowing that algorithms, connectionist, or otherwise, do not make choices. Apparently she does so for much the same reason as Pinker; computation may be inadequate to explain some mental processes, but nobody has shown her a more compelling alternative.

7.6.2 Artificial Intelligence

The claim that computer simulation can produce new information is probably a no more articulate explanation than the one offered by Simon. While admitting that a simulation is no better than the assumptions built into it, and that a computer can do only what it is programmed to do, he nevertheless says, "despite both assertions, simulation can tell us things we do not already know" [103]. One of those things turns out to be logical deductions that would be impractically long for the human to work out with a pencil and paper. The other is that we can use the computer to test various models, until we find one that best fits observed data, thereby uncovering insights into natural laws.

It is important to realize that neither of these processes leads to an algorithm to create novel information. The first process simply reveals information already present in the premises fed into the algorithm. The second process, using artificial intelligence (AI) simulations to test computational models, is simply Bayesian parameter estimation [104] done suboptimally. In both cases the algorithm does not create new information. Given novel information provided by the human user, a computer simply reveals consequences that are already present, but obscured to human sensibilities. By exploiting this power, the human can learn things faster by using a computer than by using a pencil. However, in either case, it is the human both learning and creating novel information; the computer's behavior is no more intelligent than the pencil's.

Thus, despite the overwhelming evidence that brains and minds are irreducible and nonalgorithmic, one occasionally still hears the claim that "Artificial Intelligence" programs "learn" or create new information just as brains do. Four examples are typically offered, connectionism, Markov chains, genetic algorithms (GAs), and fuzzy systems, but all of these processes simply reformat existing information without adding anything new to it.

7.6.2.1 Connectionism

It is undeniable that neural nets can do useful things [105]. What is highly deniable is that they learn. Fundamentally, neural nets are based on a curve-fitting algorithm called the perceptron. The proof that the perceptron converges in a finite number of steps shows that it is just another algorithm [106]. All it actually does is to iteratively solve a set of inequalities to discover the decision surfaces already implied by the data. The role of hidden layers in the process is nothing more than enabling the possibility that the decision surfaces to be deduced from the data might be nonlinear.

Attributing "brain-like" behavior to the connectionist paradigm fails to recognize the physical reality that brains are much more than networks of synaptically connected neurons [89,107]. Furthermore, a numerical weight purporting to describe "strength" is a superficial description of a synaptic connection [108,109]. In other words, the inferential entailment structure of algorithms inspired by the connectionist paradigm only superficially resembles the causal entailment structure of the brain, and fails to characterize most brain processes.

7.6.6.2 Markov Chains

"Intelligent" behavior is sometimes explained in terms of Markov chains. "Random variables X, Y, Z, are said to form a Markov chain in that order (denoted by $X \rightarrow Y \rightarrow Z$), if the conditional distribution of Z depends only on Y and is conditionally independent of X. Specifically, X, Y, and Z form a Markov chain $X \rightarrow Y \rightarrow Z$ if the joint probability mass function can be written as $p(x,y,z) = p(x)p(y|x)p(z|y)$" [110]. Some people argue that Markov chains behave intelligently by adding novel information to a process.

They do not. Ironically, Markov chain theory leads to a diametrically opposite conclusion. "The data processing inequality can be used to show that no clever manipulation of the data can improve the inferences that can be made from the data." After several pages of mathematical symbolism, Cover and Thomas show that if $X \rightarrow Y \rightarrow Z$ then $I(X;Y|Z) \leq I(X;Y)$. In plain language, "the dependence of X and Y is decreased (or remains unchanged) by the observation of a 'downstream' random variable Z" [110].

7.6.2.3 Genetic Algorithms

Optimization processes can be useful tools of modeling behavior of living entities [111,112]. It is also widely supposed that intelligent behavior is somehow characterized by an optimization process. Indeed, most AI algorithms have an optimizer at the heart of the process. This leads to the observation that it might be possible to produce an algorithm that "learns" directly by evolving in a way similar to a model of biological evolution.

It is well known that gradient descent is an inadequate optimizer because it is trapped at the first local minimum that it encounters. A global optimization algorithm searches for the global minimum in an objective function that is pockmarked with suboptimal local minima. Many operate on the basis of not especially well-informed guessing. This leads fairly directly to the "no free lunch" theorem, that no global optimizer is generally superior to random guessing over the entire set of objective functions [113].

Advocates of the GA claim that it makes better guesses than the free lunch theorem would lead one to expect. The schemata that are implicitly evolving during the process inform subsequent guesses of the more promising places to search in the objective function space. In fact, this gathering information from bad guesses to make better-informed subsequent guesses is thought to be a kind of "learning" [114].

However, what is interpreted as learning is simply a process of deducing information that is already present in the objective function. Global optimizers, including GAs neither learn nor add novel information. All the information is already present in the objective function and is waiting to be revealed. Like an algorithm that solves an equation, an optimizer simply uncovers information that was already there when the problem was specified. Global optimization algorithms have many useful engineering applications and in many of those applications a well-designed GA converges much faster than other optimizers.

Is emulating brains one of those applications? If an algorithm designs a patentable invention, is this not the abduction of novelty and evidence of intelligent behavior? Suppose it is not a fluke. If an algorithm designs many patentable inventions, is that not overwhelming evidence of the abduction of novelty and evidence of intelligent behavior? Genetic programming (GP), a GA that evolves computer programs does just that; it has duplicated and even improved upon the design of many patented devices [115]. In fact, this seems to settle the question once and for all that algorithms can abduct novelty.

Or so it seems if one does not understand the process of invention. Invention is based on an ambiguous description of a need, the construction of a device that fulfills the need, and the development of unambiguous specifications *after the device is constructed*. GP is an algorithm, and its input must be unambiguously specified. Going from an ambiguous need to an unambiguous specification, a task performed by the programmer and not the program, is where the inventiveness lies. Evolving an implementation from the unambiguous specification, dramatic as it seems, is still deduction based on the reshuffling of information already presented.

Edison is supposed to have said "genius is 1% inspiration and 99% perspiration." This quote overlooks the detail that without the 1% inspiration, the 99% perspiration would be wasted. A human using GP still performs the 1% that matters, the abduction of novel information by creating the unambiguous specification. The GP running on a computer merely replaces the 99% perspiration with a motorized cooling fan. It tediously infers the device implied by the novel specification, but it creates no novelty of its own.

7.6.2.4 Fuzzy Systems

The attribute of fuzziness signifies the measure of the dilution of a property. The dilution of property is sometimes confused with probabilistic uncertainty, and that supposed uncertainty is confused with the uncertainty that arises in predicting the behavior of intelligent processes. Fuzziness, the measure of the dilution of a property, is not related to probability, a measure of uncertainty about the occurrence of an event.

To appreciate the difference, imagine that you have ten wells. Suppose that there is a toxin present in the water, such that if the toxin is present in a concentration of less than 20%, it is harmless. Suppose that you have *a priori* knowledge that each of the wells has an equal probability of 0.1 of being a member of the class of lethally toxic. Is it safe to drink from any of them? The answer is no, because while nine of the wells contain no toxin at all, the tenth one is lethal; with no information as to which is which, drinking from any of them carries a one in ten risk of being killed. Suppose, on the other hand, that you have *a priori* knowledge that each of the wells has an equal fuzzy membership of 0.1 of being in the class of lethally toxic. Is it safe to drink from any of them? The answer is yes, because no matter which well is selected, it contains a harmless dose (10% concentration) of the toxin.

The rules of fuzzy set membership are defined by a straightforward appeal to crisp logic [116]. The degree of membership of a given element in a given set is determined by appeal to a membership function, but where do the membership functions come from? "Fuzzy membership functions can have different shapes depending on the designer's preference or experience" [117]. In other words, at its foundation, it refers to a lookup table of simple geometric shapes. There may be some inventiveness on the part of the designer devising the membership functions, but the algorithm merely unambiguously shuffles the unambiguous data fed into it. Like the other AI techniques discussed above, fuzzy logic algorithms have many useful engineering applications. However once again, emulating the brain's ability to deal with ambiguous information is not one of them.

7.6.3 THE SELF-REPLICATING AUTOMATION

The essence of the position of the advocates of the computational theory of mind consists of an acknowledgment of the existence of novel behaviors entailed by incomputable processes, coupled with the dismissal of these processes as irrelevant unless someone else finds a viable theory to explain them; meanwhile they will keep on computing. The fallacy underlying the AI paradigm is the mischaracterization of processes that uncover information already present as if the processes were abducting novel information. In contrast, the claim that Von Neumann made is much more substantive; he claimed to have produced an algorithm that actually does abduct something novel.

He sought to discover "What principles are involved in organizing these elementary parts into functioning organisms, what are the traits of such organisms, and what are the essential quantitative characteristics of such organisms?" [55]. Von Neumann acknowledges that growth and exact reproduction are *not* signatures of life; he observes that crystals do no less but are plainly nonliving. "A way around this difficulty is to say that self-replication includes the ability to undergo inheritable mutations as well as the ability to make an organism like the original" [118]. In other words, the whole claim that the Universal Constructor is a simulacrum of life hangs on the notion that is can pass along heritable mutation.

Von Neumann was aware of the magnitude of the problem, and acknowledged the essential abductive behavior of living processes. "Evidently, these organisms have the ability to produce something more complicated than themselves" [119]. He was probably unaware of the specific minutia of the Data Processing Inequality, but he was painfully aware of the idea and the genuine limitations that it imposes. "So, one gets a very strong impression that complication, or productive potentiality in an organization, is degenerative, that an organization which synthesizes something is necessarily more complicated, or of a higher order, than the organization it synthesizes" [119].

Although he acknowledged the principle, he believed that it had effect only below a threshold of complication. "We will stick to automata, which we know completely because we made them, either actual artificial automata or paper automata described by a finite set of logical axioms. It is possible in this domain to describe automata, which can reproduce themselves. So at least one can show that on the site where one would expect complication to be degenerative, it is not necessarily degenerative at all, and, in fact, the production of a more-complicated object from a less-complicated object is possible. The conclusion one should draw from this is that complication is degenerative below a certain minimum level" [119]. Above that threshold, the automaton should be generative, and he calls this the "completely decisive property of complexity" [120].

Had Von Neumann really been able to demonstrate the ability of the Universal Constructor to construct a thing more complicated than itself, he would have proved, by construction, the conclusions quoted above. However, his attempt to do so leads to a subtle, but fatal, contradiction in that construction. To recognize the contradiction, we must follow his construction in a bit of detail. It begins by postulating a completely artificial milieu in which float an unlimited supply of parts, of perhaps a dozen or so different kinds.

Also, found in this milieu are assorted specimens of $\phi(X)$, a chain of binary code that is a *complete and unambiguous* description of automaton X, another object floating in the milieu. Von Neumann introduces the notion of a separate $\phi(X)$ only because he does not want the description of his Universal Constructor to get sidetracked into irrelevant and nonilluminating details of how the Universal Constructor abstracts $\phi(X)$ from X. He simplifies his description of the Universal Constructor, with no loss in its replicative power, by eliminating an abstraction function, and instead postulating that for every X in the milieu there is a $\phi(X)$. However, since $\phi(X)$ is strictly an artifact of the replication process, there are no floating abstractions. This is a crucial point; there are no $\phi(X)$ chains in the milieu except those that correspond to actual objects, X, already in the milieu [121].

Into this milieu, he introduces the Universal Constructor, consisting of three components A, B, and C. A is a Universal Machine Tool, which devours a copy of $\phi(X)$ and while so doing constructs X from the parts floating in the milieu. He says that this process is a succession of unambiguous steps in formal logic. B is a duplicator, which devours a copy of $\phi(X)$ and while so doing constructs two new copies of $\phi(X)$. These two steps add nothing new. A produces X given $\phi(X)$, but X is no more complex than its complete model. B produces two copies of $\phi(X)$ given one copy of $\phi(X)$, but two copies of $\phi(X)$ is not a whit more complex than one copy. Next, he adds a control automaton, C, to A + B. C actuates A or B according to a pattern. Step 1: C causes B to duplicate $\phi(X)$. Step 2: C causes A to make X and consumes one copy of $\phi(X)$. Step 3: C ties $\phi(X)$ to X, and separates $\phi(X) + X$ from (A + B + C). The result is two entities, the Universal Constructor (A + B + C) and a new entity $\phi(X) + X$, where the new X is an exact duplicate of some other object X in the milieu [122].

He then shows that substituting A + B + C for X in $\phi(X)$ produces a perfectly legitimate recursive algorithm. To operate the algorithm, start with Universal Constructor (A + B + C). To this is attached a complete description of itself $\phi(A + B + C)$, leading to (A + B + C) + $\phi(A + B + C)$. C causes B to copy f. At the end of this step, we have (A + B + C) + $\phi(A + B + C)$ + $\phi(A + B + C)$. C causes A to make a copy of (A + B + C), given one of the two copies of $\phi(A + B + C)$ and parts from the milieu. At the end of this step, we have (A + B + C) + (A + B + C) + $\phi(A + B + C)$. Then C cuts the (A + B + C) + $\phi(A + B + C)$ loose from the agglomeration, thus resulting in two distinct entities, (A + B + C) and (A + B + C) + $\phi(A + B + C)$. Thus, the Universal Constructor has reproduced itself, with neither loss, nor gain, in complexity [122].

As a step toward producing increased complexity, he postulates an automaton D that can be temporarily tacked onto the Universal Constructor (A + B + C), just long enough to generate a copy of $\phi(A + B + C + D)$. Then (A + B + C) + $\phi(X)$ where $X \rightarrow$ A + B + C + D makes a new specimen of (A + B + C + D) + $\phi(A + B + C + D)$. This has still not added complexity to the system. The original A + B + C and the floating D have exactly the same complexity whether tacked together or not, and producing an extra copy of A + B + C + D adds no complexity to what existed before [122].

Von Neumann acknowledges that the foregoing process is no more complex than crystal growth, and is not at all the thing that he is really after. To progress from exact duplication to complexity, in this context the construction of a thing that has never existed before, the description $\phi(A + B + C + D)$ must somehow morph into $\phi(A + B + C + D')$, where D' is an object that has never existed before. Run this new description through the Universal Constructor, and you have created a new thing as a result. A + B + C + D' is an object that never existed before [123].

At this point, the construction collapses. He provides no means of obtaining $\phi(A + B + C + D')$, given $\phi(A + B + C + D)$. For this to occur, Von Neumann suggests that we need a "random change of one element" [123]. By momentarily postulating randomness, he has temporarily forgotten that he has already forbidden it in his model. The idealized world in which the Universal Constructor operates is specifically designed to preclude random change. In order to obtain exact replication, all descriptions $\phi(X)$ are complete and unambiguous. Consequently, all processes, X, are completely determined, and not subject to random behavior. The only way of obtaining $\phi(X)$ is by observation of X. There is no process for modifying $\phi(X)$. To implicitly include an additional postulate that $\phi(X)$ can fluctuate randomly contradicts the initial postulate of complete and unambiguous determination of $\phi(X)$ that is necessary to assure the stability of the Universal Constructor.

Stated simply, in order for the Universal Constructor to demonstrate a threshold of complication beyond which "the production of a more complicated object from a less complicated object is possible," it must reside in a milieu that simultaneously forbids and requires random occurrences. Rather than a proof by construction that such a thing is possible, we see instead a proof contradiction that it is not. The Universal Constructor does not provide an exception to the Data Processing Inequality. He later came to the realization that brains do not work that way. Near the end of his life, he says of processes in brains, "There exist here different logical structures from the ones we are ordinarily used to in logics and mathematics" [54]. Had he not died an untimely death, his own insights might have led him to concepts that only in recent years have begun to be discovered by others.

7.7 INCOMPUTABLE DOES NOT MEAN NONENGINEERABLE?

Because their hierarchical closed-loop inferential entailment structures have internal ambiguities, impredicatives such as hypersets afford a way to reason about natural processes whose hierarchical closed-loop endogeny of causation also entail ambiguity. Such entailment of ambiguity abounds in living processes. Because they are a means of representing ambiguity, impredicatives provide an alternative to algorithms, mathematical processes to which ambiguity is a property absolutely forbidden [53].

The notion that no algorithm describes endogeny troubles some people. The claim of incomputability is often mistaken for a claim that endogenous processes cannot be understood rationally. This error stems from two common misconceptions. One is the "Mind Projection Fallacy," the notion that our epistemological descriptions drive events in ontological reality, or more simply the inability to distinguish our models from reality [124]. The other is the incorrect assumption that mathematics is exclusively algorithmic.

Attempting to "compute the incomputable" is a fools' errand, but impredicative models afford a practical way to engineer the incomputable. Hyperset theoretic models have been used to reason about semantic problems, uncovering ambiguities not previously noticed [125]. Impredicative models, including those justified by hyperset theory, and possibly implemented by the strategies described as "situation semantics," offer a promising approach to constructing impredicative models of endogenous processes [126–129].

Rational but qualitative prediction may seem like a modest promise, but since impredicatives describe effects ignored by reductionistic methods, the implications could be dramatic. They give us a radically new strategy to rationally decide what to do next in scientific and engineering research. It is the recognition that no algorithm behaves semantically, and that leads us to the consideration that perhaps the most fruitful near-term strategy for achieving a breakthrough in sensing and control of large processes is to "wire a human into the loop" [88,89].

It is feasible with existing technology to noninvasively "wire" a stream of electronically generated symbols directly into and out of the human nervous system. At the present time, Bach-y-Rita has successfully demonstrated medical prostheses that noninvasively couple electronically generated images into the human visual cortex, enabling the blind to see [130–135]. Enough is understood about cognitive processes of unconscious integration that it is believed to be practical to train a human to experience practically any stream of data as abstracted sensations and percepts, in effect directly sensing the meaning of the data [136,137]. NASA Ames is able to noninvasively couple data from the nervous system into electronics; their research goal is to control a spacecraft "by thought" [138,139].

Combining these technologies, and using the principles of endogenous processes, it is feasible, with currently available enabling technology, to develop a process by which a human operator temporarily "becomes one" with a large system (such as a computer network, a spacecraft, or even a steelmaking process). The operator will experience input data by direct sensation and perception, assess the totality of the process by feel, and initiate control actions mentally. Rather than reading numbers or graphics from a screen and manipulating knobs and keys, the operator will mentally modulate the process such that it continually "feels right."

Slightly further in the future, the implications of endogenous systems could be quite startling. Insights into endogeny might lead to the long sought after "super-Turing" man-made artifact that operates semantically. In physics, it might enable us to make practical use of the "spooky action at a distance" effects observed in quantum entanglement, or more prosaically, but no less dramatically, it might lead to a solution to the century-old "measurement problem." In biochemistry, it could lead to an explanation of prion behavior, which, in turn, could help us to discover new principles of protein folding (we do not really know the "rules" of protein folding thus far). In medicine, understanding prion behavior is the key to dealing with "Mad Cow Disease" and possibly a whole category of degenerative brain diseases including Alzheimer's.

Longer term, perhaps, the most dramatic effects of an understanding of endogenous processes would be in psychology. If abduction could be understood mathematically, then it might serve as the foundation for understanding of human semantic processes. Starting from the question of how data are abstracted sensations, sensations into percepts, and percepts into concepts, an overarching "theory of cognition" might be discovered. This might serve as the basis of answering many of the perplexing questions of human behavior.

ACKNOWLEDGMENTS

I wish to thank Jean-Loup Risler (University of Evry, France), Laurie Fleischman of BioLingua Research, and Andy Konopka (The Editor of this volume) for careful reading of advanced drafts of this manuscript as well as for suggesting several improvements of the text.

REFERENCES

1. Lovins A, Lotspeich C. Energy surprises for the 21st century. J Int Affairs 1999; 53(1):191–208.
2. Friedman MH. Traditional engineering in the biological century: the biotraditional engineer. J Biomech Eng 2001; 123:525–527.
3. Stenger R. A smashing end for Jupiter explorer. CNN, September 21, 2003 http://www.cnn.com/2003/ TECH/space/09/21/galileo.crash/ (accessed November 17, 2003).
4. Viking Mission to Mars. Jet Propulsion Laboratory, Pasadena CA. http://www.jpl.nasa.gov/news/ fact_sheets/viking.pdf (accessed November 17, 2003).
5. Caplinger M. Life on Mars. Malin Space Science Systems, April 1995. http://www.msss.com/http/ps/ life/life.html (accessed November 17, 2003).
6. Stenger R. Debate on Mars life rages long after Viking. CNN, July 20, 2001. http://www.cnn.com/ 2001/TECH/space/07/20/viking.anniversary/ (accessed November 17, 2003).
7. Levin RV. Overview Presentation: Odyssey gives evidence for liquid water on Mars. In: Hoover RB, Rozanov AY, Weatherwak, TI, eds. Instruments Methods and Missions for Astrobiology VII, Proceedings of SPIE Vol. 5163, 2003:128–135.
8. Nobel Assembly at the Karolinska Institute, Prions — a new biological principle of infection, Announcement of the 1997 Nobel Prize in Medicine.
9. Kocisko D, Come Priola JS, Chesebro B, Raymond G, Lansbury P, Caughey B. Cell-free formation of protease-resistant prion protein. Nature 1994; 370:471–474.
10. Jackson G, Hosszu L, Power A, Hill A, Kenney J, Saibil H. Reversible conversion of monmeric human prion protein between native and fibrilogenic conformations. Science 1999; 283:1935–1937.
11. Duda RO, Hart PE. Pattern Classification and Scene Analysis. New York: Wiley, 1973:4–5.
12. Rosen R. Life Itself: A Comprehensive Inquiry into the Nature, Origin and Fabrication of Life. New York: Columbia University Press, 1991:98.
13. Hertz H. The Principles of Mechanics, New York: Dover, 1994:1–2, Original German ed. Prinzipien Mechanik, 1894.
14. Rosen R. Life Itself: A Comprehensive Inquiry into the Nature, Origin and Fabrication of Life. New York: Columbia University Press, 1991:152.
15. Eccles PJ. An Introduction to Mathematical Reasoning. Cambridge: Cambridge University Press, 1997:264.
16. Rosen R. Essays on Life Itself. New York: Columbia University Press, 2000:374.
17. Rosen R. Essays on Life Itself. New York: Columbia University Press, 2000:158.
18. Rosen R. Essays on Life Itself. New York: Columbia University Press, 2000:161.
19. Kleene SC. Introduction to Metamathematics. 13th impression. New York: North-Holland, 2000:62.
20. Rosen R. Life Itself: A Comprehensive Inquiry into the Nature, Origin and Fabrication of Life. New York: Columbia University Press, 1991:59.
21. Rosen R. Life Itself: A Comprehensive Inquiry into the Nature, Origin and Fabrication of Life. New York: Columbia University Press, 1991:61.
22. Adler MJ. Aristotle for Everybody, New York: McMillan. 1978:39–56.

23. Aristotle. Metaphysics, translated by Ross, W.D. Book III, Part 2. Oxford: Oxford University Press, 1979. http://classics.mit.edu/Aristotle/metaphysics.html (accessed November 26, 2003).

24. Rosen R. Some epistemological issues in physics and biology. In Hiley BJ, Peat FD, eds. Quantum Implications: Essays in Honor of David Bohm. London: Routledge and Kegan Paul, 1987:314 327.

25. Rosen R. Life Itself: A Comprehensive Inquiry into the Nature, Origin and Fabrication of Life. New York: Columbia University Press, 1991:248–252.

26. Letelier JC, Marin G, Mpodozis J. Autopoietic and (M, R) systems. J Theor Biol 2003; 222:261–272.

27. Kercel SW. Endogenous causes — bizarre effects. Evol Cog 2002; 8(2):130–144.

28. Rosen R. Life Itself: A Comprehensive Inquiry into the Nature, Origin and Fabrication of Life. New York: Columbia University Press, 1991:241.

29. Bateson G. Steps to an Ecology of Mind. Chicago: University of Chicago Press, 1972:435.

30. Bateson G. Steps to an Ecology of Mind. Chicago: University of Chicago Press, 1972:464.

31. Clayman CB. The American Medical Association Encyclopedia of Medicine. New York: Random House, 1989.

32. Gavin WT, Kydland FE. Endogenous money supply and the business cycle. Rev Econ Dyn 1999; 2(2):347–369.

33. Margulis L, Sagan D. What is Life? Berkeley CA: University of California Press, 1995:17–20.

34. Freeman WJ. How Brains Make Up Their Minds. London: Weidenfeld and Nicholson, 1999:135–136.

35. Freeman WJ. How Brains Make Up Their Minds. London: Weidenfeld and Nicholson, 1999:7–8.

36. Fodor J. The Mind Doesn't Work That Way. Cambridge, MA: MIT Press, 2000:37–39.

37. Rosen KH. Discrete Mathematics and Its Applications. 4th ed, Boston: WCB McGraw-Hill, 1999:629–630, 666–668.

38. Goldrei D. Classic Set Theory. London: Chapman and Hall, 1996:67–69.

39. Aristotle. Metaphysics, translated by Ross WD. Book IX, Part 1, Oxford: Oxford University Press, 1979. http://classics.mit.edu/Aristotle/metaphysics.html (accessed November 26, 2003).

40. Kleene SC. Introduction to Metamathematics. 13th impression. New York: North-Holland, 2000:63.

41. Barwise J, Echemendy J. The Liar. Oxford: Oxford University Press, 1987:39–44.

42. Aczel P. Non-well-founded sets. Lecture Notes No. 14. Stanford CA: Center for the Study of Language and Information, 1988:3–17.

43. Kleene SC. Introduction to Metamathematics. 13th impression. New York: North-Holland, 2000:42–43.

44. Picard JRWM. Impredicativity and Turn of the Century Foundations of Mathematics: Presupposition in Poincare and Russell, Ph.D. dissertation in philosophy, MIT Cambridge, MA: 1993:142–144.

45. Akansu AN, Haddad RA. Multiresolution Signal Decomposition, San Diego: Academic Press, 1992:313–315.

46. Rosen R. Life Itself: a Comprehensive Inquiry into the Nature, Origin and Fabrication of Life. New York: Columbia University Press, 1991:81–89.

47. Goldrei D. Classic Set Theory. London: Chapman and Hall, 1996:70–71.

48. Rosen R. Life Itself: A Comprehensive Inquiry into the Nature, Origin and Fabrication of Life. New York: Columbia University Press, 1991:116–123.

49. Csete ME, Doyle JC. Reverse engineering of biological complexity. Science 2002; 295:1664–1669.

50. Schrödinger E. What is Life? First published 1944. Cambridge, UK: Cambridge University Press, 1992:80.

51. Mayr E. This is Biology. Cambridge, MA: Harvard University Press, 1997:22.

52. Silberschatz A, Galvin PB. Operating System Concepts. 5th ed. Reading, MA: Addison Wesley, 1998:224.

53. Knuth DE. The Art of Computer Programming. Vol. 1. 2nd ed. Reading, MA: Addison Wesley, 1973:5.

54. Von Neumann J. The Computer and the Brain. New Haven, CT: Yale University Press, 1958 republished in 2000:82.

55. Von Neumann J. Theory of Self-Reproducing Automata. Fifth Lecture, In: Burks AW. Urbana, IL: University of Illinois Press, 1966:77.

56. Barwise J, Echemendy J. The Liar. Oxford: Oxford University Press, 1987:139–141.

57. Barwise J, Echemendy J. The Liar. Oxford: Oxford University Press, 1987:164.

58. Korzybski A. Science and Sanity. 5th ed. New York: Institute of General Semantics, 1994, Second printing 2000:14.

59. Barwise J, Perry J. Situations and Attitudes. Stanford CA: Center for the Study of Language and Information, 1999:40.

60. Barwise J, Echemendy J. The Liar. Oxford: Oxford University Press, 1987:164–170.
61. Barwise J, Perry J. Situations and Attitudes. Stanford, CA: Center for the Study of Language and Information, 1999:27.
62. Bateson G. Steps to an Ecology of Mind. Chicago: University of Chicago Press, 1972:414.
63. Rosen R. Essays on Life Itself. New York: Columbia University Press, 2000:158–159.
64. Barwise J, Perry J. Situations and Attitudes. Stanford, CA: Center for the Study of Language and Information, 1999:229–230.
65. Von Neumann J. Die Axiomatisierung der Mengenlehre. Mathematische Zeitschrift 1928; 27:669–752.
66. Barwise J, Echemendy J. The Liar. Oxford: Oxford University Press, 1987:48–52.
67. http://www.quantum-earth.com/Anap/Lord%20Kelvin%20measurement%20quote.htm (accessed November 26, 2003).
68. Church A. An unsolvable problem of elementary number theory. Am J Mathematics 1936; 58:356.
69. Copeland BJ. The Church-Turing Thesis. In: Edward N Zalta, ed. The Stanford Encyclopedia of Philosophy. Fall 2002 ed. http://plato.stanford.edu/archives/fall2002/entries/church-turing/ (accessed November 23, 2003).
70. Church A. Review of Turing 1936. J Symbolic Logic 1937; 2:42–43.
71. Rosen KH. Discrete Mathematics and Its Applications. 4th ed. Boston: WCB McGraw-Hill, 1999:673.
72. Deutsch D. Quantum theory, the Church–Turing principle and the universal quantum computer. Proceedings of the Royal Society. Series A 1985; 400:97–117.
73. Rosen R. Essays on Life Itself. New York: Columbia University Press, 2000:126–131.
74. Rosen R. Essays on Life Itself. New York: Columbia University Press, 2000:128–129.
75. Rosen R. Essays on Life Itself. New York: Columbia University Press, 2000:131.
76. Schrödinger E. What is Life? First published 1944. Cambridge, UK: Cambridge University Press, 1992:78.
77. Schrödinger E. What is Life? First published 1944. Cambridge, UK: Cambridge University Press, 1992:77.
78. Rosen R. Essays on Life Itself. New York: Columbia University Press, 2000:136.
79. Rosen R. Essays on Life Itself. New York: Columbia University Press, 2000:137.
80. Rosen R. Essays on Life Itself. New York: Columbia University Press, 2000:138.
81. Chomsky N. A review of B.F. Skinner's verbal behavior. Language 1959; 35:26–58.
82. Miller G, Galanter E, Pribram K. Plans and the Structure of Behavior. New York: Henry Holt and Co., 1960.
83. Damasio AR. The Feeling of What Happens. New York: Harcourt, Brace and Company, 1999:67–71.
84. Sacks O. An Anthropologist on Mars. New York: Vintage, 1995:253–255.
85. Grandin T. Thinking in Pictures, New York: Vintage, 1995:19.
86. Sacks O. An Anthropologist on Mars. New York: Vintage, 1995:269–271.
87. Bach-y-Rita P. Nonsynaptic Diffusion Neurotransmission and Late Brain Reorganization. New York: Demos Publications, 1995:21.
88. Kercel SW, Caulfield HJ, Bach-y-Rita P. Bizarre hierarchy of brain function. In: Priddy KL, Angeline PJ, eds. Intelligent Computing: Theory and Applications, Proceedings of SPIE. Vol. 5103, 2003:150–161.
89. Kercel SW. Softer than soft computing. In: Kercel SW, ed. Proceedings of the 2003 IEEE International Workshop on Soft Computing in Industrial Applications, Piscataway, NJ: IEEE, 2003:27–32.
90. Kercel SW. The endogenous brain. J Integr Neurosci 2004; 3(1):61–84.
91. Damasio AR. Descartes' Error. New York: Avon Books, 1994:250.
92. Rosen R. Essays on Life Itself. New York: Columbia University Press, 2000:166–170.
93. Rosen R. Life Itself: A Comprehensive Inquiry into the Nature, Origin and Fabrication of Life. New York: Columbia University Press, 1991:242.
94. Jaynes ET. Probability Theory: The Logic of Science. Cambridge, UK: Cambridge University Press, 2003:116.
95. Bateson G. Steps to an Ecology of Mind. Chicago: University of Chicago Press, 1972:405–416.
96. Chomsky N. Language and Mind. New York: Harcourt, Brace and Jovanovich, 1972:100–114.
97. Kosslyn SM, Koenig O. Wet Mind. New York: The Free Press, 1992:401.
98. Kosslyn SM, Koenig O. Wet Mind. New York: The Free Press, 1992:432.
99. Kosslyn SM, Koenig O. Wet Mind. New York: The Free Press, 1992:433–434.
100. Pinker S. How the Mind Works. New York: W.W. Norton and Company, 1997:558–559.

101. McGinn C. The Mysterious Flame. New York: Basic Books, 1997.

102. Churchland PS. Brain-Wise. Cambridge, MA: MIT Press, 2002:232.

103. Simon HA. The Sciences of the Artificial. Third Printing. Cambridge, MA: MIT Press, 1999:14.

104. Bretthorst L. An introduction to parameter estimation using Bayesian probability theory. In: Fougère PF, ed. Maximum Entropy and Bayesian Methods. Dordrecht, The Netherlands: Kluwer Academic Publishers, 1989:53–79.

105. Fogel DB. BLONDIE24. San Francisco: Morgan Kaufmann, 2002:69–83.

106. Tou JT, Gonzalez RC. Pattern Recognition Principles. Reading, MA: Addison-Wesley, 1974:165–168.

107. Mitterauer B, Kopp K. The self-composing brain: Towards a glial — neuronal brain theory. Brain Cogn 2003; 51:357–367.

108. Miles GB, Parkis MA, Lipski J, Funk GD. Modulation of phrenic motorneuron excitability by ATP: consequences for respitory-related output in vitro. J Appl Physiol 2002; 92:1899–1910.

109. Nusbaum MP. Regulating peptidergic modulation of rhythmically active neural circuits. Brain Behav Evol 2002; 60:378–387.

110. Cover TM, Thomas JA. Elements of Information Theory. New York: Wiley, 1991:32–33.

111. Rosen R. Essays on Life Itself. New York: Columbia University Press, 2000:219.

112. Rosen R. Essays on Life Itself. New York: Columbia University Press, 2000:271.

113. Macready WG, Wolpert DH. The no free lunch theorems. IEEE Trans on Evolutionary Computation 1997; 1(1):67–82.

114. Goldberg DE. Genetic Algorithms in Search, Optimization and Machine Learning. Boston: Addison Wesley, 1989:218–221.

115. Koza JR, Bennett III FH, Andre D, Keane MA, Dunlap F. Automated Synthesis of Analog Electrical Grants by Means of Genetic Programming, IEEE Trans on Evolutionary Computation 1997; 1(2):109–1282.

116. Kosko B. Neural Networks and Fuzzy Systems. Englewood Cliffs, NJ: Prentice-Hall, 1992:263–294.

117. Kosko B. Neural Networks and Fuzzy Systems. Englewood Cliffs, NJ: Prentice-Hall, 1992:341.

118. Von Neumann J. Theory of Self-Reproducing Automata. Fifth Lecture, In: Burks AW, ed. Urbana, IL: University of Illinois Press, 1966:86.

119. Von Neumann J. Theory of Self-Reproducing Automata. Fifth Lecture, edited and completed by Burks AW, ed. Urbana IL: University of Illinois Press, 1966:79.

120. Von Neumann J. Theory of Self-Reproducing Automata. Fifth Lecture, In: Burks AW, ed. Urbana IL: University of Illinois Press, 1966:80.

121. Von Neumann J. Theory of Self-Reproducing Automata. Fifth Lecture, In: Burks AW, ed. Urbana IL: University of Illinois Press, 1966:84.

122. Von Neumann J. Theory of Self-Reproducing Automata. Fifth Lecture, In: Burks AW, ed. Urbana IL: University of Illinois Press, 1966:85.

123. Von Neumann J. Theory of Self-Reproducing Automata. Fifth Lecture, In: Burks AW, ed. Urbana IL: University of Illinois Press, 1966:86.

124. Jaynes ET. Clearing up the mysteries — the original goal. In: Skilling J, ed. Maximum Entropy and Bayesian Methods. Dordrecht, The Netherlands: Kluwer Academic Publishers, 1989:1–27.

125. Barwise J, Echemendy J. The Liar. Oxford: Oxford University Press, 1987:131–138, 164–170.

126. Barwise J. The Situation in Logic. Lecture Notes No. 17. Stanford CA: Center for the Study of Language and Information, 1989.

127. Cooper R, Mukai K, Perry J. Situation Theory and Its Applications. Vol. 1. Lecture Notes No. 22. Stanford CA: Center for the Study of Language and Information, 1990.

128. Barwise J, Gawron JM, Plotkin G, Tutiya S. Situation Theory and Its Applications. Vol. 2. Lecture Notes No. 26. Stanford CA: Center for the Study of Language and Information, 1991.

129. Aczel P, Israel D, Katagiri Y, Peters S. Situation Theory and Its Applications. Vol. 3. Lecture Notes No. 37. Stanford CA: Center for the Study of Language and Information, 1993.

130. Bach-y-Rita P, Tyler ME, Kaczmarek KA. Seeing with the brain. Int J Human-Comput Interaction 2003, 15:287–297.

131. Bach-y-Rita P, Kercel SW. Sensori-'motor' coupling by observed and imagined movement. Intellectica 2002/2, 35:287–297.

132. Bach-y-Rita P, Kaczmarek KA. Tongue Placed Tactile Output Device, US Pat. 6,430,459, 2002.

133. Bach-y-Rita P, Collins CC, Saunders F, White B, Scadden L. Vision substitution by tactile image projection. Nature 1969, 221:963–964.
134. Bach-y-Rita P, Kercel SW. Sensory substitution and the human-machine interface, Trends in Cognitive Sciences December 2003, 7(12):541–546.
135. Kercel SW, Bach-y-Rita P. Noninvasive coupling of electronically generated data into the human nervous system. In: Akay M, ed. Wiley Encyclopedia of Biomedical Engineering. New York: Wiley, 2006:1960–1974.
136. Manza L, Reber AS. Representing artificial grammars: Transfer across stimulus forms and modalities. In: Berry DC, ed. How Implicit is Implicit Learning? London: Oxford University Press, 1997.
137. Reber AS, Allen R, Reber PJ. Implicit and explicit learning. In: Sternberg R, ed. The Nature of Cognition. Cambridge, MA: MIT Press, 1999.
138. Wheeler KR, Chang MH, Knuth KH, Gesture-Based Control and EMG Decomposition, IEEE Trans on Systems Main and Cybernetics Part C 2006, 36(4):503–514.
139. Wheeler K, Jorgensen C, Trejo L. Neuro-electric Virtual Devices. NASA Ames, Computational Sciences Division, Papers and Publications Archive, Report # 284, November 15, 2000.

8 The von Neumann's Self-Replicator and a Critique of Its Misconceptions

Stephen W. Kercel

CONTENTS

8.1 INTRODUCTION

A frequent conventional mischaracterization of biology is its presentation as a mere derivative, nothing more than the cumulative effect, of processes traditionally studied by physicists and believed to belong to either physics or chemistry. According to some opinions, this misrepresentation of life science finds support in the notion that von Neumann's self-reproducing automaton provides an adequate paradigm for living processes. For instance Poundstone [1] writes:

"Not only can a machine manufacture itself, but von Neumann was able to show that a machine can build a machine more complex than itself. As it happens, these facts have been of almost no use (thus far, at least) to the designers of machinery. von Neumann's hypothetical automatons have probably had their greatest impact in biology."

Despite the realization that "von Neumann's work made no specific assertions about biology [2], some authors [3] are quick to suggest that the definition of life is that which "behaves" like von Neumann's models of a self-replicating automaton.

Extravagant claims of far more significance for von Neumann's models than he claimed for them himself are by no means limited to Poundstone [1,3]. For instance, in their foreword to one of von Neumann's works [4], the Churchlands see in von Neumann's work not only an adequate characterization of life itself but also so powerful a description of cognition that they offer as how a case can be made that von Neumann is the "Newton of the Mind." It is unlikely that von Neumann saw himself as the "Newton" of anything. Newton's significance was in discovering a brief set of general principles that account for a wide range of mechanical phenomena. In contrast, von Neumann articulated the urgent need for a brief set of general principles that account for a wide range of automatic phenomena, and offered some background and suggestions as to how such a theory might be discovered. However, he never claimed to have reached nor even approached the goal as he writes:

"Let me note, in passing, that it would be very satisfactory if one could talk about a "theory" of such automata. Regrettably, what at this moment exists — and to what I must appeal — can as yet be described only as an imperfectly articulated and hardly formalized "body of experience" [5].

There are several substantive claims made about von Neumann's models that he himself did not make for the compelling reason that the claims are groundless. The first is the claim that von Neumann's complete and unambiguous description of an automaton is a paradigm for genetic description in an organism. "Biologists have adopted von Neumann's view that the essence of self-reproduction is organization — the ability of a system to contain a complete description of itself and use that information to create new copies [6]." What von Neumann saw was that a complete and unambiguous description of itself is an absolutely *indispensable* requirement for his self-reproducing automata. In contrast, speaking of the gene in the organism, he observed that "it probably need not contain a complete description of what is to happen, but only a few cues for a few alternatives [7]." Because completeness of description is indispensable to the operation of the automaton, and organisms lack that completeness, it is not reasonable to suppose that the behavior of the automaton is a description of the distinguishing features of the organism.

Another claim is that von Neumann's 29-state model is a mathematical description of biological reproduction, and stands in contrast to the mechanistic reproduction of crystal growth [1]. The fact is that von Neumann regarded the capacity to pass on heritable mutations as the property that distinguishes biological reproduction from mechanistic reproduction. "One of the difficulties in defining what one means by self-reproduction is that certain organizations, such as growing crystals, are self-reproductive by any naive definition of self-reproduction, yet nobody is willing to award them the distinction of being self-reproductive. A way around this difficulty is to say that self-reproduction includes the ability to undergo inheritable mutations as well as the ability to make another organism like the original [8]." A frequently unnoticed fact is that reproduction in the 29-state automaton *is* just like crystal growth; von Neumann specifically *excluded* the capacity to pass on heritable mutations from the 29-state model. He did so for the specific reason that he could not devise an unambiguous description of the capacity to evolve [9]. (*Note*: In his kinematic model, von Neumann did envision a machine that, given a description of itself, would produce offspring more complicated than itself. However, as demonstrated later in this chapter, the development of that model was based on contradictory axioms.)

The confusion that von Neumann's models cause in biology seems to stem from the error of supposing that a key distinguishing feature of life is the capacity of an entity to make an exact copy of itself from an internally held complete description of itself, and then attributing this concept to von Neumann [3]. As noted above, von Neumann did not see biological organisms as containing complete descriptions of themselves. Furthermore, to distinguish a complex process from a simple mechanism, he believed that the key property was the capacity of a process, having been given a complete description of itself, to produce a process more complicated than itself. "There is thus this completely decisive property of complexity, that there exists a critical size below which the process of synthesis is degenerative, but above which the phenomenon of synthesis, if properly arranged, can become explosive, in other words, where syntheses of automata can proceed in such a manner that each automaton will produce other automata, which are more complex and of higher potentialities than itself [10]." Again, this is a capacity that he specifically excluded from the 29-state model.

The task that von Neumann set for himself was not to seek a mathematical theory of either life or mind, but rather to look for some rough guidelines about how to resolve some perplexing questions in computation by asking questions about organisms and their nervous systems. He first presented the overall strategy in his 1948 lecture to the Hixson symposium, titled "The General and Logical Theory of Automata [11]." It must be appreciated that despite the title, the lecture was not the recitation of an actual theory, but rather was a statement of the conditions that any such theory would need to meet. He provided more details in a series of lectures delivered at the University of Illinois the following year [12]. He then spent a number of years designing a construct known as the 29-state automaton [12]. Finally, he articulated what he saw as fundamental differences between the computer and the biological brain [13].

8.2 THE GENERAL AND LOGICAL THEORY

Where von Neumann stood out from the crowd was in his ability to pose questions. He saw an objective of a general and logical theory of automata as answering two basic questions. "How can reliable systems be constructed from unreliable components? What kind of logical organization is sufficient for an automaton to be able to reproduce itself [14]?" He hoped that the process of constructing answers to these questions would point the way to general theoretical principles of automata.

8.2.1 THE FIRST QUESTION: RELIABILITY FROM UNRELIABILITY

Regarding the first question, he explored it, but not in detail [15]. The concern was that if a system were constructed of many unreliable components, it seems intuitively inevitable that their probabilities of error should accumulate such that the probability of error of the total system approaches 50:50, thus rendering the system useless. He found that for "single line" automata, he could put several unreliable automata in parallel and use a voting scheme that is mediated by several other unreliable "majority organs." If conditions were contrived in just the right way, the probability of error of the overall network would be a Taylor expansion of the error in any one of the parallel automata. If the probability of error was small, the high order terms would become insignificant, and the probability of error of *a system of many components* would be approximately proportional to the probability of error of *one of its components*. Thus, there exists a counterexample to the hypothesis that large systems of unreliable components are inevitably useless. However, the counterexample does not point to a strategy for reliability. The number of components necessary for implementing the strategy in practice, even for modest computational problems, turns out to be greater than the number of electrons in the Universe.

He then considered a multiplexing scheme. The idea here is that data are carried on a large bundle of multiple lines, and the automaton determines the state of the bundle by a voting scheme that leads to one of three assessments, true, false, or "I don't know." With the multiplexing scheme that von Neumann designed, "using large enough bundles of lines, any desired degree of accuracy (i.e., as small a probability of malfunction of the ultimate output of the network as desired) can be obtained by a multiplexed automaton" [16]. Remarkably, increasing the number of lines in the bundle by one order of magnitude can decrease the probability of error by *eight* orders of magnitude. The problem with the scheme is that it requires bundles on the order of 20,000 lines to achieve reasonable reliability, beyond the scope of the enabling technologies of the time.

He mentions in passing that the numbers are "not necessarily unreasonable" for what he calls "the microcomponentry of the human nervous system [17]." It is important to recognize that he makes no claim that brains are actually wired this way. He merely notes that neurons are small enough and abundant enough that brains could be wired this way.

In investigating the issue of how reliable systems could be constructed from unreliable components, von Neumann suffered from two severe handicaps. First, there was no theory of "probability as extended logic" by which reliability could be studied in principle. Second, the scheme he identified for implementing reliability was unfeasible with the technology of his time, and he had no practical means of testing any theoretical speculations that he might develop. Given the apparent hopelessness of the task, von Neumann took the eminently sensible approach of briefly saying what he could say on the subject, and then moved on.

8.2.2 THE SECOND QUESTION: SELF-REPLICATION

Another answer to the problem, suggested by biology, is that in the face of a failed component organisms either regrow the component or grow a novel work-around to compensate for the lost function. If replacement parts could be grown, in process, it might solve the reliability problem, and von Neumann sought an explanation of how it might happen. He suspected that the process was the result of replication with heritable mutation. He hypothesized that if an automaton process passes a "threshold of complication," then the offspring of the automaton can be more complicated than the parent. He then attempted to prove the hypothesis by construction. Unfortunately, in his construction, the development of the kinematic model of self-replication, he obtained this propensity to evolve by implicitly (and unknowingly) requiring contradictory hypotheses. As is discussed in some detail below, he simultaneously required and forbade noise in the kinematic model. In the models of self-replication that he developed subsequent to the kinematic model, he strengthened the prohibition on noise, focused strictly upon exact replication, and *ignored* evolution.

8.3 THE ILLINOIS LECTURES

The year after the Hixson lecture, von Neumann discussed his ideas on self-replication in much greater detail in a series of five lectures at the University of Illinois [12]. These lectures culminated in what von Neumann would later call the kinematic model of self-reproducing automaton. It is the interpretation of this model that has caused endless confusion in subsequent generations.

8.3.1 COMPUTING MACHINES IN GENERAL

It is important to note von Neumann's practical perspective. He was constructing the most complicated devices ever built by the Hand of Man. In principle, if any part failed, the whole might fail. The practical problem was how to get reliable operation out of such a big collection of parts? The strategy he pursued was to ask how organisms do it. However, because he retained a strictly reductionistic bias, presuming that the behavior of the whole is completely accounted for by the cumulative effect of local interactions of the parts, he chose to ignore much of what organisms actually do. Thus, he saw the problem as "the behavior of very complicated automata and the very specific difficulties caused by high complication," where high complication was entailed by the extreme number of parts in the system [18].

He presupposes that brains are automata, and that a general theory of automata would yield insights about brains as well as computers [19]. He focuses on computers noting that "of all automata of high complexity, computing machines are the ones which we have the best chance of understanding [18]." On its face, it seems a sensible enough strategy. We ask questions about computers because we have a serious possibility answering them. There is a reason why this is so. The computer is the closest we have come to physically realizing the reductionistic ideal, and we have nearly

400 yr of spectacularly fruitful experience asking and answering questions about reductionistic idealizations of real-world processes. In later work, von Neumann would conclude that the brain differs substantially from the computer [20]. However he never questioned his presupposition that brains are automata, nor did he consider the possibility that the difference between brains and computers might arise from the fact that the computer is a close approximation of the reductionistic ideal, while the brain departs substantially from the reductionistic paradigm [21].

He next enters into a long digression of the benefits that would accrue if computers were widely used to solve nonlinear partial differential equations. He seemed to think that most, if not all, of the open questions of science are fully characterized by nonlinear partial differential equations, and would be decisively settled by an effective strategy of solving them. He notes that computational simulation of phenomena should be less costly and dangerous than experimentation.

He then turns his attention to the idea of complexity. In a world of strict reductionism, the whole is an epiphenomenon fully captured by accounting for all the parts. Complexity is an indication of many parts. How does one measure the complexity of a system? Count the parts. There is a problem as to what constitutes a part. There is an obvious, if not quite reducible, solution. Count the parts that matter. However, the key point is that in this context, von Neumann's notion of complexity is "lots of parts."

Later in the lecture he identifies a profound problem that was ignored in his time, and, for the most part, continues to receive short shrift. "The reason for using a fast computing machine is not that you want to produce a lot of information. After all, the mere fact that you want some information means that you somehow imagine that you can absorb it, and, therefore, wherever there may be bottlenecks in the automatic arrangement, which produces and processes this information, there is a worse bottleneck at the human intellect into which the information ultimately seeps [22]." The weakest point in any computer system is the human–machine interface.

He concludes the lecture with a few practical observations about computing. A computation may require billions of invisible intermediate values; the user is interested in a very short list of results. Memory must be large enough to hold the intermediate values. Although memory can be implemented in the same switching elements used to perform the logical operations, it is prohibitively costly to do so, and memory can be implemented with far less costly technologies than are used for processing. These continue to be constraints on computer design even to the present day.

8.3.2 RIGOROUS THEORIES OF CONTROL AND INFORMATION

What does von Neumann means by "the rigorous part of information theory is just a different way of dealing with formal logics [23]." Formal logics are systems of logical variables, the traditional operators and connectives, and the quantifiers "some" and "all," and are currently known as first-order logic [24]. According to von Neumann, "If you have this machinery you can express anything that is dealt with in mathematics; or that is dealt with in any subject, for that matter, as long as it is dealt with sharply [23]." The quality of sharpness is reflected in a complete and unambiguous description.

It is at this point that von Neumann reveals a major insight in logical representation. He proves that the McCulloch and Pitts neural net and the Turing machine are equivalent. The neural net is built up from parts, whereas the Turing machine is based on a "black box" automaton, but both express the full range of formal logic. Any behavior that can be fully and *unambiguously* described can be represented in formal logic, a Turing machine, or a McCulloch and Pitts neural net.

This was an unintended consequence of the work of McCulloch and Pitts. They merely wanted to axiomatize what they considered the essential features of neuron behavior so that they could reason about neural structures without endless diversions into neuron physiology. They had no expectation that a full representation of formal logic could be built up from their axiomatized neurons. Von Neumann saw the result as potentially profound. Any behavior that could be unambiguously described could be implemented on a network of these simplified neurons. From the reductionistic perspective, it is supposed that all real-world behaviors can be unambiguously described. Although von Neumann clearly believed this, he was remarkably reluctant to present it as a foregone conclusion. He noted that even a seemingly simple concept like triangularity might lead to an ever-increasing cascade of description.

He expected that beyond a threshold level of complication, a finite neural network would in some novel way encompass a full description of any concept or behavior, even if the direct description of the behavior were absurdly large. "They may easily be in this condition already, where doing a thing is quicker than describing it, where the circuit is more quickly enumerated than a total description of all its functions in all conceivable conditions [25]." Gödel's theorem indicates that for a system of propositions, there exists a true proposition consistent with the system, but unprovable from the propositions within the system. von Neumann cited Gödel and it is clear that he saw this unprovable proposition as the novelty that would be abducted by a system that exceeds the threshold of complication. He admitted to twisting the theorem, but perhaps he twisted it a bit too much. Gödel's theorem is not limited to systems of "high complication"; it applies when the complication passes the rather undemanding level of nontriviality.

von Neumann also sought theoretical justification for his notion of a threshold of complexity in the Turing machine. "The importance of Turing's research is just this: that if you construct an automaton right, then any additional requirements about the automaton can be handled by sufficiently elaborate instructions. This is only true if A is sufficiently complicated, if it has reached a certain minimum level of complexity [26]." Again he misses a key point. There is a threshold of complexity for the automata portion of a Turing machine below which it will fail to do some things, but that threshold is *low*. The automaton need only be sufficiently complicated encompass the rather short list of relationships and operators that define formal logic.

He noted that as long as the automaton is beyond the "certain minimum level of complexity," it will perform exactly the same functions as a more complicated automaton, provided its tape includes a complete and unambiguous description of the more complicated automaton. However, this function is not to be confused with abduction of novel behavior. A Turing machine is not merely an automaton; it is an automaton *and* a tape. In the absence of either, the process collapses. Provided the state set of the simpler automaton captures all the relationships of formal logic, a Turing machine with an automaton with a sparse state set and a long program on the tape, and another Turing machine with an automaton with a rich state set and a short program on the tape might well represent exactly the same information. However, the two configurations represent nothing but alternative distributions of information between the automaton and the tape; nothing new has been abducted because one automaton is richer than the other.

In an addendum to the second lecture, Burks refers to a problem that was obscure in his day, but is widely known at present, the "halting problem [27]." The problem is this: Given any computer program as an input data set, does there exist an algorithm that can tell whether the program will execute to completion and halt, or become caught in an endless loop and execute forever? Clearly an algorithm to detect endless loops would have enormous practical value in compiler design [28]. This problem is especially perplexing because it cannot even be tested by exhaustive search. If one tries to test by letting the program run, no matter how long we wait before we intervene to abort the run, it remains possible that the program only needs an increment longer before it would halt on its own. Turing proved by contradiction that the halting problem is undecidable. There exists no proof that the problem can be solved, and there exists no proof that it cannot.

von Neumann also suspected that he saw emergent complexity of automata in this phenomenon. "Turing proved that you cannot construct an automaton, which can predict in how many steps another automaton, which can solve a certain problem will actually solve it [29]." However, all that the halting problem does is to illustrate Gödel's theorem. The property of a program to either halt or loop exclusively is consistent with the propositions implied by the program, but is patently unprovable from those propositions. There is nothing novel or emergent in this. The property of either halting or looping is unconditionally present in any program. The fact that the presence of the property cannot be uncovered by the techniques of formal logic has no effect on the fact that it is always present and immutable; it does not suddenly emerge given the right conditions.

8.3.3 STATISTICAL THEORIES OF INFORMATION

The discussion of abstractions such as the halting problem is typical of the tasks that fall within the traditional purview of formal logic. "Throughout all modern logic, the only thing that is important is whether a result can be achieved in a finite number of elementary steps or not. The size of the number of steps, which are required, on the other hand, is hardly ever a concern of formal logic. Any finite sequence of correct steps is, as a matter of principle, as good as any other. It is a matter of no consequence whether the number is small or large, or even so large that it could not possibly be carried out in a lifetime or in the presumptive lifetime of the stellar universe as we know it [30]." Clearly, the value of formal logic is in helping to avoid the undertaking of fools' errands such as attempting to include an "endless loop detector" in a compiler.

However, in practical computer design, there is another problem. Discovering that a problem is not a fools' errand, that it indeed has a solution in a finite number of steps, can we estimate that number? "There are two reasons. First, automata are constructed in order to reach certain results in certain preassigned durations, or at least in preassigned orders of magnitude of duration. Second, the componentry employed has in every individual operation a small but nevertheless nonzero probability of failing. In a sufficiently long chain of operations, the cumulative effect of these individual probabilities of failure may (if unchecked) reach the order of magnitude of unity at which point it produces, in effect, complete unreliability [30]."

It was for these reasons that von Neumann saw the need to go beyond formal logic, or "strict and rigorous questions of information," to a generalization of logic that he characterized as "statistical considerations involving information [31]." "The axioms are not of the form: if A and B happen, C will follow. The axioms are always of this variety: if A and B happen, C will follow with a certain specified probability, D will follow with another specified probability, and so on [32]."

Much of the third lecture is concerned with defending the concept rather than detailing it. He notes the shortcomings of the "frequency interpretation" of probability theory that was popular in his time. He mentions that the economist Keynes had devised an axiomatic interpretation of probability, but did not develop a comprehensive theory of probability as extended logic. "The only tie to strict logics is that when the probability is one you have an implication, when the probability is zero you have an exclusion, and when the probability is close to one or close to zero you can still make those inferences in a less rigorous domain [33]."

von Neumann did some work on a theory of probability as extended logic. In the *Collected Works*, there is a related paper [34]. In his papers he attempted to identify logical propositions about physical systems as comments about the outcomes of measurements. He developed an axiomatic definition of the logic that arises from this paradigm. He then defined a probability function, which can provide the foundations of "probability logics," in which "strict logic" is the degenerate case for probabilities of 1 and 0. He saw probability as an extension or generalization of formal logic rather than as a discipline that could be reduced to formal logic.

Clearly, he was not satisfied that his insights on probability as extended logic were sufficient to serve as a foundation for a probabilistic theory of automata. The remainder of the third lecture was devoted to observations in thermodynamics and information theory that might point the way to such a theory. He noted that in a thermodynamically closed system, entropy is a measure of how energy is degraded as it participates in a process. It is found to be proportional to the logarithm of what is *not known* about the system. By analogy to thermodynamic entropy, information entropy can be used to measure information in dealing with problems such as the carrying capacity of telephone circuits.

That the thermodynamically inspired strategy is useful can be seen in the fact that it shows the efficacy of digital representation. "Digitalization is just a very clever trick to produce extreme precision out of poor precision. By writing down 30 binary digits with 30 instruments, each of which is only good enough that you can distinguish two states of it (with intrinsic errors maybe on the 10% level), you can represent a number to approximately one part in a billion. The main virtue of the digital system is that we know no other trick, which can achieve this. From the information point of view, it is clear that this can be done, because the entropy in 30 binary instruments is 30 units, and something, which is known to one part in a billion has an entropy of the logarithm of a billion (to the base two), or about 30 units [35]."

Although he could point to various rudiments, in both his work and that of others, he noted that no comprehensive theory of extended logic existed at that point. "I have been trying to justify the suspicion that a theory of information is needed and that very little of what is needed exists yet [35]." He anticipated that the new theory would be more based on analysis than counting. "There is reason to believe that the kind of formal logical machinery we will have to use here will be closer to ordinary mathematics than present day logics is. Specifically, it will be closer to analysis, because all axioms are likely to be of a probabilistic and not of a rigorous character. Such a phenomenon has taken place in the foundations of quantum mechanics [35]." Remarkably, he thought that the new system of logic would require some sort of context dependency. "It is likely that you cannot define the function of an automaton, or its efficiency, without characterizing the milieu in which it works by means of statistical traits like the ones used to characterize a milieu in thermodynamics. The statistical variables of the automaton's milieu will, of course, be somewhat more involved than the standard thermodynamical variable of temperature, but they will probably be similar in character [35]."

8.3.4 THE ROLE OF HIGH AND OF EXTREMELY HIGH COMPLICATION

The problem that von Neumann faced was daunting. He had a rigorous theory of formal logic that would protect him from embarking on fools' errands. He had elements of a statistical theory of automata that gave him some indication of the limits of what was practically feasible. He was endowed with a virtually endless supply of 1940s vintage electronic parts. With no more than these sparse resources, he was undertaking to construct devices orders of magnitude more complicated than anything ever attempted before. Not surprisingly, he turned to the one additional source of guidance available to him. How did nature solve the problem of operability in the face of extreme complication in the design of the nervous system?

It is in no sense a criticism of von Neumann that rather than use his outdated performance numbers, it is more instructive to use contemporary numbers. Following a bit of sage advice from von Neumann whose value has not been diminished by the passage of time, modern computer designs do not use active switching elements for memory; instead, memory is typically provided by a bank of perhaps 10^9 capacitors. A modern microprocessor has on the order of 10^8 transistor switches. In brains, a synapse is triggered by calcium ions to release neurotransmitters [36], and its switching ability is at least as sophisticated as a transistor in a microprocessor. The conservative estimate is that there are 10^{14} synapses in the adult brain [37]. Thus, although the basis for counting is different from the one used by von Neumann, if we assess complication by his strategy of counting switching elements, we still get the same answer that he got. "So the human nervous system is roughly a million times more complicated than these large computing machines [38]." However, in addition to as many switches as a million Pentium IV microprocessors, the brain has processes of nonsynaptic diffusion neurotransmission, and glial transmission and interaction between the glial and neural networks that were completely unknown to von Neumann [21].

As von Neumann points out, the rate of neural operations is limited by a fatigue recovery rate. His observation that if a brain were a computer, then its clock rate would be about 2×10^2 Hz, is probably about right. Clock rates of 2×10^9 Hz are unremarkable in desktop computers. Thus, on some level, it is fair to say that a computer is ten million times faster than the brain. In terms of switching operations per hour, it might be argued that a microprocessor does ten times as many as a brain. Given the unduplicated cognitive power of the brain, all this shows is that there is much more happening in brains than the cumulative effect of switching operations.

The differences in size that von Neumann discussed have largely become irrelevant. Synapse sizes are on the order of one micron [39]. Transistor sizes in microprocessors are on the order of 0.2 μm. A difference that matters more than size is that the computer has a static morphology and the brain has a dynamic morphology.

The comparison of energy consumption is also instructive. A typical microprocessor has 10^8 switches and a clock rate of 10^9 Hz. Supposing that every switch operates on every clock tick, it performs 10^{17} switching operations per sec. At a typical power consumption of 100 W, it is consuming 10^9 ergs per sec, or 10^{-8} ergs per switching operation. Assume von Neumann is right that the brain consumes 25 W, or about one fourth as much as a Pentium IV, and that the brain does one tenth as many switching operations per second. If so, the brain requires about 2.5×10^{-8} ergs per switching operation. Given the informal basis on which the estimate is derived, the energy per switching operation can be fairly said to be about the same for microprocessors as for brains. In either case, it is about a million times the energy that von Neumann estimated to be the thermo-dynamical minimum for a switching operation. However, the more significant fact is that the 25 W brain is doing much more with its energy than the 100 W computer.

In his discussion of both computers and brains, von Neumann seemed to worry excessively about the distinction between digital and analog processing. As he notes, both kinds of systems arguably do both kinds of processing. Analog-to-digital and digital-to-analog conversion are minor engineering issues; while they may have been a bit less minor in von Neumann's time, clearly they were not a fundamental problem even then. He seems to have regarded analog and digital information as two fundamentally different kinds of information. "All computing automata fall into two great classes in a way, which is immediately obvious and, which, as you will see in a moment, carries over to living organisms. This classification is into analogy and digital machines [40]." There is a detail that was available to him whose significance he seems to have failed to notice. There is no fundamental difference between analog and digital data. According to the Shannon sampling theorem, under band-limited sampling (which is always feasible in principle because all real-world signals have finite bandwidth), they are two different representations of identically the same information [41].

The other distinction he saw was in the area of robustness. He first attributed the capacity of both computers and brains to withstand environmental insult to a generic property of soundness of the constituent materials. "Thus the natural materials have some sort of mechanical stability and are well balanced with respect to mechanical properties, electrical properties, and reliability require-ments. Our artificial systems are patchworks in which we achieve desirable electrical traits at the price of mechanically unsound things [42]." He sees the distinction in robustness as a function of size, and size as a function of material properties. "And so the differences in size between artificial and natural automata are probably connected essentially with quite radical differences in materials."

Nevertheless he does note that, "if a membrane is damaged it will reconstruct, but if a vacuum tube develops a short between its grid and cathode it will not reconstruct itself." The essential idea is that an organism will spontaneously find a way to work around any local failures of its components. In that property, he does see a hint of an organizational difference between organisms and simple mechanisms. "The fact that natural organisms have such a radically different attitude about errors and behave so differently when an error occurs is probably connected with some other traits of natural organisms, which are entirely absent from our automata. The ability of a natural organism to survive in spite of a high incidence of error (which our artificial automata are incapable of) probably requires a very high flexibility and ability of the automaton to watch itself and reorganize itself [43]."

In pondering the reason for the robustness of organisms, he notes a property that more recent observers interpret as the tendency of a living process to act to preserve its identity. Freeman sees neural dynamics as entailing a circular causation that leads to internally determined goals and goal-seeking behavior; he defines intentionality as a behavior that seeks the goal of preserving the integrity of the hierarchical closed loop of causal entailment [44]. Freeman also sees intentionality as a necessary condition for intelligent behavior. While necessary, it may not be sufficient. It is such a necessary consequence of the entailment structure, that Damasio observes it even in organisms without brains. "But the form of an intention is there, nonetheless, expressed by the manner in which the little creature manages to keep the chemical profile of its internal milieu in balance while around it, in the environment external to it, all hell may be breaking loose [45]."

The *intentionality* in Freeman and Damasio is very close to the notion of *autonomy* in von Neumann. In fact, von Neumann notes an even more remarkable detail, that if the context changes severely enough, the autonomous process, in the interest of preserving its own identity, might find it necessary to dispense with some of its functional components. In this connection, it turns out that each of the functional components has its own identity, and what is in the self-determined self-interest of the one may no longer be in the self-determined self-interests of the others. "When parts are autonomous and able to reorganize themselves, when there are several organs each capable of taking control in an emergency, an antagonistic relation can develop between the parts so that they are no longer friendly and cooperative. It is quite likely that all these phenomena are connected [43]."

8.4 REEVALUATION OF THE PROBLEMS OF COMPLICATED AUTOMATA — PROBLEMS OF HIERARCHY AND EVOLUTION

The questions that von Neumann raised in the first four lectures led to some awkward difficulties. First, there is the issue of complexity as a measure of complication beyond a threshold. Counting the parts seems like a reasonable measurement strategy, but some parts have more value to the process than others, and identifying the valuable parts, or even identifying the concept of value, are ambiguous problems in their own right. Value depends on the end to which the parts add value. Nevertheless, supposing the problem of measuring complication could be solved, there was another problem. Formal logic, Turing machines and McCulloch and Pitts neural nets could provably perform any behavior that could be completely described by a finite list of unambiguous statements. He recognized that nonambiguity was an absolute necessity for any of the three versions of the strategy. However, when von Neumann addressed the problem of trying to unambiguously identify the seemingly straightforward concept of triangularity, he found that it depended on an endlessly changing context, and that as he considered the necessary qualifying statements, they seemed to increase without bound. Again, supposing the problem of disambiguation could be solved, formal logic still remained inadequate to the task of accounting for automatic behavior. He hypothesized that an extension or generalization of logic was needed, but noted that among its properties, it must take context dependency into account. This becomes especially crucial in the case of organisms, because in the face of a changing context they appear to develop self-determined autonomous behaviors.

The deeper problem is that the problems of measuring complication, ambiguity, context dependency, and autonomic behavior seem utterly inconsistent with the reductionist paradigm in which von Neumann was steeped. He was aware of that and perplexed by it. He also thought that he had an elegant solution that finessed all four problems in a single stroke, and did so within the reductionist paradigm. He thought that the solution was an automaton whose complication exceeds a threshold of complexity.

The solution is based on the hypothesis that there exists an automaton whose degree of complication passes a threshold of complexity, and that in consequence, it is endowed with the capacity to make other automata like itself, but more complicated than itself. This would eliminate the problem of complexity as a seemingly ambiguous measure of complication; complexity would be quality of an automaton that could produce another automaton that contains more information than itself. It would do away with the infinite cascade of definition needed to disambiguate even simple concepts; the complex automaton would be the whole to which the cascade of description would converge, without the need for writing down the whole cascade. Its capacity for producing more information than it contains could lead to novel responses to novel changes in context. If that capacity expanded to the ability abduct enough novelty to preserve itself (or at least its offspring) in the face of a disruptive environment such a capacity would seem to account for the self-preserving autonomy observed in organisms.

In the earlier lectures, he indicated that he thought that he could find support for the idea of a threshold of complexity in the work of Turing and Gödel. There is a threshold implied in the work of both. However, in both cases, it is a threshold of simplicity rather than a threshold of complexity. Both a Turing machine, and a logical system that meets the necessary conditions of Gödel's theorem, have, at most, the capacity to find all the conclusions deducible from a group of hypothetical propositions. In effect, they perform the simple function of uncovering the information already present. The threshold minimum of description that must be inherent in a system to perform this simple function is nothing more than the minimum amount of description necessary to assure that the system performs all the operations of formal or first-order logic. As the sparse set of state transformations in the automaton in the Turing machine indicates, this is not a high threshold of complexity. As the behavior of the Turing machine indicates, no new information is created. This threshold of simplicity is nothing like von Neumann's threshold of complexity, a level of extreme complication beyond which an automaton creates novel information.

It might be fair to ask if the idea of a threshold of simplicity might perhaps be indicative of a thresholding principle. If there is a threshold of complication required for an automaton to uncover old information, might there not be a higher threshold of complication required for an automaton to abduct new information? It turns out that no such higher threshold exists. The idea is contradicted by the data processing inequality in information theory. If X, Y, and Z form a Markov chain, $X \to Y \to Z$, then $I(X;Y) \geq I(X;Z)$, or, "no processing of Y, deterministic or random can increase the information that Y contains about X." In other words, "no clever manipulation of the data can improve the inferences that can be made from the data [46]."

Although he was probably unaware of a rigorous demonstration of data processing inequality, von Neumann was intuitively aware of the concept, and the genuine limitations that it imposes. "So, one gets a very strong impression that complication, or productive potentiality in an organization, is degenerative, that an organization which synthesizes something is necessarily more complicated, or of a higher order, than the organization it synthesizes [7]." Although he acknowledged the principle, he believed that it had effect only below a threshold of complication. "We will stick to automata which we know completely because we made them, either actual artificial automata or paper automata described by a finite set of logical axioms. It is possible in this domain to describe automata, which can reproduce themselves. So at least one can show that on the site where one would expect complication to be degenerative, it is not necessarily degenerative at all, and, in fact, the production of a more complicated object from a less complicated object is possible. The conclusion one should draw from this is that complication is degenerative below a certain minimum level [7]." Above that threshold, the automaton should be generative, and he calls this the "completely decisive property of complexity [10]."

The problem is that we have two contradictory claims. The data processing inequality says unconditionally that we *cannot* improve the inferences that can be made from a set of data. The essence of von Neumann's fifth lecture is the demonstration of a proof by construction that there is a condition under which we *can* improve the inferences that can be made from a set of data. In the presence of such a conflict, one must consider the premises, including those implied but not explicitly stated.

As a logician, von Neumann was painfully aware that his premises might lead him astray, and he saw that defining the premises from which he would start this construction was a major difficulty.

He called his premises axioms, and called the process of identifying them "axiomatizing the automaton." He saw the problem as one of defining the parts of the automaton at just the right scale, and then forming axioms in terms of the descriptions of the parts. He was rightly concerned that if he started at too fine a scale, then most of his development would be bogged down in minute proofs of propositions that could reasonably be treated as intuitively obvious. "We are interested here in organizational questions about complicated organisms, and not in questions about the structure of matter or the quantum mechanical background of valency chemistry [47]." He was also rightly concerned that if he started at too coarse a scale, he would be in effect hypothesizing the very properties he sought to prove, or in logician's terms, begging the question. "So, it is clear that one has to use some common sense criteria about choosing the parts neither too large nor too small [47]." There is no algorithm for setting the premises; he chooses their scale by common sense.

In fact, this is only the beginning of his application of common sense. "Even if one chooses the parts in the right order of magnitude, there are many ways of choosing them, none of which is intrinsically much better than any other [47]." Here he admits the reason that the definition of his elements must be done on the basis of common sense judgment. Judgment is necessary because the answer to the question is inherently ambiguous. "Even if the axioms are chosen within the common sense area, it is usually very difficult to achieve an agreement between two people who have done this independently [48]." Indeed, he even explicitly admits that the problem is inherently ambiguous. "So, while the choice of notations, of the elements, is enormously important and absolutely basic for an application of the axiomatic method, this choice is neither rigorously justifiable nor humanly unambiguously justifiable. All one can do is to try to submit a system, which will stand up under common sense criteria [48]." There is one other criterion that a set of axioms must meet, it must be noncontradictory.

He used a common sense strategy to select his axioms, at a "relative" level in the hierarchy of logic, rather than trying to derive them as implications of axioms lower in the hierarchy. "I will give an indication of how one system can be constructed, but I want to emphasize very strongly how relatively I state this system [48]." *He recognized that the strategy is risky.* "By axiomatizing automata in this manner, one has thrown half of the problem out the window, and it may be the more important half." The "important half" that he might be throwing out the window would be a proposition that undermines his construction, but that he fails to notice because he started his proof at too high a relative level.

He begins the construction by postulating *a completely artificial milieu* in which floats an unlimited supply of parts, of perhaps a dozen or so different kinds. "We will simply assume that elementary parts with certain properties exist. The question that one can then hope to answer, or at least investigate, is: What principles are involved in organizing these elementary parts into functioning organisms, what are the traits of such organisms, and what are the essential quantitative characteristics of such organisms? I will discuss the matter entirely from this limited point of view [48]." Note most crucially that there is nothing in the milieu except the elementary parts, each of which is exactly described by a complete and unambiguous axiom that permits no variation. This is an abstract world in which there is *no randomness.* It is worth noting that von Neumann considered this detail to be crucial. In subsequent models of the self-replication, he strengthens this requirement, postulating "much more rigid determinations as to what constitutes an automaton, namely the imposition of what is best described as a *crystalline regularity*" [49].

Also, found in this milieu are assorted specimens of $\phi(X)$, an object implementing a chain of binary code that is a *complete and unambiguous* description of automaton X, another object floating in the milieu. von Neumann introduces the notion of a separate $\phi(X)$ for the perfectly legitimate reason he does not want the description of his universal constructor to get sidetracked into irrelevant and nonilluminating details of how the universal constructor abstracts $\phi(X)$ from X. He simplifies his description of the universal constructor, with no loss in its replicative power, by eliminating the abstraction function, and instead postulating that for every X in the milieu there is a $\phi(X)$. However, because $\phi(X)$ is strictly an artifact of the replication process, there are no floating abstractions of objects, not in the milieu. This is a crucial point; there are no $\phi(X)$ chains in the milieu except those that correspond to actual objects, X, already in the milieu. "All this design is laborious, but it is not difficult in principle, for it's a succession of steps in formal logics. It is not qualitatively different from the type of argumentation with which Turing constructed his universal automaton [50]." $\phi(X)$ is a chain of code that constitutes a complete and unambiguous description of automaton X. Notice that by his description he implicitly disallows randomness in the production of the chain. Being a succession of steps in formal logics, $\phi(X)$ has probability = 1 of being the description of X.

Into this milieu he introduces the universal constructor, consisting of three components A, B, and C. A is a universal machine tool, which devours a copy of $\phi(X)$ and while doing so constructs X from the parts floating in the milieu. He says that this process is a succession of unambiguous steps in formal logic. B is a duplicator, which devours a copy of $\phi(X)$ and while doing so constructs two new copies of $\phi(X)$. These two steps add nothing new. A produces X given $\phi(X)$, but X is no more complex than its complete model. B produces two copies of $\phi(X)$ given one copy of $\phi(X)$ but two copies of $\phi(X)$ is no more complex than one copy. Next he adds a control automaton, C, to A + B. C actuates A or B according to a pattern. Step 1: C causes B to duplicate $\phi(X)$. Step 2: C causes A to make X and consume one copy of $\phi(X)$. Step 3: C ties $\phi(X)$ to X, and separates $\phi(X)$ + X from (A + B + C). The result is two entities, the universal constructor (A + B + C) and a new entity $\phi(X)$ + X, where the new X is an exact duplicate of some other object X in the milieu [51].

He then shows that substituting A + B + C for X in $\phi(X)$ produces a perfectly legitimate recursive algorithm. To operate the algorithm, start with universal constructor (A + B + C). To this is attached a complete description of itself $\phi(A + B + C)$, leading to (A + B + C) + $\phi(A + B + C)$. C causes B to copy f. At the end of this step we have (A + B + C) + $\phi(A + B + C)$ + $\phi(A + B + C)$. C causes A to make a copy of (A + B + C) given one of the two copies of $\phi(A + B + C)$ and parts from the milieu. At the end of this step, we have (A + B + C) + (A + B + C) + $\phi(A + B + C)$. Then C cuts the (A + B + C) + $\phi(A + B + C)$ loose from the agglomeration thus resulting in two distinct entities, (A + B + C) and (A + B + C) + $\phi(A + B + C)$. Thus, the universal constructor has reproduced itself, with neither loss, nor gain, in complexity [51].

As a step toward producing increased complexity, he postulates an automaton D that can be temporarily tacked onto the universal constructor (A + B + C), just long enough to generate a copy of $\phi(A + B + C + D)$. Then (A + B + C) + $\phi(X)$ where $X \rightarrow A + B + C + D$ makes a new specimen of (A + B + C + D) + $\phi(A + B + C + D)$. This has still not added complexity to the system. The original A + B + C and the floating D have exactly the same complexity whether tacked together or not, and producing an extra copy of A + B + C + D adds no complexity to what existed before [8].

von Neumann acknowledges that the foregoing process is no more complex than crystal growth, and it is not at all the thing that he is really after. To progress from exact duplication to complexity, in this context the construction of a thing that has never existed before, the description $\phi(A + B + C + D)$ must somehow morph into $\phi(A + B + C + D')$, where D' is an object that has never existed before. Run this new description through the universal constructor, and you have created a new thing as a result. A + B + C + D' is an object that never existed before [52].

At this point the construction collapses. He provides no means of obtaining $\phi(A + B + C + D')$, given $\phi(A + B + C + D)$. For D' to arise, von Neumann suggests that we need a "random change of one element [8]." By postulating an axiom of randomness just at the moment that he needs it, he has temporarily forgotten that he has already forbidden it as a consequence of his prior axioms. In order to obtain exact replication, all descriptions, $\phi(X)$, are complete and unambiguous; they are the result of a succession of steps in formal logic. Consequently, all processes, X, are completely determined, and not subject to random behavior. The only way of obtaining $\phi(X)$ is by observation of X. There is no process for modifying $\phi(X)$. To implicitly include an additional postulate, that $\phi(X)$ can fluctuate randomly, contradicts the initial postulate of complete and unambiguous determination of $\phi(X)$ that is necessary to assure the stability of the universal constructor, and which von Neumann strengthens to "crystalline regularity" in subsequent models.

Stated simply, in order for the universal constructor to demonstrate a threshold of complication beyond which "the production of a more complicated object from a less complicated object is possible," it must reside in a milieu that simultaneously forbids and requires random occurrences. The universal constructor does not provide an exception to the data processing inequality. It contradicts the data processing inequality. One of the reasons for this incoherence is that the universal constructor is based on contradictory axioms. Von Neumann's own concern is confirmed. He did throw the more important half of the problem out the window. From the ambiguous range of possibilities available to him, he chose his axioms in such a way as to imply a fatal but difficult-to-notice flaw in his development. As a consequence of the data processing inequality, there is no axiomatization that he could have chosen that would have led to an automaton that would produce an automaton more complicated than itself.

8.5 THE 29-STATE AUTOMATON

8.5.1 Clearing up Some Confusion

In making sense of von Neumann's subsequent work, there are several logical details that must be appreciated. First, in a noiseless ambience, a complete and unambiguous list of descriptors and a static object described by the list are equivalent; they are two different objects representing exactly the same information. In this context, "static" includes the property of emulating "dynamics" by ratcheting through a succession of static states.

This equivalence answers a question that is relevant to the discussion of von Neumann's subsequent models. Can an automaton reside within another automaton? Recall that in von Neumann's development, an automaton is not an ontological object, but rather a simplified epistemological description of certain properties of certain objects. In the development of Turing's universal computer, von Neumann notes that essentially, a simple automaton could emulate in every detail the behavior of a complicated automaton running a program on its own tape, provided the simple automaton has a longer tape that includes a complete and unambiguous description of the more complicated automaton and its program. In principle, there is no reason that multiple layers of such automata could not be nested. The only condition is that the simplest automaton, the one at the bottom of the nesting hierarchy, meets the undemanding requirement of having a rich enough state set to capture all the operations of first-order logic. Beyond that minimal limit, the only difference in principle between rich and poor computers is the amount of description. Whether the description in the state set of the automaton or in the data on the tape is irrelevant. By analogy with Turing machines, von Neumann legitimately justifies the nesting of one automaton within another [53].

A fundamentally more important detail is that, as a consequence of the data processing inequality, no amount of processing of an unambiguous list of descriptors by an automaton will add information to the list. To appreciate what that means, suppose there exists an object X' and that object X is X' with some of its parts missing. Suppose that Y is a complete and unambiguous description of X. It follows that Y is also an unambiguous, albeit incomplete, description of X'. Suppose that Z is an automaton that can process Y. In the Markov chain, $X \rightarrow Y \rightarrow Z$, no amount of manipulation of description Y by automaton Z will increase the amount of information in Y about X; this is trivially true because Y is *a priori*, the complete description of X. More significantly, in the Markov chain, $X' \rightarrow Y \rightarrow Z$, where Y is an incomplete description of X', no amount of manipulation of description Y by automaton Z will increase the amount of information in Y about X'; this is a nontrivial consequence of the data processing inequality. This is what is meant by the expression $I(X';Y|Z) \leq I(X';Y)$, or the amount of information in Y about X' after Y is processed by Z is less than or equal to the amount of information in Y about X' with no such processing.

Contrast this with the claim made by von Neumann about the threshold of complexity. His D is equivalent to X in the Markov chain development. His $\phi(D)$ is equivalent to the Y in the Markov chain development. His $A + B + C$ is equivalent to the Z in the Markov chain development. We can also think of D' as an enriched version D, in von Neumann's development or X' as an enriched version X, in the Markov chain development. However, in his subsequent models of self-replicating automata, he considers replicative and generative capacities to be properties of the same kind. His claim is that the signature of complexity, expressed in information theoretic notation, becomes $I(X';Y|Z) \geq I(X';Y)$.

Note the crucial error in the placement of the equality. In the world of unambiguous automata, the equality $I(X';Y|Z) = I(X';Y)$ does not represent a process that is simple in some contexts and complex in others. In equating exact self-replication with crystal growth, von Neumann admitted this very point. Consistent with the data processing inequality, $I(X';Y|Z) \leq I(X';Y)$, the equality $I(X';Y|Z) = I(X';Y)$ is a signature of simplicity and *not* complexity. If one ignores the contradiction in von Neumann's development, and suppose as he did, that the "completely decisive property of complexity" is the capacity for an automaton to be generative (as opposed to being merely exactly replicative), then the information theoretic signature of complexity would be $I(X';Y|Z) > (X';Y)$, *without* the equality.

8.5.2 Five Questions and Five Models

As a general principle of scientific inquiry, posing questions is actually far more difficult than answering them. It was in this realm that von Neumann was outstanding. Von Neumann posed five questions that he regarded as central to a theory of automata. People are still trying to answer some of them to this day.

1. "Logical universality. When is a class of automata logically universal, i.e., able to perform all those logical operations that are at all performable with finite (but arbitrarily extensive) means?"
2. "Constructibility. Can any automaton be constructed, i.e., assembled and built from appropriately defined 'raw materials,' by another automaton?"
3. "Construction-universality." Can any automaton construct "every other automaton?"
4. "Self-reproduction." "Can any automaton construct other automata that are exactly like it?"
5. "Evolution." "Can the construction of automata by automata progress from simpler types to increasingly complicated types [49]?"

He showed that the Universal Turing Machine provided the answer to question (1). Regarding the next three questions he said, "We will establish affirmative answers to questions (2) to (4) as well." Then he makes what is probably the most remarkable comment to be found anywhere in his work. "An important limitation to the relevance of a similar answer to question (5) lies in the need for a more unambiguous formulation of the question." After having said in the fifth lecture that he regards an affirmative answer to question (5) to be the "completely decisive property of complexity," his subsequent models did not even address the question. What is most remarkable is his reason for not making the attempt. Although he failed to notice the specific contradiction in his development in the fifth lecture, he did see what lay at the heart of the difficulty. *He could not find a way to unambiguously pose the question.* He also recognized the need to strengthen his axioms forbidding randomness. "In addition, we will be able to treat questions (1) to (5) in this sense with much more rigid determinations as to what constitutes an automaton, namely with the imposition of what is best described as a *crystalline regularity* [54]."

To answer these five questions, he proposed five models, any one of which should be able to demonstrate answers to all five questions. The kinematic model is the automaton he described in the fifth lecture. "The *kinematic model* deals with the geometric–kinematic problems of movement, contact, positioning, fusing, and cutting, but ignores problems of force and energy [55]." The most notable model was the cellular model, or the 29-state automaton. As a further refinement, he also proposed the *excitation-threshold-fatigue model*. "von Neumann's idea was to construct this 29-state automaton out of a neuron-like element, which had a fatigue mechanism as well as a threshold [56]." "The fourth model of self-reproduction, which von Neumann considered was a *continuous model*. He planned to base this on a system of nonlinear partial differential equations of the type, which governs diffusion processes in a fluid [57]." He also proposed a probabilistic model.

It might be wondered if the probabilistic model might demonstrate a property of the evolution of increasing complexity. At first glance, this might appear to solve the problem. Unlike the kinematic model it would not simultaneously prohibit and require randomness in the process. Nor would the "crystalline regularity" of the other three models be permitted. However, von Neumann never actually produced the probabilistic model, and thus its properties cannot be explored. Even if he had, it is unlikely that it would have had the creative property such as he claimed for the kinematic model. The data processing inequality, $I(X;Y|Z) \leq I(X;Y)$, applies whether the processor Z is random or deterministic [46].

8.5.3 Implementation

von Neumann was interested in testing his ideas. He wanted an implementable version of his self-reproducing automaton. Clearly, the kinematic model was unfeasible to implement in the hardware available from 1940s technology, and remains unimplementable even today. However, it is imaginable that it might be feasible to implement in nanotechnology in a few decades in the future. Nevertheless, even if it is implemented, it will not make copies of automata more complicated than the inputs fed into it. The basis on which von Neumann predicted that it would do so is patently contradictory, and such behavior violates the data processing inequality.

The latter three models are also unavailable. von Neumann only left rough guidelines for how he would approach them. He died before he could do more.

He also failed to complete the 29-state model before his death. However, he left enough design details that others were able to complete the work. It has been implemented both as a program on other computers, and in dedicated hardware [58–60].

The scheme is actually a nested three-level cellular automaton. The milieu, or the automaton that provides the nest for the other levels, is an infinite two-dimensional array of cells. Within that largest automaton resides a large but finite automaton, originally envisioned as a collection of approximately 300,000 of the cells of the nesting automaton. These cells form self-replicating automaton, consisting of the computing unit (a rectangular box of approximately 32,000 cells) and the tape unit (a one-dimensional tail extending rightward from the lower right-hand corner of the box, and consisting of something over 200,000 cells). Also, at that same nesting level resides any offspring that the automaton might produce. Each of the 300,000 or so cells of each self-replicating automaton cells contains a 29-state finite state machine. (For that matter, each of the infinitude of cells contains a 29-state finite state machine. However, except for those used in the self-replicating automaton, all are in an "unexcitable" state.)

The automaton replicates in the following manner. From available cells in the milieu, it forms (i.e., by activating the quiescent finite state machine in the cell) a construction arm as an extension of itself reaching rightward from its upper right-hand corner. The construction arm then converts cells in the milieu into a copy of then computing unit, and then the tape. Then it activates the child automaton, which has remained inactive during construction. Then it withdraws and dissolves the constructing arm [61]. The result is two identical copies of the original self-replicating automaton.

Much of the second half of *Theory of Self-Reproducing Automata* is devoted to proving the following claim. "We now reformulate questions (1) to (4) so they apply to von Neumann's 29-state cellular structure, at the same time modifying them somewhat.

1. *Logical universality:* Can an initially quiescent automaton, which performs the computations of a universal Turing machine, be embedded in von Neumann's 29-state cellular structure?
2. *Constructibility:* Can an automaton be constructed by another automaton within von Neumann's 29-state cellular structure
3. *Construction-universality:* Can there be embedded in von Neumann's 29-state cellular structure a universal constructor M_c with this property: for each initially quiescent automaton M, there is a coded description D (M) of M such that, when D (M) is placed on a tape L attached to M_c, M_c will construct M?
4. *Self-reproduction:*
   4.1. Can a self-reproducing automaton be embedded in von Neumann's 29-state cellular structure?
   4.2. Can there be embedded in von Neumann's 29-state cellular structure an automaton, which can perform the computations of a universal Turing machine and can also reproduce itself?

All these questions are answered affirmatively in the present work [62]."

It is significant that the 29-state automaton differs substantially from the proliferation of toy programs that make copies of themselves. It also incidentally implements a Universal Turing Machine. As such, it can do anything that can be completely and unambiguously described. It does not merely consume resources and spawn. It does work.

It is also significant that it is fully compliant with the data processing inequality, $I(X;Y|Z) \leq I(X;Y)$. The 29-state automaton is characterized by $I(X;Y|Z) = I(X;Y)$, producing an exact copy of whatever is completely and unambiguously described to it, *and nothing more complicated*. As such it is completely lacking in the capacity to abduct novelty, the property that von Neumann calls "completely decisive property of complexity." The 29-state automaton is nondescriptive of the biological paradigm of evolution.

It might be objected that the proliferation of offspring in the milieu is evolution of a kind; the milieu is an automaton, and *it* is evolving. However, as von Neumann noted, "the juxtaposition of two copies of the same thing is in no sense of higher order than the thing itself [51]." In other words, two identical copies of the same automaton reflect the same degree of complication as one copy, likewise for a pair of pairs of copies, and so on. As von Neumann identifies complication, a milieu containing any finite number of identical copies of the self-reproducing automaton has the same degree of complication as a milieu that contains only one. In a superficial sense, the milieu is evolving, but its complication stays fixed.

In contrast, the similarity between the process of implementing the 29-state automaton and Special Creation is inescapable. von Neumann himself characterized the appearance of life in a lifeless milieu as a miracle. "That they should occur in the world at all is a miracle of the first magnitude; the only thing which removes, or mitigates, this miracle is that they reproduce themselves [63]." The starting configuration, or configuration of states that cannot be reached from any other configuration of the system by the operation of the automaton, is called the "Garden-of-Eden configuration [64]." The 29-state automaton is created by an external agent who uses the resources within the milieu to create the system in an inanimate form, and animates it subsequently. "And the Lord God formed man *of* the dust of the ground, and breathed into his nostrils the breath of life; and man became a living soul" (Genesis 2:7). The 29-state automaton, being a Universal Turing Machine, is designed to earn its keep. "And the LORD God took the man, and put him into the garden of Eden to dress it and keep it" (Genesis 2:15). Everything in the milieu (i.e., the set of all the cells in an unexcitable state) is available to the 29-state automaton to make copies of itself without limit. "And God blessed them, and God said unto them, Be fruitful and multiply, and replenish the earth, and subdue it: and have dominion over the fish of the sea, and over the fowl of the air, and over every living thing that moveth upon the earth" (Genesis 1:28). When the self-replicating automaton replicates, it makes copies just like itself. "And God created great whales, and every living creature that moveth, which the waters brought forth abundantly, after their kind, and every winged fowl after his kind: and God saw that *it was* good (Genesis 1:21).

This is meant neither to denigrate von Neumann nor to defend Creationism. The point is that anyone looking for either a mathematical description or a philosophical defense of biological evolution will not find it in the 29-state automaton. The 29-state automaton is neither a computational theory of mind, nor a computational theory of life, nor even a computational theory of complexity. It is a computational theory for guiding the realization of computers.

8.6 PRESUPPOSITIONS AND INSIGHTS

8.6.1 THE OBJECTIVE

The objective of von Neumann's work, a "theory of automata," was that it would describe both computers and brains [65]. The scope of automata theory was to include the structure and organization of natural and artificial systems, the role of language and information in both kinds of systems, and the principles of programming and control of both kinds of systems [66]. The big problems limiting the capabilities of computers in von Neumann's time were (1) component size, (2) reliable systems from unreliable components, and (3) the lack of a theoretical principle of computation to guide the design. The first problem has turned out to be more an issue of technology than principle. The second problem requires a principle by which the reliability of the whole exceeds the reliability of the parts. Nature seemed to solve this problem by a process of self-repair in the face of a hostile ambience. The third problem requires a coherent concept of complexity. The idea of a self-replicator that can evolve into something more complicated than itself might serve as a foundation in principle to deal with both problems.

Although von Neumann was steeped in the tradition of reductionism, and the principles of reductionism guided his attempt to characterize complexity, he was capable of looking beyond its limitations. Although the discussion of *telos* or purposeful behavior is excluded from reductionism, and was virtually never discussed in von Neumann's time, he could not disregard its existence. "There is a concept which will be quite useful here, of which we have a certain intuitive idea, but which is vague, unscientific, and imperfect. This concept clearly belongs to the subject of information, and quasithermodynamical considerations are relevant to it. I know no adequate name for it, but it is best described by calling it 'complication.' It is effectivity in complication, or the potentiality to do things. I am not thinking about how involved the object is, but how involved its purposive operations are [65]."

von Neumann's other key presupposition was the notion that natural phenomena are automata, or in other places in his writings, that they could be adequately characterized as automata. He assumed that engineering systems such as communications networks and natural processes such as nervous systems are qualitatively organized the same way. "Automata theory clearly overlaps communications and control engineering on one hand, and biology on the other. In fact, artificial and natural automata are so broadly defined that one can legitimately wonder what keeps automata theory from embracing both these subjects. von Neumann never discussed this question, but there are limits to automata theory implicit in what he said [65]." Burks, commenting on von Neumann, sees the presupposition as true by definition. He wonders how it could be otherwise. However, it must be appreciated that this is a presupposition and not a conclusion of von Neumann's work.

8.6.2 AMBIGUITY

One of the major conceptual points in von Neumann's work on automata is the demonstration that formal logic, the Turing machine, the McCulloch, and Pitts neural network are essentially three different versions of the same concept. Any proposition that could be proved by unambiguous application of formal logic could be proved by a Turing machine or a McCulloch and Pitts neural network. von Neumann saw this as having a profound implication. "It proves that anything that can be exhaustively and unambiguously described, anything that can be completely and unambiguously put into words, is *ipso facto* realizable by a suitable finite neural network. Because the converse statement is obvious, we can therefore say that there is no difference between the possibility of describing a real or imagined mode of behavior completely and unambiguously in words, and the possibility of realizing it by a finite formal neural network. The two concepts are coextensive [67]."

Of course, this leads to another problem. "A difficulty of principle embodying any mode of behavior in such a network can exist only if we are also unable to describe that behavior completely [67]." As remains the conventional wisdom to the present day, von Neumann recognized that the problem of disambiguation is thorny, but supposed that it was tractable. "This description may be lengthy, but it is always possible. To deny it would amount to adhering to a form of logical mysticism, which is surely far from most of us. It is, however, an important limitation, that this applies only to every element separately, and it is far from clear how it will apply to the entire syndrome of behavior [67]." In other words, there are elements whose ambiguity would imply logical mysticism. On the other hand, when he tries to disambiguate the seemingly straightforward concept of triangularity, the more disambiguating statements he makes, the more ambiguities flow from them. "We may have a vague and uncomfortable feeling that a complete catalogue along such lines would not only be exceedingly long, but also unavoidably indefinite at its boundaries. Nevertheless, this may be a possible operation [67]." At the beginning of the paragraph, he saw disambiguation as "always possible"; by the end he comes to the chilling awareness that the problem is potentially intractable, and can promise no more than the hope that a solution may be possible.

This was not the only instance in which von Neumann found inherent ambiguity to be a major difficulty. In the development of the kinematic model, he was very concerned over the issue of choosing his axioms. "There is in formal logics a very similar difficulty, that the whole system requires an agreement on axioms, and that there are no rigorous rules on how axioms should be chosen, just the common sense rules that one would like to get the system one is interested in and would not like to state in his axioms either things, which are really terminal theorems of his theory or things, which belong to vastly anterior fields [47]." Why should axiomatization be so difficult? "So, while the choice of notations, of the elements, is enormously important and absolutely basic for an application of the axiomatic method, this choice is neither rigorously justifiable nor humanly unambiguously justifiable [48]." He himself has identified a process that is inherently ambiguous, but is in no sense mystical. As many other logicians have done, von Neumann admits that the foundation of logic itself is inherently ambiguous.

Remarkably, he saw that the ambiguity arose from context dependency. The same proposition might be true in one context and false in another. "I mention this because when you consider automata whose normal function is to synthesize other automata from elementary parts (living organisms and such familiar artificial automata as machine tools), you find the following remarkable thing. There are two states of mind, in each of which one can put himself in a minute, and in each of which we feel that a certain statement is obvious. But each of these two statements is the opposite or negation of the other [63]!" Although he does not mention it, one is tempted to wonder if von Neumann was familiar with the Austin's context-dependent theory of truth [68]. One reason that the problem of disambiguation is not generally tractable is that if the truth of a proposition is context dependent, and if the context can vary without limit, then the necessity for disambiguating statements grows without limit, just as von Neumann found in his discussion of triangularity.

Ambiguity had a major impact on von Neumann's later work. In the kinematic model he was interested in the problems of self-reproduction, and the reproduction of offspring more complicated than the parent. In the 29-state model he focused on the problem of self-reproduction, and ignored the reproduction of offspring more complicated than the parent. He did so despite the fact that he considered this property to be the completely decisive property of complexity. Why did he leave the completely decisive property out of the 29-state automaton? He could not find an unambiguous formulation of the problem [49].

8.6.3 IMPREDICATIVITY

Although von Neumann showed that the Turing machine, the McCulloch, and Pitts neural network are essentially three different versions of the same concept, there was one exception. Both the Turing machine and neural network have a sequential property not present in formal logic. The sequential character of these two constructs implicitly bans the impredicativity that can arise in logic, and von Neumann considered himself well rid of it. "It prevents the occurrence of various kinds of more or less overt vicious circles (related to 'nonconstructivity,' 'impredicativity,' and the like, which represent a major class of dangers in modern logical systems [69]." At various places in his work he is at pains to explain why it does not lead to vicious circles.

Remarkably, despite his disavowal of impredicativity, von Neumann did reason impredicatively when he could not avoid it. This is most apparent in the strategy that he adopted in axiomatizing his models. He notes that the choice of axioms "is neither rigorously justifiable nor humanly unambiguously justifiable. All one can do is to try to submit a system, which will stand up under common sense criteria. I will give an indication of how one system can be constructed, but I want to emphasize very strongly how relatively I state this system [48]." The impredicativity is seen in the mutual dependency between the axioms and conclusions. He subjectively hypothesizes a set of axioms that seem sensible, and draws a conclusion from the axioms. Being unsatisfied with the conclusion, he subjectively modifies the axioms in the hope of reaching a more satisfying conclusion. He continues the process until the combination feels about right. The process has deductive rigor, but only in the fact that the current conclusion follows rigorously from the current axioms. If he later develops a bad feeling, he might once again change both the axioms and the conclusion to something that seems more satisfactory. This is exactly what von Neumann did in going from the axioms that led to of the unrealizable kinematic model to the crystalline regularity of the 29-state model, and deciding to do so at the cost of the capacity of the automaton to produce offspring more complicated than itself.

Again, this was not the only instance in which von Neumann saw that impredicativity is indispensable to creative reasoning. Even more remarkably, he sees an analog to the process in autocatalysis. Also, he sees the big drawback, it depends on individual genius. "In pure mathematics, the really powerful methods are only effective when one already has some intuitive connection, with the subject, when one already has, before a proof has been carried out, some intuitive insight, some expectation, which, in a majority of cases, proves to be right. In this case, one is already ahead of the game and suspects the direction in which the result lies. A very great difficulty in any new kind of mathematics is that there is a vicious circle: *You* are at a terrible disadvantage in applying the proper pure mathematical methods unless *you* already have a reasonably intuitive heuristic relation to the subject and unless *you* have had some substantive mathematical successes in it already. In the early stages of any discipline, this is an enormous difficulty; progress has an autocatalytic feature. This difficulty may be overcome by some exceptionally lucky or exceptionally ingenious performance, but there are several outstanding instances where this has failed to happen for two, three, or four generations [70]."

8.6.4 PROBABILITY

von Neumann's skills at problem identification are perhaps best demonstrated in description of the new form of logic that he thought was necessary to understand computational reliability and brain function. "The axioms are not of the form: if A and B happen, C will follow. The axioms are always of this variety: if *A* and *B* happen, *C* will follow with a certain specified probability, *D* will follow with another specified probability, and so on [32]." It would be an alternative to the "frequency interpretation." "Strict logic" is the degenerate case for probabilities of 1 and 0. It would be generalization of formal logic rather than as a discipline that could be reduced to formal logic. It would be information-theoretic, and largely inspired by the analogy between thermodynamic entropy and information. It would be based on analysis rather than counting.

Had he lived a normal lifetime, he would have witnessed the development of just such a theory. Jaynes was concerned with slightly different issues that von Neumann. How do we prevent our reasoning from becoming confused by hidden assumptions? How do we use what we already know to maximum advantage? Can we reason without getting hopelessly bogged down in irrelevancies? To keep our thinking clear, we must keep the proper perspective on both our need and capacity to make decisions on the basis of incomplete information. "We are hardly able to get through one waking hour without facing some situation (e.g., will it rain or won't it?) where we do not have enough information to permit deductive reasoning; but still we must decide immediately what to do [71]." Jaynes says that we reach such necessary but partially informed decisions by *plausible reasoning*.

Most crucially, he sees the proper role of probability as a tool for facilitating the process. He starts with three desiderata of plausible reasoning: (1) degree of plausibility is represented by an irrational number. (2) Plausible reasoning should correspond qualitatively to common sense. (3) It should be consistent. From these desiderata he derives the principles of probability theory. Speaking of the quantities *p*, in his calculations, he says, "They define a particular scale on which degrees of plausibility can be measured." In other words, a probability is a measure of what we know [72].

Plausible reasoning is a generalization of logic. He notes that "Aristotelian deductive logic is the limiting form of our rules for plausible reasoning [73]." That limit is complete information. In the absence of complete information, plausible reasoning served as an algorithm for making the best possible guess with whatever information is available.

For simple problems, he gets the same formulas as arise from counting balls in urns. However, how he arrived at those formulas was significantly different; it was not based on counting ontological events. "The important new feature was that these rules were now seen as uniquely valid principles of logic in general, making no reference to "chance" or "random variables [74]."

Jaynes' theory of "Probability as Extended Logic," began to be developed shortly after Shannon's discoveries in information theory and was published in a substantially complete form in the 1980s. Remarkably, it meets all of von Neumann's conditions for a probabilistic logic. As an interpretation of probability, it is more general than previous interpretations; the frequency and Bayesian interpretations are degenerate cases of this more general theory. As a theory of logic, it is more general than formal logic, which turns out to be the degenerate case for probabilities of 1 and 0. It is based on analysis rather than counting. It is information-theoretic, and thermodynamically inspired. The underlying principle is entropy maximization.

The theory of probability of extended logic affords a strategy for guiding the decision of what to do next in the face of incomplete information. This has a consequence that is considered quite radical by many. It is epistemological; it interprets probability as a measure of our ignorance of a phenomenon. In contrast, the "frequency interpretation" is ontological; it interprets probability as a measure of the tendency of a process in reality to gyrate in an uncaused manner. Importantly, probability of extended logic does not specifically forbid the occurrence of such "random processes." However, it does make the point that most of our uses of probability are for estimating outcomes that are causally determined, but whose causes are unknown to us. In such cases, we are tempted to interpret our ignorance of the cause of a phenomenon as inherent randomness of the phenomenon. That error is reinforced by the fact that if we use the model of randomness to make guesses about what to do next with a phenomenon whose causes we do not understand, the guesses often turn out to be correct.

von Neumann was neither ignorant of nor hostile to the interpretation of probability as a measure of what we do not know about a determined phenomenon. He sees this very problem in the cumulative rounding error that arises due to fixed word length in a computer. "This error is, of course, not a random variable like the noise in an analogy machine. It is, arithmetically, completely determined in every particular instance. Yet its mode of determination is so complicated, and its variations throughout the number of instances of its occurrence in a problem so irregular, that it usually can be treated to a high degree of approximation as a random variable [66]." He does consider this to be a legitimate, albeit pathological, use of probability theory.

Where Jaynes radically departs from von Neumann is precisely in this interpretation. Jaynes sees the use of probability to account for our ignorance of a determined situation as the norm, and not the pathological exception. If there is pathology, perhaps it is the occurrence of ontological noise, or a genuine random process in reality.

A result of probability as extended logic that was anticipated by neither Jaynes nor von Neumann is that it gives us a new insight into impredicativity and ambiguity. Impredicativity is not precluded by formal logic. As with all the logicians of his time, von Neumann saw it as a dangerous pitfall that could easily lead to paradox. What he did not realize was that the paradoxes could be precluded by a proper axiomatization. He would still have considered such axiomatization to be practically useless because the resulting logic is inherently ambiguous.

Nevertheless, von Neumann was able to see that some situations are ether inherently ambiguous, or at least hopelessly impossible to disambiguate. Although he wished he could deny the existence of ambiguity in real-world situations, he knew it when he saw it, and proceeded as best he could in the face of it. What he failed to appreciate was that a logic that includes inherent ambiguity might serve as an effective model of processes whose causal structures inherently entail ambiguous behaviors. In one situation, the process would produce one outcome, and in another virtually identical situation it would produce another. In either case, the outcome is fully determined by the causes that led to it, but no matter how much we know about those causes, we would not be able to predict the outcome with certainty [75]. Impredicativity can provide an ambiguous epistemological model of an ambiguous ontological situation; we can gain insight about the one by asking questions about the other. Where probability as extended logic can give us a new insight into both ontological and epistemological ambiguity is that it can serve as an auxiliary model that it allows us to make informed guesses about an entailed situation, when the entailments do not give us enough information to give a complete description of the situation, as would always be the case if the entailments are inherently ambiguous.

8.6.5 COMPUTERS VS. BRAINS

From this perspective on ambiguity, impredicativity, and probability, we have the background to make an informed evaluation of von Neumann's about the differences in principle between the computer and the brain. It must first be recognized that von Neumann's comparison of size and energy requirements of neurons and computing elements were really issues of enabling technology rather than fundamental organizing principle. He also considered system organization. "These differences influence the organization of the system: natural automata are more parallel in operation, digital computers more serial [76]." However, this turns out not to be a difference in principle. Any parallel computing process can be converted to a serial process [77]. In addition, in parallel processing, while the number of processors grows linearly, the amount of interprocess communications grows combinatorially, and the interprocess communications can actually slow the overall process down as more processors are added [78].

Although von Neumann hoped that parallel architecture would solve the problem that he called "logical depth," he was aware that the brain probably solved the problem some other way. "Logical depth" is the number of sequential operations that are required to obtain a particular result. He considers large "logical depth" to be bad because each step has a probability of error, and, except in unusual cases, these accumulate from step to step. He is concerned that the feedback in analog biological neural circuits may be especially bad about amplifying such errors, and wonders how the process can produce any useful results. In accounting for this conundrum, he forms a most remarkable hypothesis. "'Depth' introduced by feedback in the human brain may be overcome by some kind of self-stabilization [79]."

He also saw that more than internal stability is required. Brains differ from computers in having active repair processes. "The very design and construction of such large automata would result in many mistakes. Moreover, the large number of components would result in a very short mean free path between errors and make localization of failures too difficult. Natural automata are clearly superior to artificial ones in this regard, for they have strong powers of self-diagnosis and self-repair [80]."

Given these two properties, stability of some unusual character, and self-repair, it is surprising that he held on to a presupposition for which he had no real justification. He presumed that the brain possesses a static architecture. "It is true, that the animal nervous systems too, obey some rigid 'architectural' patterns in their large-scale construction, and those variations that make one suspect a merely statistical design, seem to occur only in the finer detail on the microlevel [17]." Certainly, the artificial automata that he uses to reason about brain function have this sort of rigidity. Nevertheless, von Neumann suspects that there must be more going on. "Yet the nervous system seems to be somewhat more flexibly designed [81]." In fact, he notes that instead of brains being "somewhat more flexibly designed" than artificial automata, "deep differences in the basic organizational principles are probably present [81]."

von Neumann recognized that brain could not possibly be organized by the same principle as a computer. "The statistical character of message system used in the arithmetics of the central nervous system and its low precision also indicate that the degeneration of precision, described earlier, cannot proceed very far in the message systems involved. Consequently, there exist here different logical structures from the ones we are ordinarily used to in logics and mathematics [82]." He expected that logic to be probabilistic in character, and expected that this would be necessary because the brain is somehow inherently stochastic. This was not an unreasonable interpretation given what was available to him at the time. He had little reason to expect that when the theory of probability as extended logic was finally developed, it would turn out to be primarily a description ignorance of cause rather than a measure of uncaused behavior. He had very little reason to anticipate the development of an impredicative logic, in which formal logic would be encompassed as the special case that occurs when the amount of constraint on an ambiguous entailment structure limits the number of possible outcomes to one. He had no reason to expect that the dynamic architecture of the brain would be observed experimentally. In the absence of these insights, he arrived at what was probably the best guess that was allowed the information available to him.

Given his predisposition to disallow ambiguity and impredicativity, he was not receptive to the possibility of just how deep those differences in organizational principle might go. The brain has a dynamic morphological architecture and not a rigid architecture [21]. That architecture results in a hierarchical loop of causation and the inherent stability of the dynamic process is the only property that stays fixed. Self-repair is a consequences of the dynamic architecture [44,45]. Inherently ambiguous behavior is also a consequence of the dynamic architecture [75]. That ambiguity implies that brain behavior is not fully predictable, but as with any process for which we lack full description, we can make educated guesses about it with a probabilistic theory of logic [74].

8.7 CONCLUSIONS

The work of von Neumann is often cited by advocates both of the computational theory of mind and of the computational theory of life. However, it turns out that his work provides no support for either of these theories. His notion of complexity was an automaton that could make copies like itself, but that are more complicated than itself. The only place in his work in which he claimed to have demonstrated such a construct was in the kinematic model of the self-reproducing automaton. Even here von Neumann expressed his misgivings that perhaps he had not taken a sharp enough look at his axiomatization. His concern was well founded. A sharper look reveals that the kinematic model had contradictory axioms. *By design* the 29-state model avoids the issue of complexity. Quite simply put, nothing in von Neumann's discussion of self-reproducing automata actually serves to model evolution from the simple to the complicated.

An error that von Neumann made in connection with the 29-state automaton has been propagated by commentators on his work [83]. That error is interpreting the capacity of a process to make an exact copy of itself to be the signature of complexity. As the data processing inequality and von Neumann's own comments on crystal growth indicate, the capacity of a process to make an exact copy of itself, particularly in a milieu whose crystalline regularity precludes the possibility of noise, is the signature of simplicity.

The automata in von Neumann's work are strictly unambiguous devices that process strictly unambiguous data streams. The idea that ambiguous streams of signs could be effectively processed in brains, or that the description of brain dynamics would be inherently ambiguous, seems never to have occurred to him.

This does not change the fact that he left several clues that pointed to the possibility. He recognized certain situations as being beyond his power to disambiguate, and in those situations he resorted to a qualitative approach very similar to the strategies of impredicative logic to decide what to do next. He recognized that there were processes of stability and repair in biological brains that are completely beyond the power of computation. He recognized that first-order logic is inadequate to describe biological brain behavior.

von Neumann's automata describe computation, and not biology. They do not model genetic behavior; genetic behavior does not require complete description, but without complete description von Neumann's automata collapse. His automata do not model biological reproduction; von Neumann regarded the propagation of exact copies to be no more significant than crystal growth; nevertheless this is exactly the property of the 29-state automaton. His automata do not model biological evolution; designing the 29-state automaton to make only an exact copy of itself from an exact description of itself, von Neumann specifically forestalled any capacity for it to evolve. It is true that in the kinematic model he saw the capacity to for an automaton to produce something more complicated than itself from an exact description of itself, and so to facilitate evolution. However, that model is based on contradictory axioms, and von Neumann himself not only expressed misgivings about its axiomatization, but also changed the axiomatization when he formulated the 29-state model. When one considers von Neumann's actual results, and his own opinions about them, the assertion that "von Neumann's hypothetical automatons have probably had their greatest impact in biology [84]" is difficult to comprehend. In reality very few biologists have even remotely been familiar with von Neumann's work and thereby an alleged impact of self-replicating automata on biology would be minimal at best. Moreover, for these biologists who were familiar with von Neumann models (primarily theoretical biologists of 1970s through present) the obvious misgivings and mistakes described in this chapter invalidated automata theories as the sole tools to explain the phenomenon of life itself.

ACKNOWLEDGMENTS

I wish to thank Jean-Loup Risler (University of Evry, France) and Andy Konopka (The Editor of this volume) for careful reading of the advanced drafts of this manuscript, as well as for suggesting several improvements of the text.

REFERENCES

1. Poundstone, W. The Recursive Universe: Cosmic Complexity and the Limits of Scientific Knowledge. Oxford: Oxford University Press, 1984:16–17.
2. Poundstone, W. The Recursive Universe: Cosmic Complexity and the Limits of Scientific Knowledge. Oxford: Oxford University Press, 1984:190.
3. Poundstone, W. The Recursive Universe: Cosmic Complexity and the Limits of Scientific Knowledge. Oxford: Oxford University Press, 1984:191.
4. von Neumann, J. The Computer and the Brain, 2nd ed. New Haven: Yale University Press, 2000:22.
5. von Neumann, J. The Computer and the Brain, 2nd ed. New Haven: Yale University Press, 2000:2.
6. Poundstone, W. The Recursive Universe: Cosmic Complexity and the Limits of Scientific Knowledge. Oxford: Oxford University Press, 1984: 18.
7. von Neumann, J. Theory of Self-Reproducing Automata. Burks, A.W., ed. Urbana: University of Illinois Press, 1966:79.
8. von Neumann, J. Theory of Self-Reproducing Automata. Burks, A.W., ed. Urbana: University of Illinois Press, 1966:86.
9. von Neumann, J. Theory of Self-Reproducing Automata. Burks, A.W., ed. Urbana: University of Illinois Press, 1966:92–93.
10. von Neumann, J. Theory of Self-Reproducing Automata. Burks, A.W., ed. Urbana: University of Illinois Press, 1966:80.
11. von Neumann, J. "The general and logical theory of automata," (from the 1948 Hixson Lecture). In: Taub, A.H., ed. John von Neumann Collected Works, Vol.5. New York: Pergammon, 1963:288–328.
12. von Neumann, J. Theory of Self-Reproducing automata. In: Burks, A.W., ed. Urbana: University of Illinois Press, 1966.
13. von Neumann, J. The Computer and the Brain, 2nd ed. New Haven: Yale University Press, 2000:80–82.
14. von Neumann, J., Theory of Self-Reproducing Automata. Burks, A.W., ed. Urbana: University of Illinois Press, 1966:19.
15. von Neumann, J. "Probabilistic logics and the synthesis of reliable organisms from unreliable components." In: Shannon, C.E., McCarthy, J., eds. Automata Studies. Princeton: Princeton University Press, 1956:43–98.
16. von Neumann, J. "Probabilistic logics and the synthesis of reliable organisms from unreliable components." In: Shannon, C.E., McCarthy, J., eds. Automata Studies. Princeton: Princeton University Press, 1956:360.
17. von Neumann, J. "Probabilistic logics and the synthesis of reliable organisms from unreliable components." In: Shannon, C.E., McCarthy, J., eds. Automata Studies. Princeton: Princeton University Press, 1956:368.
18. von Neumann, J., Theory of Self-Reproducing Automata. Burks, A.W., ed. Urbana: University of Illinois Press, 1966:32.
19. von Neumann, J., Theory of Self-Reproducing Automata. Burks, A.W., ed. Urbana: University of Illinois Press, 1966:21.
20. von Nuemann, J. The Computer and the Brain, 2nd ed. New Haven: Yale University Press, 2000:80–82.
21. Kercel, S.W. "The endogenous brain." J Integr Neurosci 2004; 3:47–72.
22. von Neumann, J., Theory of Self-Reproducing Automata. Burks, A.W., ed. Urbana: University of Illinois Press, 1966:38.
23. von Neumann, J., Theory of Self-Reproducing Automata. Burks, A.W., ed. Urbana: University of Illinois Press, 1966:42.
24. Barwise, J. The Situation in Logic, CSLI Lecture Notes Number 17. Stanford, CA: Center for the Study of Language and Information, 1989:293.
25. von Neumann, J., Theory of Self-Reproducing Automata. Burks, A.W., ed. Urbana: University of Illinois Press, 1966:48.
26. von Neumann, J., Theory of Self-Reproducing Automata. Burks, A.W., ed. Urbana: University of Illinois Press, 1966:50.
27. von Neumann, J., Theory of Self-Reproducing Automata. Burks, A.W., ed. Urbana: University of Illinois Press, 1966:52.

28. Rosen, K.H. Discrete Mathematics and its Applications, 4th ed. Boston: McGraw-Hill, 1999:181.
29. von Neumann, J., Theory of Self-Reproducing Automata. Burks, A.W., ed. Urbana: University of Illinois Press, 1966:52–53.
30. von Neumann, J. "The general and logical theory of automata," (from the 1948 Hixson Lecture). In: Taub, A.H., ed. John von Neumann Collected Works, Vol. 5. New York: Pergammon, 1963:303.
31. von Neumann, J., Theory of Self-Reproducing Automata. Burks, A.W., ed. Urbana: University of Illinois Press, 1966:57.
32. von Neumann, J., Theory of Self-Reproducing Automata. Burks, A.W., ed. Urbana: University of Illinois Press, 1966:58.
33. von Neumann, J., Theory of Self-Reproducing Automata. Burks, A.W., ed. Urbana: University of Illinois Press, 1966:59.
34. von Neumann, J. "Quantum Logics (Strict- and Probability-Logics)," (from an unfinished manuscript written about 1937). In: Taub, A.H., ed. John von Neumann Collected Works, Vol. 4. New York: Pergammon, 1963:195–197.
35. von Neumann, J., Theory of Self-Reproducing Automata. Burks, A.W., ed. Urbana: University of Illinois Press, 1966:61–62.
36. Schwartz, J.M., Begley, S. The Mind and the Brain. New York: HarperCollins, 2002:362.
37. Schwartz, J.M., Begley, S. The Mind and the Brain. New York: HarperCollins, 2002:111.
38. von Neumann, J., Theory of Self-Reproducing Automata. Burks, A.W., ed. Urbana: University of Illinois Press, 1966:65.
39. Mackenzie, P.J., Kenner, G.S., Prange, O., Shayan, H., Umemiya, M., Murphy, T.M. "Ultra structural correlates of quintal synaptic function at single CNS synapses." J Neurosci 1999; 19:13.
40. von Neumann, J. "The general and logical theory of automata," (from the 1948 Hixson Lecture). In: Taub, A.H., ed. John von Neumann Collected Works, Vol. 5. New York: Pergammon, 1963:292.
41. Shannon, C.E. "Communication in the presence of noise." Proceedings of IRE, 1949; 37:10–21.
42. von Neumann, J., Theory of Self-Reproducing Automata. Burks, A.W., ed. Urbana: University of Illinois Press, 1966:70.
43. von Neumann, J., Theory of Self-Reproducing Automata. Burks, A.W., ed. Urbana: University of Illinois Press, 1966:73.
44. Freeman, W.J. How Brains Make Up Their Minds. London: Weidenfeld and Nicholson, 1999:7–8.
45. Damasio, A. The Feeling of What Happens. New York: Harcourt Brace, 1999:136.
46. Cover, T.M., Thomas, J.A. Elements of Information Theory. New York: Wiley, 1991:32.
47. von Neumann, J., Theory of Self-Reproducing Automata. Burks, A.W., ed. Urbana: University of Illinois Press, 1966:76.
48. von Neumann, J., Theory of Self-Reproducing Automata. Burks, A.W., ed. Urbana: University of Illinois Press, 1966:77.
49. von Neumann, J., Theory of Self-Reproducing Automata. Burks, A.W., ed. Urbana: University of Illinois Press, 1966:93.
50. von Neumann, J., Theory of Self-Reproducing Automata. Burks, A.W., ed. Urbana: University of Illinois Press, 1966:84.
51. von Neumann, J., Theory of Self-Reproducing Automata. Burks, A.W., ed. Urbana: University of Illinois Press, 1966:85.
52. von Neumann, J., Theory of Self-Reproducing Automata. Burks, A.W., ed. Urbana: University of Illinois Press, 1966:86–87.
53. von Neumann, J., Theory of Self-Reproducing Automata. Burks, A.W., ed. Urbana: University of Illinois Press, 1966:107.
54. von Neumann, J., Theory of Self-Reproducing Automata. Burks, A.W., ed. Urbana: University of Illinois Press, 1966:93–94.
55. von Neumann, J., Theory of Self-Reproducing Automata. Burks, A.W., ed. Urbana: University of Illinois Press, 1966:94.
56. von Neumann, J., Theory of Self-Reproducing Automata. Burks, A.W., ed. Urbana: University of Illinois Press, 1966:96.
57. von Neumann, J., Theory of Self-Reproducing Automata. Burks, A.W., ed. Urbana: University of Illinois Press, 1966:97.

58. Signorini, J. "How a SIMD machine can implement a complex cellular automaton? A case study: von Neumann's 29-state cellular automaton." IEEE Proc Supercomput 1989:175–186.

59. Pesavento, U. "An implementation of von Neumann's self-reproducing machine." Artif Life J 1995; 2(4):337–354.

60. Beuchat, J., Haenni, J. "Von Neumann's 29-state cellular automaton: A hardware implementation." IEEE Trans Educ 2000; 43(3):300–308.

61. Freitas, R.A., Gilbreadth, W.P. "Advanced automaton for space missions," Chapter 5, Proceedings of the 1980 NASA/ASEE Summer Study, Santa Clara, California June 23–August 29, 1980, NASA Conference Publication 2255.

62. von Neumann, J., Theory of Self-Reproducing Automata. Burks, A.W., ed. Urbana: University of Illinois Press, 1966:292.

63. von Neumann, J., Theory of Self-Reproducing Automata. Burks, A.W., ed. Urbana: University of Illinois Press, 1966:78.

64. Moore, E.F. "Machine models of self-reproduction." Proc Symp Appl Math 1962; 14:17–33.

65. Moore, E.F. "Machine models of self-reproduction." Proc Symp Appl Math 1962; 14:18.

66. Moore, E.F. "Machine models of self-reproduction." Proc Symp Appl Math 1962; 14:20.

67. von Neumann, J. "The general and logical theory of automata," (from the 1948 Hixson Lecture). In: Taub, A.H., ed. John von Neumann Collected Works, Vol. 5. New York: Pergammon, 1963:310.

68. Austin, J.L. "Truth," In Proceedings of the Aristotelian Society. Supp. Vol. 24 (1950). Reprinted in Philosophical Papers. In: Urmson, J.O., Warnock, G.J., eds. Oxford: Oxford University Press, 1961:117–133.

69. von Neumann, J. "Probabilistic logics and the synthesis of reliable organisms from unreliable components." In: Shannon, C.E., McCarthy, J., eds. Automata Studies. Princeton: Princeton University Press, 1956:330.

70. von Neumann, J., Theory of Self-Reproducing Automata. Burks, A.W., ed. Urbana: University of Illinois Press, 1966:33.

71. Jaynes, E.T. Probability Theory: The Logic of Science. Cambridge: Cambridge University Press, 2003:1.

72. Jaynes, E.T. Probability Theory: The Logic of Science. Cambridge: Cambridge University Press, 2003:37.

73. Jaynes, E.T. Probability Theory: The Logic of Science. Cambridge: Cambridge University Press, 2003:31.

74. Jaynes, E.T. Probability Theory: The Logic of Science. Cambridge: Cambridge University Press, 2003:10.

75. Kercel, S.W. "Endogenous causes — bizarre effects." Evol Cognition 2002; 8(2):130–144.

76. von Neumann, J., Theory of Self-Reproducing Automata. Burks, A.W., ed. Urbana: University of Illinois Press, 1966:23.

77. Almasi, G.G., Gottlieb, A. Highly Parallel Computing. Redwood City, CA: Benjamin Cummings, 1989:431.

78. Almasi, G.G., Gottlieb, A. Highly Parallel Computing. Redwood City, CA: Benjamin Cummings, 1989:126.

79. von Neumann, J. "Probabilistic logics and the synthesis of reliable organisms from unreliable components." In: Shannon, C.E., McCarthy, J., eds. Automata Studies. Princeton: Princeton University Press, 1956:375.

80. von Neumann, J., Theory of Self-Reproducing Automata. Burks, A.W., ed. Urbana: University of Illinois Press, 1966:24.

81. von Neumann, J., Theory of Self-Reproducing Automata. Burks, A.W., ed. Urbana: University of Illinois Press, 1966:369.

82. von Neumann, J. The Computer and the Brain, 2nd ed. New Haven: Yale University Press, 2000:81–82.

83. Poundstone, W. The Recursive Universe: Cosmic Complexity and the limits of Scientific Knowledge. Oxford: Oxford University Press, 1984: 194–195.

84. Poundstone, W. The Recursive Universe: Cosmic Complexity and the limits of Scientific Knowledge. Oxford: Oxford University Press, 1984: 17.

9 The Mathematical Structure of Thermodynamics

Peter Salamon, Bjarne Andresen, James Nulton, and Andrzej K. Konopka

CONTENTS

9.1 INTRODUCTION

Thermodynamics is unique among physical and chemical descriptions of our surroundings in that it does not rely on a detailed knowledge of any interior structure of the systems* to which it pertains but rather treats such systems as "black boxes" whose equilibrium states are determined by the surroundings with which they can coexist and which can be described by a few parameters. This feature assures that the theory holds true when the system is a collection of molecules, a beaker of water, or a black hole. Einstein expressed this feature of thermodynamic theory in his classic quote:

> Thermodynamics is the only physical theory of universal content which, within the framework of the applicability of its basic concepts, I am convinced will never be overthrown. — Albert Einstein

Foremost among these basic concepts is the notion of equilibrium, the situation where the state of the system does not vary noticeably in time. The "noticeably" in the previous sentence has two complications. The first is that if this system were to be cut off from its surroundings the state would remain the same. This distinguishes equilibria from steady states. The second is the fact that the notion of equilibrium is associated with a particular time scale. Over larger periods of time,

---

* In standard presentations of thermodynamics [2,5], the term *thermodynamic system* is universally used. In order not to deviate from this established usage, we have retained the expression here in spite of the collision of meanings of the overloaded term system in this handbook. Every occurrence of the term system in this chapter is to be taken in the thermodynamic sense.

any system will eventually evolve until the final dead state of ^{56}Fe is reached through nuclear transformations.

As an example, consider a carbon filament suspended in air. On a human timescale, that carbon filament will be in equilibrium thermally with the air surrounding it while retaining its chemical integrity as free carbon. On a much longer timescale, it would also achieve chemical equilibrium with the oxygen of the surroundings to form CO_2. If we were to pass an electric current through the filament, it would heat up and glow and appear not to change over many minutes. While this would be a steady state on this timescale, it is not at equilibrium because when we stop the current, the state changes to a lower temperature.

9.2 A HISTORICAL INTRODUCTION TO THERMODYNAMICS

Thermodynamics began with the development of early theories of heat and mechanics. The analysis was primarily motivated by an economic impetus: the newly invented steam engine. Carnot's [1] great accomplishment was to show that the conversion of heat into work has limitations set by what we today view as a "no free lunch" principle: one cannot extract more work from an initial configuration of states of a collection of thermodynamic systems than it would take to restore those systems to such initial states after the work was extracted [2]. This is often stated as the impossibility of a perpetual motion machine of the second kind [2] because it is the essence of what we today call the second law of thermodynamics. It predates the first law by about 25 yr.

The first law of thermodynamics also can be stated as a "no free lunch" principle: we cannot get more work out of a system than the change in its internal energy. More commonly, however, the first law is stated as a conservation law, which, historically, combined the two separate conservation laws for heat (caloric) and mechanical energy [3] into one conservation law for a quantity called internal energy whose change is defined by

$$\Delta U = W + Q, \tag{9.1}$$

where U is the internal energy, W is the work done on the system, and Q is the heat added to the system. The fact that this defines a *conserved* quantity is a consequence of an empirical observation — a given amount of work turned into heat by friction always produces the same amount of heat. (Here as well as in the following we only consider the exchange of heat and work with the surroundings except where explicitly stated otherwise for simplicity.) Because heat and work are conserved in processes where no interconversion takes place, this means that internal energy U will be conserved even in processes including such conversions. The fact that U is a *function of state* is not obvious but requires an assumption to this effect. It is physically stated in terms of cyclic processes, i.e., processes in which a system starts and ends in the same state. For such processes, the work produced by the system to the surroundings must equal the heat withdrawn from the surroundings. A Carnot cycle and a Stirling cycle [4] are examples of such cycles.

Energy has become firmly rooted in our language and intuition, and the conservation of this energy, as stated by the first law of thermodynamics, is so widely applied that the subtleties associated with it have faded from the collective consciousness. The second law on the other hand carries a great deal of subtlety. It is associated with a far lesser known quantity called entropy that most people, even many scientists, find difficult and abstruse. The deeper meaning of the second law is a unidirectionality associated with physical processes and as such occupies a unique and important position among physical laws. All other physical laws view a process and its time-reversed version as equally acceptable or unacceptable on physical grounds. The second law of thermodynamics asserts that only one direction is physically possible — unless of course the process is reversible, which is only an ideal abstraction. The second law is associated with the increase of entropy whose changes are defined by

$$\Delta S = \Delta Q_{rev}/T. \tag{9.2}$$

Here ΔS is the increase of entropy of a thermodynamic system at temperature T when the amount of heat ΔQ_{rev} is added reversibly. Intuitively, the construction of entropy [5] expresses the unidirectionality of transport, notably of heat, between two systems, from higher to lower temperature. The mathematical theory of thermodynamic systems focuses on one such system, a thermodynamic system, and describes geometrically the set of equilibrium states it can have. The allowed modes of interaction with the surroundings define the equilibrium. Important for this perspective was a far subtler form of the second law introduced by Caratheodory [6], which asserts that *arbitrarily close to any equilibrium state of a system there exist states that are not accessible without the transport of heat out of the system*. Exchange of other quantities like work and mass can occur freely.

9.3 DEFINITIONS AND AXIOMS

A *simple thermodynamic system* is a homogeneous macroscopic collection of components. The system is treated as a black box and its *state* is describable by a small number of macroscopic parameters, typically its energy, entropy, volume, and particle number, dictated by the surroundings with which it coexists. However, not all of these are necessarily independent. A simple single-phase system consisting of $n - 1$ components has n independent parameters called thermodynamic degrees of freedom. Parameters in excess of this will be interdependent. Simple thermodynamic systems may interact with one another to form a nonuniform thermodynamic system.

Such interactions between systems and between systems and the surroundings occur through *walls*, which are constructed to allow passage of certain quantities. All other exchange is blocked. An isolated system has no exchange with its surroundings, i.e., its volume, energy, particle number, etc. are fixed. An adiabatic wall allows passage of only volume (e.g., by moving a piston, i.e., exchanging work, but not particles). A diathermal wall allows in addition passage of heat. Similarly, a semipermeable wall further allows passage of specific types of particles.

By selecting a proper wall, desired standard *processes* are allowed. Thus an adiabatic wall between a system and its surroundings allows only adiabatic processes, i.e., processes where no heat is exchanged with the surroundings and production of work therefore must be accompanied by a decrease of the internal energy. A diathermal wall permits isothermal processes at the temperature of the surroundings. Here work produced is compensated by influx of heat, keeping the internal energy fixed. Semipermeable walls may allow, e.g., passage of oxygen, sodium ions, water, and/or glucose. Inside a system coupled to its surroundings through such a wall, chemical reactions may proceed while exchanging reactants and products with the surroundings.

Thermodynamic cycles are made up of sequences of such standard processes. For example, a Carnot cycle involves a system (called the working fluid), which undergoes a cyclic process by following a sequence of standard processes while connected to a corresponding sequence of surroundings. The sequence followed is isothermal (hot) — adiabatic — isothermal (cold) — adiabatic. Systems with ongoing chemical reactions may steer such reactions through the permeability of the wall and by controlling the work and heat flows (isothermal, adiabatic, isobaric, etc.).

The Carnot cycle also illustrates another important version of the second law — the fact that the conversion of heat to work is a limited affair in which only a certain fraction of the heat can be captured as work. How large a fraction can be converted is the so-called Carnot efficiency, which depends on the temperatures of heat sources and sinks available for contact during the isothermal branches of the cycle. Let us look at a simple reversible Carnot heat engine operating between a hot heat reservoir at temperature T_H and a cold reservoir at T_L. The engine absorbs the amount of heat Q_H from the hot reservoir accompanied by the entropy influx $S_H = Q_H/T_H$. Because the engine operates in a cycle and thus cannot accumulate entropy, it must somehow dispose of this much entropy. The work produced does not carry any entropy, so the only available sink is the cold heat reservoir. However, entropy and heat are transported together, so discharging S_H must be accompanied by a discharge of heat equal to $Q_L = S_H \times T_L$. The fraction of heat turned into work in this reversible machine is thus $(Q_H - Q_L)/Q_H = 1 - T_L/T_H$, the famous Carnot efficiency. A realistic irreversible engine will of course produce even less work. Note that the engine cannot convert all incoming heat into work, not for energetic reasons but due to entropy constraints.

In all cases, the quantities exchanged belong to a class of variables called extensive, i.e., variables, which are additive when systems are merged (energy, entropy, volume, and particle number). The corresponding intensive variables (temperature, pressure, and chemical potential) are not additive over subsystems but describe possible gradients. At equilibrium either between systems or between a system and its environment, these intensive variables will be the same in all the connected systems. Thus extensive variables scale with the power 1 of the size of the system considered, the intensive variables with power 0.

Using the definitions above, we are in a position to state the axioms in their traditional geometrical form [7].

Axiom 1: For any thermodynamic system, there exists an extensive function of state U called the internal energy.

Axiom 2: For any thermodynamic system, there exists an extensive function of state S called the entropy. The entropy is a concave function of any set of complete independent extensive parameters of the system $S = f(U, V, N_1, N_2, ..., N_k, P, M, ...)$, where the arguments are the internal energy U, volume V, particle numbers N of the k molecular species, polarization P, magnetization M, as well as any other relevant extensive quantity.

These two axioms are essentially the first and second laws expressed in terms of a single system. The geometrical picture that goes with this formulation is the concave surface $S = f(U,V, ...)$ in $n + 1$ dimensions for the n degree of freedom system. The sections below present a modern differential geometrical alternative to this picture including a rigorous proof of Caratheodory's principle. The presentation is perforce rather sketchy in that it provides the bare minimum of examples, although all the definitions are carefully stated and rigorous. More details can be found in any modern differential geometry book [8,9].

9.4 THERMODYNAMIC STATES, COORDINATES, AND MANIFOLDS

Roughly speaking, a manifold is a coordinatizable set. More correctly, it is a set equipped with re-valued coordinates which uniquely label the elements and whose values change in a "continuous" fashion. Historically, manifolds arose as a collection of variables subject to equations. Early examples were well studied by the founders of differential geometry as curves and surfaces [10]. Going to higher dimensions was an obvious and yet conceptually difficult leap that required a higher level of abstraction [11]. Spaces of states of dynamical systems were one strong impetus toward such abstraction. Mechanical systems, such as compound pendulums, provided ready examples. The set of equilibrium states of a thermodynamic system is yet another example. This is *the* example for the present chapter. The set of equilibrium states of a thermodynamic system was also conceptualized initially as a surface; James Clerk Maxwell had a plaster model of the equilibrium states of water constructed and sent it as a present to Josiah Willard Gibbs (Figure 9.1), the pioneer responsible for the dramatic shift in point of view of thermodynamics from a theory of processes to a theory of equilibrium states [3].

As illustration, consider an ideal gas. There are many functions of state for the gas: pressure p, temperature T, volume V, energy U, entropy S, mass M, density ρ, heat capacity C_v .... The variables on this list are not independent in the sense that for a particular ideal gas, once we know

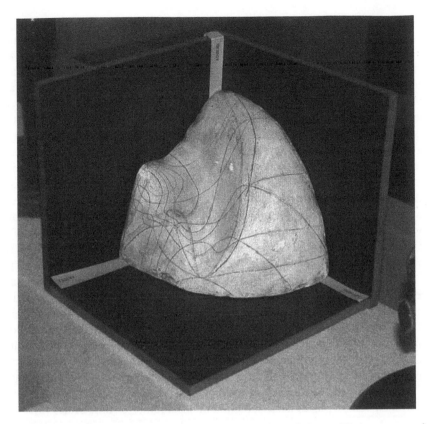

FIGURE 9.1 (See color insert following page 150.) Plaster model of the equilibrium states of water constructed by James Clerk Maxwell and sent as a present to Josiah Willard Gibbs. (©The Cavendish Laboratory, University of Cambridge. Reproduced with permission.)

two of them, the others are determined. In usual parlance, this means that the dimension of the manifold of equilibrium states is two.*

We now proceed on a more formal level. Recall that a *topological space* (M, T) is a set M and a collection T of subsets of M that are designated as open. All that is required here is that the collection T be closed under unions and intersections, the latter only over finite subcollections of T. We say that T defines a topological structure on the set M. Such structure is needed to be able to talk about the continuity of functions defined on M. In practice, the coordinates defined by our observables define this topology.

A *manifold* $(M, \{\varphi_k, k \in K\})$ is a topological space equipped with a collection of coordinate functions φ_k each of which establishes a topological isomorphism between an open set O_k in M and an open subset U_k of R^n, such that the open sets cover M, i.e., such that

$$\bigcup_{k \in K} O_k = M$$

---

* This at least is the situation described in the early chapters of thermodynamics books and traditional physical chemistry books. In fact, this list of functions of state gets expanded once the reader hits the chapters on the behavior of the entity in the presence of electric and magnetic fields. In that case, the polarization D, and magnetization M of the gas also play a role and the entity is said to have more degrees of freedom. The corresponding mathematical object, the manifold, has to have a higher dimension equal to this number of freedoms because it takes that many functions of state to uniquely specify a state.

A topological isomorphism is an invertible function, which is continuous and has a continuous inverse. In simple terms, this assumes that, at least locally, we can coordinatize the set M and that nearness in the sense of approximately equal coordinate values implies nearness in M. One final condition is needed: wherever there are two or more possible sets of coordinates, the transition between the two sets of coordinates must be well behaved. Formally, if for some j and k in K, $O_j \cap O_k \neq \varnothing$, then the function $\varphi_j \circ \varphi_k^{-1}$ is smooth on $\varphi_k(O_j \cap O_k) \subset R^n$. "Smooth" is a nebulous word and serves to define the type of manifold under consideration; for example, smooth can mean "continuous," "differentiable," "twice differentiable," or "infinitely differentiable," or "analytic." The standard meaning of smooth for the manifold of equilibrium states of a thermodynamic system is piecewise analytic.* Note that if the manifold is a connected set, the overlap condition requires that the dimension of the images of all the coordinate charts be the same value n. This number is called the dimension of the manifold.

The manifold that we concentrate on below is the manifold of equilibrium states of a thermo-dynamic system. The dimension of this manifold is what is known as the number of degrees of freedom of the system. This is the number of independent parameters that need to be specified in order to reproduce the experimental realization of the system. This often depends on the number of external (environmental) degrees of freedom we are able to vary. If we only vary pressure and temperature, we only get two degrees of freedom. If we also vary (say) the magnetic field surrounding the system, we get a third degree of freedom, the magnetization M. The number of degrees of freedom also depends on the time scale on which we desire to view the system. For example, on a certain time scale we can take the amount of oxygen and iron in the system as independent variables. On another (slower) time scale we could assume this degree of freedom to be set by chemical equilibration to form iron oxide. On intermediate time scales comparable to the relaxation time of this degree of freedom, thermodynamic arguments strictly speaking do not apply. This state of affairs is usually referred to as the assumption of separability of time scales [5].

9.5 MANIFOLDS AND DIFFERENTIAL FORMS

The abstract formulation of differential geometry via the theory of manifolds [8,9] gives an ideal tool for studying the structure of any theory and this has been one of its primary roles. Such structure is typically specified by additional properties beyond "coordinatizability." To make this possible, we need a number of concepts, which comprise the standard baggage of the theory: tangent vectors, differential forms, wedge products, and submanifolds.

A *path*** in a manifold is a differentiable one-parameter family of points defined by a continuous function γ that maps an interval in the real numbers to points in the manifold. In classical thermodynamics books, such paths are called quasistatic loci of states because every point on the path is an equilibrium state. Finite rate processes do not quite proceed along such paths because equilibrium is only approached asymptotically. Again the notion of separability of time scales comes to the rescue. A quasistatic locus is a good representation of a process that occurs on time scales that are slow compared to the equilibration time of the system.

A *tangent vector* at a point is an equivalence class of *paths* that "go in the same direction at the same speed." We may think of a tangent vector *at a particular state* as an n-tuple of time derivatives of the coordinate functions at the point. Thus a tangent vector represents any path that has the same instantaneous values of all these derivatives. The set of tangent vectors at a point is a vector space. This comes naturally through the identification with n-tuples of time derivatives. Note that by the chain rule, a tangent vector assigns a time derivative to any function of state not

---

* Recall that a function is analytic if it has a convergent power series. Piecewise analytic is needed here because different analytic forms correspond to different phases (e.g., liquid, solid, and gas) of the entity.
** Sometimes this is called a parameterized path.

just the coordinate functions. For example, if f is any function of state of a fixed quantity of some ideal gas, then the tangent vector $(dp/dt, dT/dt)$ assigns to f the time derivative

$$\frac{df}{dt} = \frac{\partial f}{\partial p}\frac{dp}{dt} + \frac{\partial f}{\partial T}\frac{dT}{dt}. \tag{9.3}$$

An equivalence class of paths corresponding to one tangent vector are exactly those paths along which a set of coordinate functions change at the given rates, e.g., $(dp/dt, dT/dt)$ in the above example.

A *cotangent vector* at a point (p,T) is an equivalence class of *functions* at the point where now two functions are deemed equivalent if their rates of change are the same along every tangent vector at the point. The coordinate expression of a cotangent vector is what we would normally associate with the gradient of any one of the functions in the equivalence class, i.e., each equivalence class is just the set of functions whose gradient vectors at the point are equal. There is good reason to identify this at a particular point (p_0, T_0) with the differential of any one of the functions in the equivalence class and write

$$dF = fdp + gdT \tag{9.4}$$

for the cotangent vector corresponding to the function F. Such functions exist for every pair of *numbers* f and g, e.g., $F = fp + gT$. It follows that cotangent vectors also form a vector space of dimension n and in fact this vector space is the dual of the tangent space — hence the name. The duality means that each cotangent vector may be thought of as a linear map of tangent vectors to real numbers. In coordinate form, this means that each tangent vector is identified with a row n-tuple and each tangent vector with a column n-tuple. The assignment of a real number to a vector and a covector at a point is by means of the chain rule (9.3) where f is any function in the equivalence class represented by the cotangent vector and $(dp/dt, dT/dt)$ is taken along any path in the equivalence class represented by the tangent vector.

In summary, we have defined the tangent space and the cotangent space of a manifold at a point. The *tangent space* is the set of tangent vectors, which we may think of as infinitesimal displacements. Formally, we defined them as an equivalence class of curves that go in the same direction at the same speed. Dual to the tangent space we have the *cotangent space*, the set of all covectors at the point. These were defined as an equivalence class of functions whose differentials are equal at the point of tangency.

With the definition of tangent and cotangent vectors come the notions of *vector field* and *differential form*. These are just smooth choices of a tangent vector or, respectively, a cotangent vector at each point on the manifold. We choose to follow the long-standing tradition in thermodynamics, which focuses the development on differential forms. Because of the duality, most things can be done with either differential forms or vector fields.

The reader should pause here to note that our definition of a differential form is merely a modern statement of the traditional notion of a differential form. In coordinates, such forms all look like

$$\omega = f(p,T)dp + g(p,T)dT \tag{9.5}$$

with f and g now functions of state. Examples of important differential forms in classical thermodynamics are heat Q and work W. Note that while the differential of a function is a differential form, not all differential forms are differentials of functions although any form may be written as a linear combinations of differentials of state functions.*

9.6 PFAFFIAN EQUATIONS

The first law of thermodynamics in its familiar form asserts that there exists a function of state U = internal energy such that its differential is equal to

$$dU = W + Q \tag{9.6}$$

When coordinate expressions for the differential forms of heat and work are included, this equation becomes a Pfaffian [6] partial differential equation in any coordinate system. The solutions of such a Pfaffian equation are submanifolds. Because we will be needing solutions of such equations, we sketch the main results concerning such equations: the theorems of Frobenius and Darboux. To motivate the machinery needed, consider the following.

It turns out that once we require one equation among the differential forms on our (sub)manifold, other equations logically follow. In particular, taking the differential of both sides of such an equation must also hold. As an example, consider Equation 9.6 with the usual elementary form of the coordinate expressions substituted in for heat Q and work W. Then it follows that

$$d(dU) = d(W + Q) = d(TdS - pdV) = dT \wedge dS - dp \wedge dV \tag{9.7}$$

In fact, this equation, though hardly recognizable as such, is equivalent to the Maxwell relations [7]. To make sense of this equation, we need definitions of the exterior derivative operator $d(.)$ and the wedge product $\wedge$.

The product of differential forms is indispensable for multiple integrations and the reader likely saw such products in a calculus course. Alas, these products are all too often handled without comment and by mere juxtaposition. This ignores the orientation implied by the order of the factors. We thus adopt the symbol $\wedge$ (wedge) for the product and add the requirement of antisymmetry

$$dx \wedge dy = -dy \wedge dx \tag{9.8}$$

The natural thing to do with a differential one-form is to integrate it along a path to get a number

$$W = \int_\gamma \omega = \int_{(p_1,T_1)}^{(p_2,T_2)} \left(f(p(t),T(t))\frac{dp}{dt} + g(p(t),T(t))\frac{dT}{dt} \right) dt, \tag{9.9}$$

where ω is the one-form in Equation 9.5. Similarly, the natural thing to do with two-forms is to integrate them along a two-dimensional region and so on for higher forms. Note that the set of k-forms again forms a vector space at any point. The dimension of this vector space is $\binom{n}{k}$. To get a feel for the concept just introduced, consider the first two-form in Equation 9.7 above. Expanding dT in the coordinates (S,V) results in

$$dT \wedge dS = \left(\left(\frac{\partial T}{\partial S} \right)_V dS + \left(\frac{\partial T}{\partial V} \right)_S dV \right) \wedge dS = \left(\frac{\partial T}{\partial V} \right)_S dV \wedge dS \tag{9.10}$$

* Note that while any given cotangent vector (necessarily at one point) is equal to the differential of many functions, a differential form specifies a cotangent vector at each point. Thus for a differential form to be the differential of a function is asking the same function to match a smoothly defined cotangent at all points and this in general is not possible.

where we have used one consequence of antisymmetry: $dS \wedge dS = 0$. Performing a similar expansion of the second two-form in Equation 9.7 gives

$$d(dU) = dT \wedge dS - dp \wedge dV = \left(\left(\frac{\partial T}{\partial V} \right)_S + \left(\frac{\partial p}{\partial S} \right)_V \right) dV \wedge dS. \qquad (9.11)$$

As a second illustration of what the mathematical machinery of differential forms and wedge products can do for us, consider the product of two differential forms du and dv where u and v are functions of x and y. It then follows that

$$du \wedge dv = \left(\left(\frac{\partial u}{\partial x} \right)_y dx + \left(\frac{\partial u}{\partial y} \right)_x dy \right) \wedge \left(\left(\frac{\partial v}{\partial x} \right)_y dx + \left(\frac{\partial v}{\partial y} \right)_x dy \right)$$

$$= \left(\left(\frac{\partial u}{\partial x} \right)_y \left(\frac{\partial v}{\partial y} \right)_x - \left(\frac{\partial u}{\partial y} \right)_x \left(\frac{\partial v}{\partial x} \right)_y \right) dx \wedge dy \qquad (9.12)$$

Note that the coefficient of $dx \wedge dy$ is exactly the Jacobian determinant of the coordinate change from (x,y) to (u,v). It follows that the usual change of variable formula for multiple integrals is just a consequence of the fact that we are really integrating a wedge product. The machinery of k-forms also gives a definition of *functional independence*. We say that k functions $f_1, f_2, \ldots, f_k$ are independent in a region if $df_1 \wedge df_2 \wedge \ldots \wedge df_k \neq 0$.

Equipped with the wedge product, the set of differential forms on a manifold have an algebraic structure known as a ring. The interesting subsets in rings are *ideals* — subrings such that the product of any element in the ring with an element of the ideal is an element of the ideal. The standard elementary example of a ring is the set of integers. Ideals in this ring are of the form "all multiples of k" for some integer k. The notion of ideal turns out to be central to characterizing, which differential forms can be solutions of systems of Pfaffian equations. By rearranging the equation so all terms are on one side, we may view each equation as a condition that a differential one-form vanishes on the solution submanifold. For example, instead of writing the first law as in Equation 9.6, we could write

$$dU - W - Q = 0 \qquad (9.13)$$

The advantage of writing it this way comes about from the fact that zero times anything will still give zero. It follows that any form that has a factor that should vanish on our solution must still vanish on our solution. In algebraic jargon, this means that the set of differential forms that vanish on a submanifold comprise an ideal. Not all ideals work, however; we are missing the condition that these be differential ideals, i.e., that they be closed under the action of taking differentials. To make sense of this, we need the extension of the exterior derivative operator $d(.)$ to higher forms. This is obtained by the following three requirements:

- for functions (0-forms) d gives precisely the one-form which is the differential of the function
- $d[d(\text{anything})] = 0$
- d obeys the product rule

$$d(\omega \wedge \eta) = d\omega \wedge \eta + (-1)^K \omega \wedge d\eta \qquad (9.14)$$

where ω is any K-form and η is any L-form. To illustrate this definition, we calculate the exterior derivative of a general differential one-form in two variables

$$d\big(f(x,y)dx + g(x,y)dy\big) = dfdx + fd^2x + dgdy + gd^2y = dfdx + dgdy$$

$$= \left(\frac{\partial f}{\partial y} - \frac{\partial g}{\partial x}\right)dx \wedge dy. \tag{9.15}$$

As a check, note that this automatically vanishes as required by condition A, if the form we start with happens to be exact.

As a second illustration, we note that applying condition B above in Equation 9.11 implies one of the Maxwell relations

$$\left(\frac{\partial T}{\partial V}\right)_S = -\left(\frac{\partial p}{\partial S}\right)_V \tag{9.16}$$

Note that this is followed from Equation 9.7 by expanding all the terms in the coordinates dS and dV. Similar expansions of Equation 9.7 in other coordinates give the other Maxwell relations, establishing our claim that Equation 9.7 is really all of the Maxwell relations combined into one coordinate-free expression!

At last we are in a position to state the definitive theorem concerning the solution of Pfaffian systems: Frobenius's theorem. The theorem says that associated with any given system of Pfaffian equations

$$\omega_j = 0, \qquad j = 1, ..., J,$$

where for each ω_j a differential one-form on a manifold M, there is a differential ideal of forms generated by the ω_j which must vanish on any solution of this Pfaffian system.

Specifically, it is an ideal in which the exterior derivative of any form in the ideal is still in the ideal.

The computational implications for a single Pfaffian equation $\omega = 0$ are the following. Examine the sequence

$$\begin{aligned}
&\omega, \\
&d\omega, \\
&\omega \wedge d\omega, \\
&d\omega \wedge d\omega, \\
&\omega \wedge d\omega \wedge d\omega, \\
&d\omega \wedge d\omega \wedge d\omega, \\
&\omega \wedge d\omega \wedge d\omega \wedge d\omega, \\
&\cdots
\end{aligned} \tag{9.17}$$

On any solution of $\omega = 0$, every one of these differential forms must vanish. Some of them are identically zero on the entire manifold. Once one is identically zero, all subsequent terms are identically zero. Suppose that the first identically zero term occurs at position r. The number r expresses something fundamental about the form ω: r is the minimum number of variables that can be used to express ω. This is the classic theorem of Darboux. To state it carefully, we need to distinguish the cases where r is even and where r is odd. The theorem states that there exist smooth independent functions $x_1, x_2, ..., x_m, y_1, y_2, ..., y_m, z$ such that for $r = 2m + 1$,

$$\omega = dz + \sum_{i=1}^{m} y_i dx_i \tag{9.18}$$

while for $r = 2m$,

$$\omega = \sum_{i=1}^{m} y_i dx_i. \tag{9.19}$$

Furthermore, the dimension of the maximal solutions of $\omega = 0$ is m.

We do not here present proofs of the theorems of Frobenius and Darboux but make use of them to understand the implications for theory building. We hope that the development above makes their validity easy to accept. Our purpose here is to present the mathematical structure of thermodynamic theory and this is best understood with these facts in hand. The proofs can be found in standard texts on differential geometry [8,9] and occasionally even in books on theoretical physics [6].

9.7 THERMODYNAMICS — THE FIRST LAW

Let us examine the structure of thermodynamic theory with the machinery above. The first law, as usual, is the Pfaffian differential Equation 9.10

$$dU - W - Q = 0 \tag{9.20}$$

Let us calculate the parameter r for this form in the case when $W = -pdV$ and $Q = TdS$. Viewing for the moment p and T as independent variables, the sequence above becomes

$$
\begin{aligned}
&dU + pdV - TdS \\
&dp \wedge dV - dT \wedge dS \\
&-TdS \wedge dp \wedge dV - pdV \wedge dT \wedge dS \\
&-2dp \wedge dV \wedge dT \wedge dS \\
&0
\end{aligned}
\tag{9.21}
$$

We thus conclude that $r = 5$ and thus the maximal solutions of (9.20) will be two-dimensional. Robert Hermann turned these facts into a mathematical definition [12]. Hermann defined an n degree of freedom *thermodynamic system* as a maximal submanifold of a $2n + 1$ dimensional manifold equipped with a differential form Ω such that $r(\Omega) = 2n + 1$, i.e., such that $\Omega \wedge (d\Omega)^n \neq 0$.

A $2n + 1$ dimensional manifold equipped with a differential form Ω such that $\Omega \wedge (d\Omega)^n \neq 0$ is called a contact manifold and the form Ω the contact form. Hermann's definition can be restated in the following form: a thermodynamic system is a maximal solution of $\Omega = 0$, where Ω is a contact form. The name "contact" has some significance; contact forms arose in mechanics to deal with surfaces rolling on each other [13]. Also in thermodynamics we can interpret the form Ω as coming from the coexistence of a system with its environment. By Darboux's theorem, there exist coordinates that make Ω assume the canonical form

$$\Omega = dz + \sum_{i=1}^{m} y_i dx_i.$$

For simplicity, we will carry out our discussion for the explicit two-dimensional case $\Omega = dU + pdV - TdS$. The parameters (p,T) can be thought of as parameters describing the environment of the system. At equilibrium, the system chooses its state to coexist with this (p,T). In the geometrical picture introduced by Gibbs wherein we view the system as the surface of the function $U = U(V, S)$, the normal vector describing the tangent plane to the surface is $(-1, -p, T)$. As p and T are changed, this tangent plane rolls on the surface in much the same way that mechanical cogs roll on each other. Moving our perspective to the space of the five variables (U, p, V, T, S) reveals the essential nature of this coexistence.

The functions (U, p, V, T, S) are by no means the only *contact coordinates*, i.e., the only coordinates, which make Ω assume the form in Equation 9.18. For example, the classical Legendre transformation always result in contact coordinates

$$\Omega = dH - Vdp - TdS = dG - Vdp + SdT = dF + pdV + SdT \tag{9.22}$$

where H, G, and F are the enthalpy, the Gibbs free energy, and the Helmholtz free energy. Their usefulness derives from exactly those situations when the environment specifies the coefficients in front of the differentials, i.e., the "y" variables from Darboux's theorem. This further justifies the view that the "y"s describe the environment, the "x"s describe the system, and "z" characterizes the contact.

Requiring a coordinate change to preserve the appearance of Ω shown in Equation 9.18 allows many more coordinate changes. These are generalized Legendre transforms [14] and form an infinite dimensional group known as the contact group [15]. Its use to date has been limited by the paucity of exotic environments. Its use has been demonstrated for a system inside a balloon whose pressure and volume obey a definite relationship although neither pressure nor volume is constant. It is potentially useful for biological systems with complex constraints.

Besides possible uses of these generalized Legendre transforms, the first law of thermodynamics in the form $\Omega = 0$ gives a deeper perspective regarding the thermodynamic method. It shows us that this method may be thought of as a theory for "viewing" the inside of black box systems [14]. We manipulate n parameters y in the environment and observe the changes in our black box system as it moves to states of coexistence. In this way we find a *thermodynamic theory of the black box*. The form Ω chooses the particular x and z that must go with the y's describing the system's surroundings.

9.8 THERMODYNAMICS — THE SECOND LAW

In the context of differential forms, the natural choice for the second law of thermodynamics is Caratheodory's form: the heat form Q has an integrating factor. In terms of the machinery above, this is most naturally stated as $Q \wedge dQ = 0$ [16], which, according to Equation 9.19 with $m = 1$, says that Q must be of the form TdS.

This way of stating the second law completely omits the phenomenology. Caratheodory's original statement is that at any state of an equilibrium thermodynamic system has arbitrarily close states that cannot be reached by adiabatic processes, i.e., along states with $Q = 0$. Physically, these inaccessible states correspond to the states that we would reach by removing a little bit of heat from our system. Such states cannot be reached by relaxing internal degrees of freedom or adding or extracting work; these mechanisms could only add heat (through friction). Caratheodory's work on Pfaffian equations shows that this condition is equivalent to the condition of the existence of an integrating factor. For completeness, we now present a direct proof of this fact following Pauli [6]. We begin by making use of Darboux's theorem to choose coordinates, which make the differential form for heat take its simplest guise

$$Q = dz + \sum_{i=1}^{m} y_i dx_i,$$

or

$$Q = \sum_{i=1}^{m} y_i dx_i$$

where the functions x_1, x_2, ..., x_m, y_1, y_2, ..., $y_{m,z}$ (respectively, x_1, x_2, ..., x_m, y_1, y_2, ..., y_m) are independent. For convenience in the present proof, we combine the two cases by setting

$$z = x_{m+1}$$

$$y_{m+1} = 1 \tag{9.23}$$

$$k = m + 1$$

in the first case and $k = m$ in the second case, making

$$Q = \sum_{i=1}^{k} y_i dx_i. \tag{9.24}$$

Now consider any equilibrium state s^0 with coordinates $(x_1^0, x_2^0, ..., x_k^0, y_1^0, y_2^0, ..., y_k^0, w^0)$, where w represents any additional independent coordinates, which do not appear in Q. Consider any nearby state s^1 with coordinates $(x_1^1, x_2^1, ..., x_k^1, y_1^1, y_2^1, ..., y_k^1, w^1)$ and let

$$\Delta x = \left(x^1 - x^0 \right) = \left(x_1^1 - x_1^0, x_1^1 - x_1^0, ..., x_k^1 - x_k^0 \right). \tag{9.25}$$

As we will show below, there exists an adiabatic path, i.e., a path γ such that $\int_\gamma Q = 0$, from s^0 to s^1 provided there exists a nonzero k-tuple $y^* = (y_1^*, y_2^*, ..., y_k^*)$, which is orthogonal to Δx. This is indeed the case for any Δx unless $k = 1$, i.e., unless Q is of the form $y_1 dx_1$ in which case Q has an integrating factor $1/y_1$.

To see the existence of an adiabatic path for the case where a nonzero y^* exists, we splice together three partial paths (Figure 9.2)

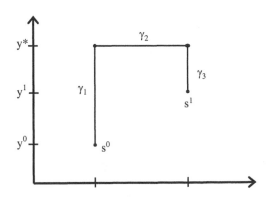

FIGURE 9.2 The adiabatic path connecting nearby points.

$$\gamma(t) = \begin{cases} \gamma_1(t) & 0 \le t < 1/3 \\ \gamma_2(t) & 1/3 \le t \le 2/3 \\ \gamma_3(t) & 2/3 < t \le 1 \end{cases} \tag{9.26}$$

with

$$\gamma_1(t) = \left(x^0, \, y^0 + 3t\left(y* - y^0\right), \, w^0 + t\left(w^1 - w^0\right) \right),$$

$$\gamma_2(t) = \left(x^0 + 3(t - 1/3)\left(x^1 - x^0\right), \, y*, \, w^0 + t\left(w^1 - w^0\right) \right), \tag{9.27}$$

$$\gamma_3(t) = \left(x^1, \, y* + 3(t - 2/3)\left(y^1 - y*\right), \, w^0 + t\left(w^1 - w^0\right) \right).$$

Portions γ_1 and γ_3 are adiabatic because $dx_i = 0$, $i = 1, \ldots, k$, while γ_2 is adiabatic, because along γ_2,

$$Q = \sum_{i=1}^{k} y_i * \frac{dx_i}{dt} dt = \left(\sum_{i=1}^{k} y_i * \left(x_i^1 - x_i^0\right) \right) dt = \left(y* \cdot \Delta x \right) dt = 0. \tag{9.28}$$

9.9   RIEMANNIAN STRUCTURE

In addition to the contact structure Ω and the heat form Q, thermodynamic systems possess another structure that can be attributed to the specialness of extensive coordinates, i.e., coordinates that scale with the size of the system. In all such coordinates, the second derivative matrix of the entropy S with respect to a set of independent extensive variables (say, U,V) defines a Riemannian structure on our manifold. This follows purely from the fact that entropy is a maximum at equilibrium and thus its second derivative is a negative definite symmetric matrix. For our simple two degree of freedom system, this matrix is

$$\frac{\partial^2 S}{\partial(U,V)} = \begin{pmatrix} -C_v/U^2 & 0 \\ 0 & -R/V^2 \end{pmatrix}, \tag{9.29}$$

where R is the gas constant and C_v is the constant volume heat capacity of the system. Such Riemannian structure associates lengths to processes. This *thermodynamic length* has been shown to be the relevant quantity in a covariant improvement to Einsteinian fluctuation theory [17,18]. It can be interpreted as the "number of fluctuations" traversed. The statistical mechanical expression for this distance shows it to be mathematically identical to Fisher's genetic distance introduced to measure genetic drift [19]. For thermodynamic processes in finite time, the square of the distance is proportional to the minimum entropy produced by traversing the process [20,21].

9.10   CONCLUSIONS FOR SYSTEMS BIOLOGY

Our presentation was an attempt to present thermodynamics in a way that starts as a theory about a set for which we can measure certain parameters — a manifold. The first law was revealed to be a statement that the theory was really applicable in any context where the system will choose a state once its environment is specified. For this context, the theory gives a constructive prescription through the postulate that the energy deficiency form $\Omega = dU - Q - W$ be a contact form. The prescription defines the corresponding variables of the system and the coexistence function — a generalized thermodynamic potential. From this perspective, we see thermodynamics as a theory of black box systems that are characterized by the surroundings to which they equilibrate.

REFERENCES

1. Carnot S. Reflexions sur la puissance motrice du feu et sur les machines propres a developper cette puissance. Bachelier, Paris, 1824.
2. Pippard, AB. Elements of classical thermodynamics for advanced students of physics. Cambridge University Press, 1957.
3. Tisza L. Generalized thermodynamics. Cambridge, Mass: M.I.T. Press, 1966.
4. Keenan JH. Thermodynamics, New York, John Wiley, 1941.
5. Salamon P, Salamon A, Konopka AK. Chapter 1, Thermostatics: a poster child of systems thinking. In Konopka, AK, ed. Handbook of Systems Biology. New York: Kluwer, 2005.
6. Carathéodory, C. Untersuchung über die Grundlagen der Thermodynamik. Math. Annalen 1909, 67, 355–386. See also Pauli W. Lectures on Physics: Vol. 3, Thermodynamics and the Kinetic Theory of Gases. Cambridge, MA: MI Press, 1973.
7. Callen, H. Thermodynamics. New York: John Wiley and Sons, 1960.
8. Warner, F. Foundations of Differentiable Manifolds and Lie Groups. New York: Springer, 1983.
9. Bishop RL, Goldberg SI. Tensor Analysis on Manifolds. New York: Dover Publications, 1980.
10. Eisenhart LP, A Treatise on the Differential Geometry of Curves and Surfaces. Boston: Ginn and Company, 1909.
11. Poincare H. Les methodes nouvelles de la mecanique celeste. Gauthier Villars, Paris, 1892.
12. Hermann R. Geometry, Physics, and Systems. New York: M. Dekker, 1973.
13. Ball RS. Treatise on the Theory of Screws. Cambridge, UK: Cambridge UP, 1900.
14. Salamon P. The Thermodynamic Legendre Transformation or How to Observe the Inside of a Black Box, PhD Thesis, Department of Chemistry, University of Chicago, 1978.
15. Salamon P, Ihrig E, Berry RS. A group of coordinate transformations preserving the metric of weinhold. J. Math. Phys. 1983; 24: 2515.
16. Edelen, D. The College Station Lectures on Thermodynamics. College Station, TX: Texas A & M University, 1993.
17. Diosi L, Lukacs B. Covariant evolution equation for the thermodynamic fluctuations. Phys. Rev. A, 1985; 31: 3415–3418.
18. Ruppeiner G. Riemannian geometry in thermodynamic fluctuation theory. Rev. Mod. Phys. 1995, 67: 605–659.
19. Salamon P, Nulton JD, Berry RS. Length in statistical thermodynamics. J. Chem. Phys. 1985; 82: 2433–2436.
20. Salamon P, Berry RS. Thermodynamic length and dissipated availability. Phys. Rev. Lett. 1983; 51: 1127–1130.
21. Nulton J, Salamon P, Andresen B, Anmin Q. Quasistatic processes as step equilibrations. J. Chem. Phys. 1985; 83: 334–338.

APPENDIX
Systems Biology:
A Dictionary of Terms

Activation energy — (1) Difference between the energy of substrates of a chemical reaction and the energy of hypothetical transition state (activated complex) of this reaction. (2) Minimum amount of energy needed to activate atoms or molecules to a condition in which it is equally likely that they will undergo chemical reaction or transport as it is that they will return to their original (nonactivated) state. The transition state theory of chemical reactions postulates the existence of a high-energy transition state between the low-energy initial conditions and the (also low-energy) product conditions of a one-step reaction (or every single step of a multistep reaction). Within this theory, the activation energy is the amount of energy required to boost the initial materials (substrates) "uphill" to form the transition state (consistent with meaning [1] above. The chemical reaction then proceeds "downhill" of the energy barrier to form the intermediate (in multistep reaction) or final (in one-step reaction) products. Catalysts (such as enzymes) lower the activation energy by altering the transition state.

Active site — A region within its molecular structure to which an appropriate substrate can bind and thereby form the so-called enzyme–substrate complex.

Adaptation — (1) A heritable feature of an individual's phenotype that improves its chances of survival and reproduction in the existing environment. (2) An activity of a system (such as, for instance, an organism, a population, or an ecosystem) that leads to better chances of its own survival in a given environment.

Algorithm — A systematic list of a finite number of step-by-step instructions for accomplishing some task that has an end (result). More precisely, an algorithm is a process of representation that can be characterized by the following five properties: (1) It has zero or more input data. (2) It has one or more output data. (3) It is resolved into unambiguous steps. (4) It halts after finitely many steps. (5) It is effective.

Allele — One of possible alternative forms of the same gene occupying a given locus (position) on a chromosome.

Allele frequency — The relative proportion of a particular allele in a specific population.

Alphabet — An ordered finite set of symbols (letters). Symbols can be elementary (not fractionable into "smaller" symbols) or composite (fractionable into smaller units). The alphabet that contains only elementary symbols is referred to as an *elementary alphabet*.

Ambiguity — Although vague or sloppy thinking may lead to an ambiguous representation of an unambiguous situation, the property of ambiguity is not to be confused with either vagueness or sloppiness. Rather, ambiguity is a process in which more than one possible outcome is consistent with the entailments that produce the outcome. Semantically, this is reflected in context dependent multiple meanings of a sign. As Quine and others have shown, some ambiguities are inherent; adding disambiguating contextual information introduces new ambiguities more quickly than it resolves the old ones. Although commonly considered a property of epistemological representations, ambiguity also encompasses ontological processes. Causal ambiguity is the situation in which the causal constraints on a process are sufficiently loose that any of several different outcomes is consistent with the entailment structure, and, having occurred, can be said to have been caused by the entailment structure.

Analogy — (1) A resemblance in function or structure, which is due to convergence in evolution but not to common ancestry. (Resemblance due to common ancestry is referred to as homology). (2) A relation between two phenomena that can be described by means of the same model. For example, a mechanical pendulum and electric condenser are analogies of each other because their behavior can be adequately described by the same mathematical formalism (differential equations).

Anticipatory system — A system that is capable of making and utilizing a predictive model of itself and/or its environment. If we would want to model such a system with a dynamic systems formalism, its current changes of states would be affected by the future (predicted) states.

Ascendency — A measure of how well a system is performing. It is the product of the total system activity and an informational measure of how well organized (see **organization** below) its connections are. Self-organization thereby is seen as an increase in a system's ascendency. Both the system activity and its organization are seen as results of autocatalytic (see **autocatalysis** below) agency at work in the system.

Attractor — A subset of a (dynamic) system's states to which the system "is attracted." When time passes, a dissipative dynamical system settles into an attractor and even then remains confined to the states belonging to the attractor unless forced to leave; mathematically — a set in the phase space that has a neighborhood in which every point stays nearby and approaches the attractor as time goes to infinity; *chaotic system* has *chaotic attractor* that displays exponentially sensitive dependence on initial conditions; strange attractor is one that has noninteger *fractal dimension;* unfortunately, the term "strange attractor" is often used for any chaotic attractor; however, there exist chaotic attractors that are not strange, as well as strange attractors that are not chaotic.

Autocatalysis — A causal circuit whereby the activity of an entity serves indirectly to enhance itself via a chain of directed, positive interactions. This is a special form of positive feedback (see below) wherein all links in the loop are enhancing. The direction of the loop possesses an inherent asymmetry in that the result of a random event upon the circuit elicits a nonrandom response.

Automaton — (1) A completely or partly self-operating machine or mechanism. (2) A system whose activity can be described by a set of subsequent transitions between states.

Binomial distribution — A discrete probability distribution of obtaining an exact number (say n) of successes out of N Bernoulli trials. Each Bernoulli trial is a success with a fixed probability p and failure with (also fixed) probability $q = 1 - \mathrm{p}$.

Bioinformatics — (1) Information technology–based trend in biology-related academic and industrial activities of computer scientists, engineers, mathematicians, chemists, physicists, medical professionals, and biologists. Activities include archiving, searching, displaying, manipulating, and otherwise integrating life-science-related data. Examples of bioinformatics tasks include mapping genomes, high-throughput contig annotation, interpretation of gene expression data, and creation of databases. (2) A subfield of computer science that is devoted to biological applications of computer programming, databases, and related activities.

In both meanings bioinformatics differs from computational biology (the field of biology in which some of the methods require use of computers). The difference is primarily in the motivation of practitioners of bioinformatics (demonstrate better software than the competition) as opposite to motivation of practitioners of computational biology (curiosity and desire to learn and understand biological systems).

Biological species (concept) — A population or group of populations within which there is a significant amount of gene flow under natural conditions, but which is genetically isolated from other populations.

Biota — All of the organisms, including animals, plants, fungi, and microorganisms, found in a given area.

Biotechnology — Application of molecular biology to produce food, medicines, and other chemicals usable for controlling living conditions of individual humans or populations thereof.

Body plan — A hypothetical spaciotemporal design that includes an entire animal, its organ systems, and the integrated functioning of its autonomous parts.

Catalyst — A chemical substance that accelerates a reaction and can be detected as unchanged after this reaction. Within the transition state theory of chemical reactions, catalysts lower the activation energy of a reaction via restructuring its transition state to a lower energetic level than the transition state would assume without catalysis.

Centripetality — The aggregation of material and energetic resources within and among system members that participate in autocatalytic causal circuits. This tendency elsewhere has been called "chemical imperialism," and it provides the drive behind the competition and striving that is only tacitly acknowledged in conventional evolutionary theory.

Chaotic system — A nonlinear (dynamic) system that displays *deterministic chaos*; chaotic systems display also extreme sensitivity to small changes in system parameters.

Chemical reaction — A chemical process in which substances are changed into different ones (with different properties) via rupture or rearrangement of the chemical bonds between atoms in a way that does not affect atomic nuclei.

Based on IUPAC recommendation: A process that results in the interconversion of *chemical species*. Chemical reactions may be *elementary reactions* (one-step reactions) or *stepwise reactions* (multistep reactions). It should be noted that this definition (meaning 2) includes experimentally observable interconversions of conformers of the same compound. Detectable chemical reactions normally involve sets of *molecular entities* but it is often conceptually convenient to also use the term to denote changes involving single molecular entities (i.e. "microscopic chemical events").

Every chemical reaction can be studied in terms of its mechanism (series of subsequent intermediary compounds between the substrates and products).

Reactions occur at a particular rate that depends on several parameters such as the temperature and concentrations of the reactants.

Many chemical reactions go to completion, i.e., attain equilibrium, over a period of minutes or hours, and can be monitored by classical techniques, such as pressure change or electrochemistry. Chemical dynamics explores the detailed behavior of molecules during the most crucial moments of reactions, for example, when bonds are being broken and new bonds being formed.

Chemical kinetics — Studies of rates of chemical reactions and inference of possible mechanism on the basis of such studies. (Knowledge of a reaction mechanism comes in part from a study of the rate of a reaction).

The fact that a reaction is thermodynamically favorable does not necessarily mean that it will take place quickly. Many reactions that do proceed are endothermic. Therefore, the enthalpy change is not the ultimate arbiter of the spontaneity of a chemical reaction, and an additional term, the change in free energy, is an important factor. In addition, the rate of the reaction — the change in concentration of a reactant or of a product with respect to time — gives information on how the reaction will proceed. Some of the factors that influence the rate of a reaction are: concentration of reactants and/or products, temperature, surface area, and pressure.

Where a catalyst is present (e.g. an enzyme), then the concentration of the catalyst also influences the rate of reaction.

Church–Turing Thesis — Copeland BJ. The Church-Turing Thesis. In: Zalta E, ed. Stanford Encyclopaedia of Philosophy. Stanford: http://plato.stanford.edu/, 1996; gives the following description that is adequate to the content of this volume:

> ... *Church's thesis: A function of positive integers is effectively calculable only if recursive.*
>
> *Turing's thesis: LCMs [logical computing machines: Turing's expression for Turing machines] can do anything that could be described as "rule of thumb" or "purely mechanical".*
>
> *It is important to distinguish between the Turing-Church thesis and the different proposition that whatever can be calculated by a machine can be calculated by a Turing machine. (The two propositions are sometimes confused.) Gandy (1980) terms the second proposition 'Thesis M'.*
>
> *Thesis M: Whatever can be calculated by a machine is Turing-machine-computable.*
>
> *Thesis M itself admits of two interpretations, according to whether the phrase 'can be calculated by a machine' is taken in the narrow sense of 'can be calculated by a machine that conforms to the physical laws (if not to the resource constraints) of the actual world', or in a wide sense that abstracts from the issue of whether or not the notional machine in question could exist in the actual world. The narrow version of thesis M is an empirical proposition whose truth-value is unknown. The wide version of thesis M is simply false....*

As far as systems biology is concerned, the narrow version of thesis M (one can call it thesis N) is believed to be true:

Thesis N: Any computation that can be completely described by an algorithm expressible in a formalized language can be translated into an equivalent set of operations of a Turing machine (a formal system).

(Formalized language has its syntax and semantics while a formal system has syntax only.)

Code — (1) A mapping from one symbolic representation (input representation) of a system into another representation (output representation). If the input and output representations can be expressed in the form of elementary symbols chosen from finite alphabets, a code can be seen as a relation between input and output alphabets of a device that does encoding (an encoder) or decoding (a decoder). (2) A transformation whereby messages are converted from one representation to another.

Codominance — A condition in which two alleles at a locus produce different phenotypic effects and both effects appear in heterozygotes.

Codon (of the genetic code) — A trinucleotide in messenger RNA that according to the genetic code corresponds to a specific amino acid.

Coenzyme — An additional substance (compound) that plays a role in catalysis by an enzyme.

Coevolution — Concurrent evolution of two or more species that are mutually affecting each other's evolution.

Communication — (1) A process of sharing knowledge or units thereof. (2) A process of sending and receiving messages.

Community — Any ecologically integrated group of species of microorganisms, plants, and animals inhabiting a given area.

Comparative analysis (in evolutionary biology) — An approach to studying evolution in which hypotheses are tested by measuring the distribution of states among a large number of species.

Complex (complexity) — (1) There are dozens, if not, hundreds of (often mutually exclusive) definitions of complexity in the scientific literature. Probably, one of the most popular is von Neumann's idea of "exceeding a threshold of complication." This definition had numerous problems, not the least of which is knowing when the threshold is exceeded. A more useful definition of complexity, as used in several chapters of this work, is that a complex process is characterized by a closed-loop hierarchical structure of entailment. In this context, complexity comes in two distinct versions. Ontological complexity is characterized by a closed loop of inherently hierarchical efficient causes, and is sometimes called an "endogenous process" or "closed to efficient cause." Epistemological complexity is characterized by a closed loop of inherently hierarchical logical relationships, and is typically characterized by an impredicative representation. A consequent property of the closed-loop hierarchical structure of entailment is that it has no complete representation or "largest model." (2) A system is complex (convoluted) if it does not a single model that would completely represent it. Instead, an unlimited number of models complementary to each other but not derivable from each other are needed to represent a convoluted (complex) system for any practical purposes. By definition, there is no (and there cannot be) largest model of a convoluted system.

Computing — Executing an algorithm by a symbol-manipulating device placed in appropriate conditions. The nature of appropriate conditions usually includes a modeling relation between the device and the formal description of algorithm.

Convergent evolution — The evolution that leads to similar features in apparently otherwise unrelated life forms.

Contingency — A chance or aleatoric event that remains beyond prediction under the assumption of continuous behavior. Whenever contingent events are generic, simple, and repeatable, a degree of prediction can be achieved using probability theory. Contingent events, however, can be complex and unique for all time, thereby eluding treatment by probability theory. The existence of contingencies frustrates efforts at arbitrarily precise determination.

Covalent bond — A chemical bond formed by sharing electrons between two atoms.

Cotangent space — The vector space at a point in a manifold, which is dual to the tangent space. A vector in this space can be defined as an equivalence class of functions that "change at the same rate in every direction." A cotangent vector specifies a rate of change along any tangent vector. If $\{X_i\}$ is a coordinate system at a point, the differentials dx_i form a basis for the cotangent space and thus define the corresponding coordinate system in the cotangent space.

Crossing over — The reciprocal exchange of corresponding segments between two homologous chromatids.

Data (datum) — Unit of representation (elementary model) of selected aspects of reality provided in a form suitable for symbolic manipulation such as intellectual reflection, discussion, or mechanical transformation (computation). Data can be characterized (described) with the help of at least three properties: (1) content, (2) medium, and (3) structure.

Set of unitary representations (1) organized in a way suitable for symbolic processing.

Synonym of "information": a unit of knowledge that can be communicated and processed.

A set of units (3) or a combination of these units plus some consequences of combining them.

Data structure (information structure) — An organized (and often codified) form in which data is available for symbolic processing with the help of specific algorithm or classes of algorithms.

Deterministic chaos — Behavior with no simple means of prediction despite the fact that it arises in a deterministic (dynamic) system because of extreme sensitivity to initial conditions due to presence of chaotic attractor; chaos arises in a dynamical system if two arbitrarily close starting points diverge exponentially, so that their future behavior is eventually unpredictable.

Development — A progression toward ever more constrained and harmonious behaviors among the elements of a living ensemble. Development begins with an increase in the number of distinguishable parts of the system. As the system develops, any activity by one member is focused (constrained) to have an effect on a single or small subset of other members. Development can be considered any increase in system organization (see below).

Differential form — A smooth choice of a cotangent vector at each point in a manifold. If $(x_1, x_2, ..., x_n)$ is a coordinate system at a point, then the differentials dx_i form a basis for the cotangent space with coefficients that are smooth functions of $(x_1, x_2, ..., x_n)$. Differential forms typically represent "infinitesimal" quantities that need to be integrated along a path to give net changes.

Downward causation — Downward causation is the idea that the properties of the parts are influenced by the whole, of which they are members. Traditionally, it was assumed that God was the source of all downward causation, and for the past several centuries, virtually all discussion of downward causation was dismissed as "unscientific." However, there is no justification for the notion that downward causation must have a mystical or supernatural origin. The admission of the possibility of a naturally occurring downward causation helps to formulate a number of potentially solvable problems in science.

Dynamic programming — An optimization technique designed primarily for search problems.

Dynamical system (dynamic system) — A system that has the capability to change with time.

Ecological niche — The functioning of a species in relation to other species and its physical environment.

Ecology — The scientific study of the interaction of organisms with their environment, including both the physical environment and the other organisms that live in it.

Ecosystem — An interacting combination of living and nonliving elements that occupy a given physical region. The living members of each ecosystem are separated (usually to some degree in space and time) into an autotrophic group that fixes energy and heterotrophic components that take their sustenance from what is fixed by the autotrophs. There is usually physical transport linking the autotrophic and heterotrophic regions of any ecosystem.

Effective — In the context of theoretical computer science, the property of being effective is the property of being fully representable by a Turing machine. In other contexts, it has very different meanings. For example, in the context of causation, an effective process is one that produces an effect.

Efficient cause — Efficient cause is an ontological influence not to be confused with the modern notion of efficiency. Furthermore, it is a different concept from the "billiard balls bouncing off each other" paradigm that inevitably arises in the discussion of causation in physics. If causation is considered as the transformation of one event into another event, efficient cause is the constraint that regulates how that transformation unfolds. Efficient cause arises from the intersection of two influences, the global constraints on the relationships between phenomena (epistemologically represented as the laws of nature) and the morphology of the process. Efficient causes in ontology are often represented epistemologically by "dynamical laws," usually derived by applying some "law of nature" to the structure of the process. A simple example of an efficient cause is the ontological constraint on the operation of a linear time-invariant system that is represented by the system's transfer function.

Elementary alphabet — An alphabet whose letters cannot be fractioned into smaller textual units.

Emergent property — A property of a complex system that is not exhibited by its individual component parts determined from a model of the system. The emergence can take place as a result of spontaneous changes in the system (such as chemical reaction or phase transitions) or as a consequence of anticipatory behavior of the anticipatory system (future, anticipated, states of the system can affect the current changes of states.)

Endergonic reaction — One for which energy must be supplied. (Contrast with exergonic reaction.)

Endo- — A prefix used to designate an innermost structure.

Endogenous (endogeny) — The term originally arises in medicine and economics. Informally, an endogenous process is one that "makes itself up as it goes along." In the context of this work, endogeny is the ontological version of Rosen's complexity; an endogenous process is a process in reality that is entailed by a closed-loop hierarchic structure of efficient causes. Endogeny is the ontological analog to the epistemological concept of impredicativity. In keeping with Rosen's principle of separation of the model from the process being modeled, there is no such thing as an endogenous formal system or an impredicative natural system.

Endosymbiosis — The living together of two species, with one living inside the body (or even the cells) of the other.

Endosymbiotic theory — An (speculative) evolutionary scenario in which the eukaryotic cell evolved from a prokaryote that contained other endosymbiotic prokaryotes.

Energy — The capacity to do work.

Entailment — (1) Generalized implication, semantic implication, or material implication. X entails Y if Y is a consequence of X. A given fact, phenomenon, process, or statement X entails another fact, phenomenon, process, or statement Y if and only if the occurrence of Y is a consequence of the occurrence of X. (2) A constraint that limits the possible outcomes of a process. Causal entailments are the causal constraints that allow particular effects. Inferential entailments (such as chains of subsequent implications) are the logical constraints that give rise to particular conclusions.

Entropy — (1) A state function of a thermodynamic system in equilibrium. Spontaneous irreversible processes in a closed system lead from an equilibrium state with lower entropy to another equilibrium state with higher entropy. Entropy is defined by how it changes in a reversible process and its definition is constructed so as to distinguish the forward process, which occurs spontaneously, from the reverse process, which can never happen. This is embodied in the second law of thermodynamics, which states that the total entropy of the universe increases in any spontaneous process. Such increase defines "time's arrow" and this is the only physical law without time reversal symmetry. As traditionally defined, the entropy of the system changes by the amount of heat entering the system divided by the temperature of the system. One of the deepest results of 20th century physics is that an equivalent definition is to define the entropy of a system to be the logarithm of the number of microstates consistent with its macroscopic description. In a classical mechanical description of the system, this is just the volume in phase space accessible to the system. (2) Shannon entropy — a mathematical function of a random variable that measures deviation of probability distribution (of this variable) from the discrete uniform distribution. Shannon entropy and its variants have been excessively used as statistical tools of choice in several fields of engineering and "hard" science.

Environment — An organism's surroundings, both living and nonliving; include temperature, light intensity, and all other species that influence the focal organism.

Enzyme — A protein that serves as a catalyst for a chemical reaction. (RNA-based catalysts are also called RNA enzymes by some scientists but, in principle, the term "enzyme" refers to a protein.)

Epi- — A prefix used to designate a structure located on top of another.

Epigenesis — A process of interaction between genes and environment, which ultimately results in the distinctive phenotype of an organism.

Epigenetic effect (phenomenon) — Any gene-regulating activity that does not involve changes to the regions in DNA coding for proteins being regulated and that can persist through one or more generations. Two known examples of epigenetic phenomena are gene silencing and imprinting.

Epigenetic hypothesis — Patterns of gene expression, not genes themselves, define each cell type. (This is the original concept of epigenetics coined by Waddington in early 1950s in order to explain differentiation.)

Epigenetic rule — A model of an alleged trend in epigenesis that could channel development in particular directions.

Epigenetics — Study of heritable changes in gene function, which occur without a change in the DNA sequence (at least not the one that encoded the studied protein).

Episome — A plasmid that may exist either free or integrated into a chromosome.

Epistasis — An interaction between genes, in which the presence of a particular allele of one gene determines whether another gene will be expressed.

Epistemology — Theory of representation of reality in the human mind (theory of knowledge). The origin, structure, methods, and validity of knowledge or specific mental representations of reality are typical epistemological subjects. The term "epistemology" appears to have first been used by J. F. Ferrier, *Institutes of Metaphysics* (1854) who distinguished two branches of metaphysics (science of being in itself) — epistemology and ontology. (*See also* Ontology; Modeling Relation; Natural System)

Equilibrium — In biochemistry, a state in which forward and reverse reactions are proceeding at counterbalancing rates, so there is no observable change in the concentrations of reactants and products.

In evolutionary genetics, a condition in which allele and genotype frequencies in a population are constant from generation to generation.

In thermodynamics, a state of the system in which all parameters describing it have constant values that do not change with time.

Estimator — A rule (a random variable) for determining the value of a population parameter based on a random sample from this population. An example of an estimator is the sample mean, which is an estimator of the population mean.

Evolution — (1) Any gradual change. Organic evolution, often referred to as evolution, is any genetic and resulting phenotypic change in organisms that can take place after the passage of many generations. (2) A temporal sequence of patterns in biotic systems over long periods of time. Charles Darwin, the commonly recognized originator of evolution in biology, suggested that biotic taxa and patterns arise somehow gradually out of preexisting taxa. The dynamic of evolution is assumed to be an overproduction of similar forms, which is pruned back by external natural factors collectively referred to as "natural selection." Evolution was once thought to be progressive, like development, but now is commonly held to be adirectional.

Exterior product — A formal, anticommuting product defined on differential forms to give two-forms, three-forms, etc. which get integrated multiple times. Exterior products of differential forms typically represent densities that need to be integrated over a region to give net quantities. The anticommuting feature means that $dx_i \wedge dx_j = -dx_j \wedge dx_i$ and is needed to force products with repeated factors to vanish and brings the added benefit of oriented values for integrated densities such as area. Also called wedge product.

Exterior forms — Linear combinations of exterior products of differential forms.

Exergonic reaction — A reaction in which free energy is released. (Contrast with endergonic reaction.)

Exergy — The work available from the disequilibrium of a system with its surroundings. When the surroundings maintain constant temperature, it reduces to the free energy.

Feedback — The situation by which the activity of a system element propagates over a causal pathway back to affect its own workings — a causal circuit. The effect upon the original element can be either positive or negative. Feedbacks are the crux of the science of "cybernetics," which explains how the behavior of a system can be controlled externally or from within. (See also "autocatalysis" above.)

Feedback control — Control of a particular step of a multistep process, induced by the presence or absence of a product of one of the later steps.

Final cause — Final cause is the causal influence by which the properties of the parts are influenced by the whole of which they are members. Traditionally, it was assumed that God was the only final cause, and for the past several centuries, virtually all discussion of final causation was dismissed as "unscientific." However, we have little difficulty admitting the concept of a linear hierarchy of causation into scientific discussion. For example, if a small robot changes the layout of a linear time-invariant system, consequently updating its transfer unction, we can say that the robot has entailed the efficient cause of the linear system. Similarly, if a big robot modifies the small robot, we can say that the big robot has entailed the efficient cause of the small robot. We can also imagine a progression of ever larger robots, each rebuilding the next smaller one. Thus, we have an example of a linear hierarchy of efficient causes. At adjacent positions in our linear hierarchy, if efficient cause A entails efficient cause B, we have only one direction for tracing the path from A to B, and this is traditionally assigned the direction of "upward" causation. In Rosen's complexity, the linear hierarchy is bent into a closed loop, and remarkable consequence ensues. We can still move upward from efficient cause A to efficient cause B. However, we can start at B and move in the same direction the long way around until we eventually arrive at A; in effect, we can as easily say that B causes A (downward causation without divine intervention) as A causes B (upward causation). As discussed by Rosen, Freeman, and Damasio, once such a loop forms, it has a property of dynamic superstability; it will do whatever it can (including making novel modifications to its own structure) to preserve the integrity of the loop. Final cause can be seen as the "downward" influence of the loop on efficient causes that are its components. It is obtained by traversing the loop of causation the long way around, and it has a self-determined goal, to preserve the integrity of the loop.

Formal cause — Formal cause is not to be confused with formalisms or formal systems. It is the influence that entails the particular form of an effect. It does so by modulating the efficient cause. In other words, formal causes change the properties of the efficient cause within its structure, but they do not update the structure of efficient causes. A simple example of formal cause can be seen in a linear time-invariant system. If the transfer function is the description of the efficient cause of the operation of the system, the parameter set of the transfer function is the description of the formal cause.

Formalism — (1) A school (academic tradition) of metamathematics developed by Hilbert, which claims derivability (provability) of every mathematical statement within a properly designed axiomatic theory. A consequence of this claim was that every true proposition about a formal axiomatic system could be derived from the axioms of the same system. This particular consequence has later been shown to be untrue for some axiomatic systems of mathematics (Gödel, 1937) and thereby untrue in general. (2) Any inferential (formal) system that can be used to model another system. In terms of Hertz–Rosen modeling relation, the modeled system can be a natural system or another formal system. (*See also* Church–Turing Thesis; Model; Modeling Relation; Formal System)

Formal system — If a Rosen modeling relation defines a congruency between a process in reality and an epistemological representation of that process, then the epistemological representation is labeled a formal system. It is not to be confused with a Hilbert's formalism. Hilbert's formalisms disallow impredicatives. In contrast, Rosen modeling relations often use impredicative structures as formal systems.

First law of thermodynamics — Total energy of the closed thermodynamic system can be neither created nor destroyed and therefore must remain constant. Arguably, an example of closed system is the entire Universe. That is why the original formulation of the first law was: Energy of the Universe cannot be either destroyed or created.

Fitness — The contribution of a genotype or phenotype to the composition of subsequent generations, relative to the contribution of other genotypes or phenotypes. (See inclusive fitness.)

Free energy — A function of state for thermodynamic systems. It is a special case of exergy for processes that take place at constant temperature and pressure (in which case it is called the Gibbs free energy) or for processes that take place at constant temperature and volume (in which case it is called the Helmhotz free energy). The change in free energy is the maximum work that could have been obtained from the net changes subject to the constraints appropriate to that type of free energy.

Gene — A unit of heredity. The term is used in biology in three alternative variants: (1) The unit of genetic function which carries information for a functionally important unit of biological function. (2) The determinant (Mendelian factor) of an observable characteristic of an organism. (3) Protein-coding region in cellular DNA (or RNA) sequence. The protein encoded by such region is referred to as "gene product." Variant (3) is favored in molecular biology and its biotechnology-related derivatives.

Gene duplication — A DNA rearrangement that generates a supernumerary copy of a gene in the genome. This would allow each gene to evolve independently to produce distinct functions. Such a set of evolutionarily related genes can be called a "gene family."

Gene family — A set of genes derived from a single parent gene via gene duplication. Individual genes from a family need not be on the same chromosomes.

Gene flow — The exchange of genes between different species (an extreme case referred to as hybridization) or between different populations of the same species caused by migration following breeding.

Gene pool — All of the genes in a population.

(the) Genetic Code (translation code; universal genetic code) — Many-to-one mapping (function) that assigns one of 20 naturally occurring amino acids to each of 61 "sense" trinucleotide code words (codons) during protein biosynthesis. Three additional trinucleotides — TAA, TAG, and TGA — serve as signals for the termination of translation during protein biosynthesis but do not encode any amino acid. The initiation of translation is often encoded by the trinucleotide ATG, which is also a codon for amino acid methionine. (Many protein sequences begin with methionine for this reason.)

Genetic drift — Changes in gene frequencies from generation to generation in a small population as a result of random processes.

Genetics — The study of heredity.

Genome — (1) Complete set of genes of an organism. (2) All the DNA contained in an organism (or a representative cell thereof), which includes both the chromosomes within the nucleus and the DNA in mitochondria. (3) Complete genetic information defining an organism.

Genotype — An exact description of the genetic constitution of an individual, either with respect to a single trait or with respect to a larger set of traits.

Genus — A group of related, similar species.

Homeostasis — The maintenance of a steady state, such as a constant temperature or a stable social structure, by means of physiological or behavioral feedback responses.

Hypothesis — (1) A tentative assumption (usually about a set of facts) that can be verified as true or false by either experiment or reasoning. (2) A statement assumed to be true for the purpose of argument or further testing. (3) An antecedent H of conditional statement "if H then P". The P is a prediction or consequence, which should be valid if H could be proven true.

Hypothesis testing — In statistics, making a decision between rejecting or not rejecting a given null hypothesis on the basis of a set of specific observations.

Impredicative — An impredicative representation is one that is characterized by the constraint on its relationship with itself. The conventional definition of impredicativity is given by Kleene. "When a set M and a particular object m are so defined that on one hand m is a member of M, and on the other hand, the definition of m depends on M, we say that the procedure (or the definition of m, or the definition of M) is impredicative. Similarly, when a property P is possessed by an object m whose definition depends on P (here M is the set of the objects which possess the property P). An impredicative definition is circular, at least on its face, as what is defined participates in its own definition." Kleene's definition is purely epistemological; it is a property of representations. In the context of this work, impredicativity is the epistemological version of Rosen's complexity. It is the epistemological analog to the ontological concept of endogeny. In keeping with Rosen's principle of separation of the model from the process being modeled, there is no such thing as an endogenous formal system or an impredicative natural system.

Information — A vague, metaphoric concept that relates what we perceive (by observing, reading, or listening) to a change in our knowledge. There is no single, universally accepted definition of the term. The two most plausible explications of the term *information* are: (1) Anything that constrains the number of options available to a systems behavior. Although one of the notions of information arose out of a communications context, where it is manifested in "atomic" form, such as a letter of a printed text over a finite alphabet, the idea pertains as well to the distributed and innate constraints that serve to channel flows in networks along preferred pathways. The same calculus used on "atomic" forms of information applies as well to its more diffuse manifestations. (2) A tacit component of functionally competent entity (i.e., of a system) that is in charge of assuring this functional competence. In Aristotle's writings (see his On Soul as well as Metaphysics), every object, phenomenon, and process has its soul in addition to its material form. The soul is inside the form but is not the same as the form. Nonetheless, the soul of things is responsible for their nature, for their being what they are. In this sense, the Aristotelian soul is nonmaterial organizing power for the matter. This concept of organizing power is intuitively the same as the notion of information contained in the functionally robust system. The same intuition is also present in all considerations of knowledge acquisition, storage, and sharing (communication).

Informatics — (1) The art of management of data and management-oriented data analysis. (2) Synonym of applied computer science that includes database-related tasks, computer programming as well as hardware-related tasks.

Inhibitor — An "anticatalyst" that prevents a catalyst from proper functioning in its catalytic capacity. When the enzyme is a catalyst, an inhibitor is a substance, which binds to the surface of the enzyme and interferes with its action on its substrates.

Internal energy — A function of state for thermodynamic systems. Internal energy is defined by how it changes: either by work being done on or by the system, or heat being added to or taken from the system. The conservation of energy is the first law of thermodynamics and asserts that the internal energy added up over all systems never changes.

Law of independent assortment — The random separation during meiosis of nonhomologous chromosomes and of genes carried on nonhomologous chromosomes. Mendel's second law.

Law of segregation (Mendel's first law) — Alleles segregate independently from one another during gamete formation.

Likelihood — A hypothetical estimate of chance that an event, which has already occurred would yield a specific outcome. The concept of likelihood differs from that of probability.

(Probability refers to the occurrence of future events with possibly unknown outcomes, while likelihood refers to past events with known outcomes.)

Manifold — (1) (of sense): The ingredients of sense experience (such as colors, sounds, etc.) considered as a multiplicity of discrete items. [*See* I. Kant, *Critique of Pure Reason*, A. 77-9-B. 102–105. — L.W.] (2) Multiplicity of options for a specific activity. (A manifold of possibilities.) (3) A set equipped with real-valued coordinates, which uniquely label the elements and whose values change in a "continuous" fashion. Historically, manifolds arose as a collection of variables subject to equations. Curves and surfaces in multidimensional Euclidean spaces were early examples of manifolds that were well studied by the founders of differential geometry. Going to higher dimensions was an obvious and yet conceptually difficult leap that required a higher level of abstraction.

Mapping — (1) In genetics, determining the order of genes on a chromosome and the distances between them. (2) In sequence analysis, older name for sequence annotation. Putative functional or structural domains in long nucleic acid or protein sequences are detected (mapped) and the sequence is labeled (annotated) with names of domains (annotations). (3) In mathematics, a name for function or transformation, which assigns every element of one set (domain) to one and only one element of another set (counter domain).

Material cause — If we think of causation as a process in which one occurrence is transformed into another, then material cause is the input to the transformation. As a degenerate example, consider the traditional concept that cause is like "billiard balls bouncing off each other." The material cause is seen as the energy of the ball doing the striking that is transformed into the effect, the energy of the ball being struck. However, the only cause admitted in the billiard ball example is material cause; it is then confused with causation in general. Likewise, the "information from past and present," also a material cause, is seen as the generalized cause in a "causal" filter.

Markov process — A probabilistic model in which the probability of the next state of a system depends solely on the probability of the previous state or a finite sequence of the previous states.

Markov source — A device that generates symbols with probabilities that could be determined from a Markov process. The terms Markov source and hidden Markov model are synonymous.

Median — "Middle value" of a sorted list of numbers. The smallest number such that at least half the numbers in the list are not greater than it.

Mean (the arithmetic mean) — Given a finite list of N real numbers $x_1, x_2, x_3, \ldots, x_N$ the mean is a real number M calculated from the formula:

$$M = \frac{1}{N} \sum_{i=1}^{N} x_i.$$

In statistics sample, mean is an estimator of the expected value of the distribution.

Meiosis — Division of a diploid nucleus to produce four haploid daughter cells. The process consists of two successive nuclear divisions with only one cycle of chromosome replication.

Meta- — A prefix used in different fields in different way, which often pertains to some concept of "beyond" or "above." (1) In general biology, the prefix can denote a change (like in metabolism) or a shift to a new form or level (like in metamorphosis). (2) In logic and systems science, the prefix "meta-'can denote description (or other form of "processing" such as formal derivation or computation) from a perspective of a next higher level of organization of thought. For instance (object) language can be described with the help of a metalanguage; logic can be discussed with the help of metalogic, mathematics with the help of metamathematics and so on.

Metabolic pathway — A sequence (succession) of enzyme-catalyzed reactions in which product(s) of one reaction is(are) the substrate(s) of the next.

Metabolism — A set of all chemical reactions that occur in an organism, or a well-defined specific subset of that set (as in "respiratory metabolism").

Model (of a system or a complex thing) — (1) A representation of selected aspects (characteristics) of the system with a simultaneous abstraction — away of all other imaginable or observable points of view and facts. (2) In the Rosen Modeling relation, a model is a particular kind of representation. The inferential entailment structure of the model within a formal system is congruent with the entailment structure of the process being modeled. The utility of the model is that novel insights can be gained about the process being modeled by asking questions about the model. (3) In logic, a specific semantic realization of a formal axiomatic system that is given by interpreting the system's basic notions. This logical notion of a model is different than the foregoing two concepts used in science. In this volume, we assume that model is defined by definition (2).

Molecular biology — Field of life sciences that aims at biologically relevant explanations of structure and function of chemical compounds (i.e., their molecules) found in living cells and tissues. It has developed from a tradition that adopted mechanistic methods of thinking from physics and applied them to integrate genetics with cell biology, evolutionary biology, embryology, immunology, and other classical fields of biology (such as systematics).

Molecular clock — An assumption that biopolymers (nucleic acids or proteins) diverge from one another over evolutionary time at an approximately constant rate which — in turn — is assumed to be well correlated with phylogenetic relationships between organisms.

Molecular evolution — An evolutionary process leading to present day DNA and protein sequences from the ancestral ones.

Moment (the kth moment) — (1) Of a sequence of N numbers (a list) — the arithmetic mean of kth powers of elements: $(x_1^k + x_2^k + x_N^k)/N$. (2) Of a random variable X — the expected value $E(X^k)$ of a random variable X^k. The mean of X is the first moment of X.

Natural system — If a Rosen Modeling Relation defines a congruency between a process in reality and an epistemological representation of that process, then the ontological process is labeled a Natural System. It is important to appreciate that Natural, signifies the property of "occurring in reality." Natural Systems can include human-made processes. (*See also* Modeling Relation; Formal System; Formalism)

Nonlocality — A nonlocal property is distributed throughout a process. It is everywhere in the process at once. If an attempt is made to localize the process by trying to observe it at a single point, as the location becomes infinitesimal, the property vanishes. An example of nonlocality is Bateson's description of semantics that meaning lies in the correspondence between a symbol and its referent. Nonlocality has one property not often appreciated; it applies everywhere inside a nonlocal process, but nowhere outside it.

Ontology — (1) The branch of metaphysics concerned with the nature of being (reality itself) as opposite of the nature of our representations of reality. (The latter is handled by another branch of metaphysics: epistemology.) (2) Name given by a group of computer scientists to an idealized integrated collection of databases of structured vocabularies that can be connected to life-sciences — relevant databases. Each vocabulary is a data structure of a computer-manageable kind such as — for instance — directed acyclic graph in which nodes are in a well-defined relation of ancestry (descend). (3) A computer-friendly representation of complex data that can be read and manipulated by computer programs.

Organic — Pertaining to any aspect of living matter, e.g., to its evolution, structure, or chemistry. The term is also applied to any chemical compound that contains carbon.

Organism — Any living creature. A necessary but not sufficient can differ for a process to be an organism is that it be closed to efficient cause.

p-Value — The probability of erroneously rejecting an acceptable null hypothesis by chance alone. For instance a *p* value < 0.05 means that the probability that the evidence supporting rejection of null hypothesis was due to chance alone is less than 5%.

Paradigm — A general methodological framework along with a set of assumptions, beliefs, and cultural biases within which questions are asked and hypotheses are formed.

Parsimony — A principle of preferring a minimal change between subsequent steps of a process. For instance, a single point mutation is a parsimonious change between two generations of evolving DNA sequence.

Pattern — Any logical, geometrical, or (broadly) factual connection between elements of a model that attracts our attention. "Pattern" is usually understood pragmatically by experienced explorers of the same model.

Phenotype — The observable properties of an individual as they have developed under the combined influences of the genetic constitution of the individual and the effects of environmental factors.

Phenotypic plasticity — The fact that the phenotype of an organism is determined by a complex series of developmental processes that are affected by both its genotype and its environment.

Population — Any group of organisms coexisting at the same time and in the same place, and capable of interbreeding with one another.

Population density — The number of individuals (or modules) of a population in a unit of area or volume.

Population genetics — The study of genetic variation and its causes within populations.

Population structure — The proportions of individuals in a population belonging to different age classes (age structure). Also, the distribution of the population in space.

Pragmatic — Practical. Dealing with facts and occurrences.

Pragmatics — In linguistics, one of three main aspects of studying languages (two others being syntax and semantics). It refers to usage of sentences in the context of other sentences as well as of real-world situations.

Pragmatic inference — (1) The art of determining sequence motifs from their instances and the knowledge context they pertain to. (2) The art of determining an alphabet (vocabulary) of function-associated motifs based on one or more individual patterns and the knowledge of structures or mechanisms that correlate well with the presence of these patterns. (3) Inference to the best explanation; abduction. (4) Reasoning that explores consilience of inductions or consilience of inductions, deductions, and abductions.

Probability — A measure of chance (possibility) that a particular event (or set of events) will occur. Values of probability measure are positive real numbers from the closed interval [0, 1]. Probability of impossible events equals 0 while probability of sure events equals 1. (The concept of probability is related to but different from the notion of likelihood.)

Proteome — Complete set of all proteins encoded in the nuclear component of a genome of a given organism.

Randomness — Logical equivalent of patternlessness. Randomness is a situation in which patterns are either undetectable or nonexistent.

Random genetic drift — Evolution (change in gene proportions) by chance processes alone.

Rate constant — Of a particular chemical reaction, a constant which, when multiplied by the concentration(s) of reactant(s), gives the rate of the reaction.

Reactant — A chemical substance that enters into a chemical reaction with another substance.

Recursion — A recursion is a form of algorithm in which the algorithm invokes nested versions of itself. However, it preserves the finite character of an algorithm in that it has defined bottom, an unambiguous test for identifying when it has reached the bottom, and an unambiguous step to take at the bottom. The step at the bottom is what defines a particular recursive algorithm. Recursion is not to be confused with impredicativity, which no *bottom*, but rather is defined by the relationship between successive levels in its hierarchy. In special instances such as the impredicatively defined Fundamental Wavelet Equation, the impredicative can be approximated by a recursion, and the resulting error is bounded. However, there is no general principle that allows for the approximation of all impredicatives by recursions.

Reduction — (1) In chemistry, gain of electrons, the reverse of oxidation. Reductions often lead to the storage of chemical energy, which can be released at any time via an oxidation reaction. (2) In methodology of science, replacement of the modeled complex system by a simpler surrogate system (a model).

Reductionism — (1) Causal — an academic doctrine according to which the only legitimate conclusion about a system can be reached from studying its parts, but one should not infer properties of parts from studying the whole system. [Downward causation is "forbidden."] (2) Methodological — an academic trend according to which complex systems should be represented by simple models (surrogate systems) which, in turn, could be studied with a variety of scientific methods.

Regular expression — (1) In sequence analysis and bioinformatics, a flexible definition of a sequence pattern allowing groups of motifs to reside in the place of single motif. (2) In information technology (IT), a valid formula of a specialized programming or scripting language dedicated to serve a software tool or database. For example, most of information retrieval systems (such as library searchable catalogues) accept queries formulated in a "language" that consists of simple regular expressions. Another example is a scripting language for textual pattern matching in UNIX operating system. Some programming languages (such as PROLOG or PERL) are entirely designed for textual pattern matching as a way of writing down programs, i.e., regular expressions actually constitute the programming language.

Regular grammar —A formal grammar that is equivalent of finite-state automaton.

Reversible process — A thermodynamic process is called reversible provided the reverse process can be achieved without expending additional exergy. In fact all spontaneous processes are irreversible and reversibility is an idealization that started with frictionless mechanics. Any friction, heat transfer between systems at different temperatures, or chemical reactions not exactly at equilibrium take place irreversibly, i.e., with the degradation of some exergy.

Semantics — In linguistics, systematic studies of meaning of linguistic expressions (such as sentences). Semantics is one of the three components of studying sentences, the two others being Syntax and Pragmatics.

Semiconservative replication — The common way in which DNA is synthesized. Each of the two partner strands in a double helix acts as a template for a new partner strand. Hence, after replication, each double helix consists of one old and one new strand.

Sequence hypothesis — First of two basic assumptions of molecular biology (the second is the central dogma).
(1) Original formulation — "…Specificity of a piece of nucleic acid is expressed solely by the sequence of its bases, and this sequence is a (simple) code for the amino acid sequence of a particular protein…." [Crick F H C (1958) On protein synthesis. Symp. Soc. Exp. Biol. 12: 138–163.] (2) In cellular protein biosynthesis, the amino acid sequence of polypeptide (primary structure of a protein) is collinear with and determined (solely) by a sequence of a specific protein-encoding region in chromosomal DNA.

Significance (Statistical significance) — (1) The probability that a test or experiment leads by chance alone to an erroneous rejection of the null hypothesis when the null hypothesis is in fact acceptable. (2) The likelihood that a statement is true. (3) The degree of deviation of an observation from its occurrence by chance alone according to a model of chance. The degree of nonconformity to a (presumed correct) model of chance in the foregoing sense.

Simulation — Imitating the behavior of a real system via using properties of its model or a class thereof. A simulation is often confused with a model. A simulation is an epistemological description of the entailed outcomes of a process. In contrast, a model is an epistemological description of the entailment structure that produces the outcomes.

Species — (1) A population or series of populations of closely related and similar organisms. (2) Biological species — A set (group) of individual organisms capable of interbreeding freely with each other and — at the same time — incapable of interbreeding with organisms from outside this set.

Spontaneous reaction — A chemical reaction that will proceed on its own, without any outside influence. A spontaneous reaction need not be rapid.

Stability — (1) The capacity for a system to remain within a nominal range of (nonextreme) behaviors. (2) Resistance to change, deterioration, or displacement. [AHD]. (3) The ability of a system (or an object) to maintain equilibrium or resume its original state after alteration (such as resuming original position after displacement.)

Stabilizing selection — Selection against the extreme phenotypes in a population, so that the intermediate types are favored. (Contrast with disruptive selection.)

Statistical inference — Finding out properties of an unknown statistical distribution from data generated by that distribution.

Standard deviation (SD) — Square root of variance of a distribution. Can be estimated by sample standard deviation, which is square root of sample variance.

Strong Church–Turing Thesis — The Strong Church–Turing Thesis is an unproven hypothesis that any physically realizable process can be fully described as a Turing machine. (*See also* Church–Turing Thesis.)

Sub- — A prefix often used to designate a structure that lies beneath another or is less than another.

Substrate — One of the chemicals (chemical entities) that enter chemical reaction. [Every reaction processes from substrates to products.]

Symmetry — (1) The property of the relation between two objects A and B, such that if A is in relation with B, then B must be in relation with A. (2) Identification of two objects regarding dislocation in space (such as translation or rotation), time, or generating a mirror image.

Syntax — In linguistics, set of rules whereby words or other elements of sentence structure are combined to form grammatically correct sentences without regard to their meaning.

System — A group of interacting, interrelated, or interdependent elements forming a complex whole, whose properties are not a simple combination of properties of elements. Specific additional meanings include: (1) A functionally related group of elements, especially: 1a. The organism regarded as a physiologcal unit. 1b. A group of physiologically or anatomically complementary organs or parts. The immune system, nervous system, or a digestive system are representative examples. 1c. A group of interacting mechanical or electrical components functioning in a robust manner within a mechanical or electrical device (machine). Exhaust system or electric system in an automobile are representative examples here. So do parts of the proverbial watch functioning toward measuring passage of time. [The watch is more the sum of parts as a proverb about the hammer and the watch clearly indicates.] 1d. A network of objects or structures with indication of connections between them. Metro or road system in a big city is a representative example here. (2) An organized set of interrelated ideas or principles such as a religion, ideology, or other belief-based general paradigm. Science system, legal system, and ethical system are representative examples here. (3) An actual social, economic, or political organizational form. The establishment. (4) A naturally occurring group of objects or phenomena. The solar system, a (specific) ecosystem, or a pack of wolfs are representative examples here. (5) A set of objects or phenomena grouped together for the purpose of classification or analysis. All life forms in the ocean or all elementary particles in the Cosmos could be representative examples here. (6) A method, a procedure, or a paradigm, which can be systematically reused without changing details each time it is used. Number system (such as binary or decimal), writing system (such as roman script), and cryptographic system are representative examples here.

Systematics — The scientific study of the diversity of organisms via appropriate classification and coding (naming).

Tangent Space — The vector space of infinitesimal displacements at a point in a manifold. A vector in this space can be defined as an equivalence class of paths that "go in the same direction at the same speed." A tangent vector specifies the rate at which any function will change as we move in the direction of the tangent vector. If it is a coordinate system at a point, then the partial differential operators form a basis for the tangent space and thus define the corresponding coordinate system in the tangent space.

Thermodynamics — A field of science devoted to studies of interconversions between different forms of energy (such as for instance heat and work).

Theory — (1a) Systematically organized knowledge applicable in a relatively wide variety of circumstances, especially a system of assumptions, accepted principles, and rules of procedure devised to analyze, predict, or otherwise explain the nature or behavior of a specified set of phenomena. (1b) Such knowledge or such a system. [AHD-1] (2) Abstract reasoning; speculation. [AHD-2] (3) A belief that guides action or assists comprehension or judgment: rose early, on the theory that morning efforts are best; the modern architectural theory that less is more. [AHD-3] (4) An assumption based on limited information or knowledge; a conjecture. [AHD-4] (5) A narrative describing a possible scenario (chain) of events that could lead to a given outcome. (6) A pejorative term meant to characterize outcomes of lunacy of an irresponsible thinker. Incomplete, second-hand opinions or speculations not always relevant to the subject matter. The opposite of "practice."

Turing machine — A Turing machine is an automaton, or a general mathematical model of computation. As given by K. Rosen, in his classic reference on discrete mathematics: "A *Turing machine* $T = (S, I, f, s_0)$ consists of a finite set S of states, an alphabet I containing a blank symbol B, a partial function f from $S \times I$ to $S \times I \times \{R, L\}$, and a starting state s_0." The Church–Turing thesis says that any process that can be described by this automaton is "effective." Von Neumann demonstrated that any process that could be described unambiguously could also be described by this automaton. The advocates of the Strong Church–Turing thesis believe that any physically realizable process can be described by this automaton. (*See also* Church–Turing Thesis; Strong Church–Turing Thesis.)

Upward causation — Upward causation is the notion that the properties of the whole are caused as a cumulative effect of the properties of the parts. Until recently "traditional" science did not admit the possibility of causation running in the other direction. The academic tradition that advocates upward causation and forbids downward causation is called causal reductionism. (*See also* Downward causation; Reductionism)

Vector — (1) An agent such as an insect that carries a pathogen affecting another species. (2) An intermediary object (such as a plasmid or a virus) that carries an inserted piece of DNA into a chromosomal DNA of a cell.

In mathematics (algebra), an element v of a vector space V over a field F. A one-dimensional array. For a fixed natural number k, any sequence of k real (or complex) numbers can be considered a vector in a k-dimensional vector space. In particular, a sequence of three numbers can be considered a vector in three-dimensional space provided that the appropriate algebraic operations are defined.

In physics, a mathematical object characterized by magnitude and direction that can be used to quantitatively represent properties of physical systems such as velocity or force.

Working fluid — A thermodynamic system used as a temporary medium for storing heat and work during a thermodynamic process, typically the operation of a heat engine.

z-**Value (*z*-Score)** — The observed value of a Z statistic, which is constructed by standardizing some other statistic. The Z statistic is related to the original statistic by measuring number of standard deviation by which a given data point differs from expected value: $Z =$ (observed − expected value of original)/Standard Deviation (observed).

Index